R. W. JENSEN is associated with the technical staff of the Hughes Aircraft Company, Culver City, California. A practicing automated circuit analysis consultant, Mr. Jensen has also published widely on ECAP in technical journals.

M. D. LIEBERMAN is the Assistant Director, Campus Facility, Stanford University Computation Center. The recipient of an award for his role in the implementation of ECAP while associated with the Hughes Aircraft Company, Mr. Lieberman was author of the original *Hughes ECAP User's Manual*.

IBM

ELECTRONIC

CIRCUIT

ANALYSIS

PROGRAM

Techniques

and

Applications

Courtesy of International Business Machines Corporation

RANDALL W. JENSEN

Department of Electrical Engineering
Utah State University
(Formerly of Hughes Aircraft Company)

MARK D. LIEBERMAN

Assistant Director, Campus Facility
Stanford Computation Center
Stanford University
(Formerly of Hughes Aircraft Company)

PRENTICE-HALL, INC., Englewood Cliffs, New Jersey

IBM

ELECTRONIC

CIRCUIT

ANALYSIS

PROGRAM

Techniques

and

Applications

to our wives, Marjorie and Adele,
without whom we might have finished this book
three months earlier—
or might not have finished it at all
and
to the IBM team of
Tyson, Wall, Nisewanger, Hogsett, Peirce, and Mitchel
for its role in the development of ECAP

SERIES IN COMPUTER APPLICATIONS IN ELECTRICAL ENGINEERING
Franklin F. Kuo, *editor*

PRENTICE-HALL INTERNATIONAL, INC., *London*
PRENTICE-HALL OF AUSTRALIA, PTY. LTD., *Sydney*
PRENTICE-HALL OF CANADA, LTD., *Toronto*
PRENTICE-HALL OF INDIA PRIVATE LTD., *New Delhi*
PRENTICE-HALL OF JAPAN, INC., *Tokyo*

IBM ELECTRONIC CIRCUIT ANALYSIS PROGRAM:
Techniques and Applications
Randall W. Jensen and Mark D. Lieberman

© 1968 by Prentice-Hall, Inc., Englewood Cliffs, N.J.

Current printing (last digit):
10 9 8 7 6 5 4

Library of Congress Catalog Card Number 68–18515

Printed in the United States of America

PREFACE

This book provides the student, as well as the working engineer, with the information necessary to facilitate the optimum use of the IBM Electronic Circuit Analysis Program, known as ECAP. The book is offered both as an introduction to ECAP, even for the reader with no previous computer experience, and as an easy-to-use reference for the experienced user. In each case, the book is a complete source of information for the application of any of the standard ECAP versions to which it is assumed the reader has access. The "standard ECAP versions" are the program tapes for the 1620, 7094, and 360 computers. Although this book is primarily directed to the 360 version, appendices describe the differences between the 1620 and 7094 versions and the 360 program.

The purpose of this book is not to teach circuit theory, although the use of ECAP is likely to give additional insight into the subject. The techniques employed internal to the program are not emphasized, unless the information is necessary in the development or the understanding of the material. However, a brief description of some of the internal operations is included in the appendices; this description should convey an appreciation of the solution techniques used in the program.

The book has three sections. The first, chapters 1 through 4, explains the basic techniques of the program. The second section, chapters 5 through 7, explains advanced techniques and the practical application of ECAP. The third section, the appendices, provides reference material.

Chapter 1 contains introductory material to acquaint the reader with the capabilities of ECAP and the background of the program. The chapter also describes the basic procedures used in most data processing facilities.

Chapters 2, 3, and 4 develop the basic skills for use in the DC analysis, AC analysis, and transient analysis programs. In these chapters the essential program techniques are presented and profusely illustrated. Since each new

v

feature of the program is introduced as quickly as possible in an example, the reader should keep in mind that a Comments section generally follows the example and clarifies any points not self-evident in the example.

Chapter 5 develops techniques for deriving equivalent circuits of elements that are not standard in the program. This chapter contains several equivalent circuits, such as a large-signal charge-control transistor model and the non-linear capacitance, which are developed as examples.

Chapter 6 contains several complete analysis examples which demonstrate the capabilities and the potential of the program in the solution of practical electronic circuit analysis problems.

Chapter 7 explains the use of ECAP in the solution of engineering problems in fields other than electrical by the application of physical analogies. Examples include mechanical, heat-transfer, and feedback control system problems.

The appendices include reference information, such as commonly used transistor equivalent circuits, brief descriptions of pertinent program-solution techniques, summaries of program capacities, and a list of the error messages produced by ECAP in response to user errors in input data.

Each chapter concludes with suggested problems which enhance the usefulness of the book as a text in the increasing number of college and industrial courses giving attention to ECAP. Answers to all problems are included in the back of the book, and complete solutions manuals are available.

We believe that the book can be used at various levels in the electrical engineering curriculum. In a beginning course in circuits, after traditional analysis techniques have been presented, teaching the use of ECAP itself as a solution technique is very worthwhile. Thereafter, in many courses, students will choose ECAP as the simplest and most effective tool for obtaining solutions to various problems. Instructors will find that ECAP reduces some of the drudgery of obtaining practical solutions, thereby allowing more time for contemplation of concepts and results.

As to the computer facilities available, it should be noted that, although most users probably will be using a facility set up for the batch processing of input card decks (such as described in Chapter 1), the techniques for using ECAP on a remote terminal (such as a teletypewriter or visual display device) are virtually identical to those described here. This text, therefore, is of value in all these computing environments.

Although the style of the book is relatively unorthodox in its insistent use of illustrative examples, we feel—particularly for a tutorial or a practical, problem-oriented topic such as ECAP—that this approach is particularly effective. To be sure, any "unorthodoxy" is present only as a by-product of what we hope is a better-than-standard presentation. The examples in chapters 2, 3, and 4 are contrived solely to optimally illustrate ECAP techniques; they are purposely made as easy as possible to allow the reader to focus his attention on the ECAP techniques rather than on the circuit operation. The simplicities and sizes of the circuits permit the easy checking of results. Repeated use of the same circuits further supports these ideas. The examples in chapters 5, 6, and 7 are necessarily more detailed to present more advanced and complex applications of the program.

A practical suggestion to the ECAP user is that, unless much time is available to "experiment," the quickest way of doing the job correctly is to follow closely the procedures described in this book. When a question arises, the fastest way to find an answer is to study a pertinent example. We feel that the teach-by-example approach (employed in presenting the basic rules), the practical device-modeling techniques, and the multidisciplined applications will stand out as the most important features of the text.

We wish to acknowledge the contributions of Howell N. Tyson, Jr., Herb M. Wall, Donald A. Nisewanger, Gerald R. Hogsett, Rosalie Peirce, and Charles E. Mitchel of IBM Corporation, who constituted the major ECAP development team and who have always been most cooperative and encouraging in our various ECAP endeavors. We also express our gratitude to Mrs. Sally Chuck for her assistance in the preparation of the original manuscript; to the Computer Applications Department and the Digital Circuits and Support Department of Hughes Aircraft Company, and to the Stanford University Computation Center for their cooperation and support.

RANDALL W. JENSEN
MARK D. LIEBERMAN

CONTENTS

1

INTRODUCTION

2

DC ANALYSIS

3

AC ANALYSIS

4

TRANSIENT ANALYSIS

5

MODELING TECHNIQUES

6

PRACTICAL EXAMPLES

7

ANALYSIS OF PHYSICAL SYSTEMS

APPENDICES

TERMINOLOGY

A	network topology matrix	C_{te}	emitter transition capacitance
A_t	transpose of A matrix		
a	amperes	CV_{from}	element voltage in the *from*-branch (a branch number is not assigned to this term)
BAi	branch current in branch i		
Bi	branch i or ith B card		
BETA	current gain of dependent current source	CVi	element voltage in branch i
$\beta_{i,j}$	current gain from branch j to the dependent current source in branch i	Δt	time between solutions in a transient analysis
		Dxx	floating point representation of 10 to the xx power (used in printouts)
BPi	element power dissipation in branch i		
CA$_{from}$	element current in the *from*-branch (a branch number is not assigned to this term)	e	node voltage vector
		E	voltage source vector
		E_i	independent voltage source in branch i
CAi	element current in branch i	$Ei(t)$, Ei	time dependent voltage source in branch i
C_i	capacitance element in branch i	EO, E0	initial condition voltage across capacitance
$C_{b'e}$	emitter junction capacitance	EQ	EQUILIBRIUM solution request statement
$C_{b'c}$	hybrid-π transistor model parameter; $C_{b'c} \simeq C_{ob}$	EX	EXECUTE statement
		Exx	floating point representation of 10 to the xx power (used in input data and printouts)
C_{ob}	collector base junction capacitance		

f	farads	Mi	mutual inductance i or ith M card
FI	FINAL TIME statement		
FR	FREQUENCY statement	MO	MODIFY command statement
f_T	transistor gain-bandwidth product (frequency where $\lvert h_{fe} \rvert = 1$)	NVi	voltage at node i with respect to the datum node (ground)
f_β	transistor beta cutoff (frequency where $h_{fe} = 0.707\, h_{FE}$)	OU	OUTPUT INTERVAL statement
		$\mho$	mhos
G_i	conductance in branch i	Ω	ohms
GM	transconductance of dependent current source	PR	PRINT statement
		$r_{bb'}$	transistor base resistance
		R_i	resistance in branch i
h	henries	sec	seconds
h_{FE}	transistor low frequency current gain (DC)	SE	SENSITIVITY calculation request statement
h_{fe}	transistor AC current gain	Si	switch i or ith s card
		ST	STANDARD DEVIATION calculation request statement
Hz	Hertz (1 cycle per second)		
I	independent current source vector	TI	TIME STEP statement
		TR	TRANSIENT analysis command statement
I_i	independent current source in branch i	Ti	dependent current source i or ith T card
$I_i(t)$, Ii	time dependent current source in branch i		
		v	volts
IN	INITIAL TIME statement	w	watts
IO, I0	initial condition current in inductance	ω	angular frequency (radians per second)
k_{ij}	mutual inductance coupling coefficient between branches i and j	ω_β	transistor beta cutoff frequency
		WO	WORST CASE calculation request statement
L_i	inductance in branch i	Y	admittance (mhos)
MI	MISCELLANEOUS output request statement	**Y**	admittance matrix
		Z	impedance (ohms)

1

INTRODUCTION

Circuit designers face the problem of guaranteeing reliable electronic circuit performance at minimum cost. The demand for extremely high reliability in circuit performance is greatest in the aerospace industry because of space and military requirements. In these areas where high reliability is desired the performance of electronic circuits is most critical. In other industries, such as consumer electronics, the problem becomes one of achieving a reasonable degree of reliability—one that will exceed the minimum performance acceptable to the customer and still be economical enough to be competitive. In the design of electronic circuits it is thus necessary to determine the nominal value and tolerance of each component so that circuit performance will meet the required specification within certain tolerances. For example, in designing a circuit to attain a worst case tolerance of ± 1 per cent on an output parameter, it is uneconomical and unrealistic to assume that all components used in the circuit must have a tolerance of arbitrarily less than ± 0.1 per cent so that the sum of the tolerances of components in the circuit does not exceed the specified ± 1 per cent. A circuit design using such a philosophy is generally unreliable and often includes additional costly parts with stringent tolerances. Actually, to achieve an overall tolerance of ± 1 per cent, some components may be required to have a tolerance of ± 0.01 per cent, while for others a ± 10 per cent tolerance may be adequate.

The process of designing an electronic circuit usually follows a relatively standard procedure (Fig. 1.1). To obtain some insight into the usefulness of the digital computer in the circuit design process, consider the steps that an engineer might follow in designing a circuit. Once a project assignment has been made, the problem is divided into sections, and an engineer is assigned to design and test one or more of the sections. For the system to operate as a unit, each section must meet a set of specifications which defines its operating characteristics. The engineer must design the circuitry either analytically or experimentally so it will meet specifications and perform its function when incorporated into the system. The engineer must analyze the circuit to determine the effects of element value tolerances, parameter variations, and power dissipation. The circuit is then tested to check the validity of the calculations or to measure its response or both. If the results of the analysis or the tests show that the design will not meet circuit specifications, the engineer must modify the original design (or the latest modification). The design process is iterative; it is here that circuit analysis programs are very useful.

One might consider the circuit analysis program as an electronic circuit simulator, although it can also work effectively in analyzing other physical systems which have corresponding electrical network analogs. The circuit analysis program permits the designer to use the capabilities of the digital computer in several areas. First, the effects of a parameter variation can be much more easily observed using a computer program than using breadboard evaluation. For example, the effects of changing only the collector capacitance or the alpha cutoff frequency of a transistor are very difficult to determine experimen-

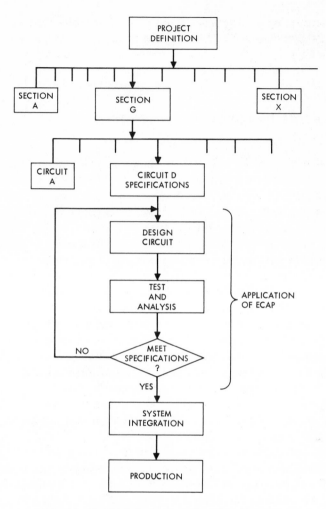

Fig. 1.1. Standard design procedure flow diagram.

tally, but they are easy to ascertain when using the digital computer. Second, the circuit designer can analyze the effects of a component failure on circuit performance without destroying the component involved. Third, the computer allows the designer to study combinations of component parameters, such as worst case, which are difficult if not almost impossible to achieve in the laboratory. Fourth, the program makes it possible to simulate the effects of expensive or hard-to-obtain components. Finally, the program makes it possible to perform measurements on a circuit model that would be difficult to make or time-consuming and expensive to instrument in the laboratory.

The electronic computer has not eliminated the need for experimental testing, but it definitely can reduce the time and tedium involved in circuit development by performing a large amount of analysis and breadboarding. A good example of the savings gained by using the digital computer in circuit analysis is in the design of integrated circuits, for which breadboard circuits are both time-consuming and expensive to build.

1.1 REQUIREMENTS FOR A CIRCUIT ANALYSIS PROGRAM

Before investigating ECAP in detail, we shall discuss the capabilities of a good general network analysis program and the features that should be incorporated into it for maximum usefulness of the program.

Briefly, the general network analysis program should be capable of obtaining (1) a steady state or DC solution, (2) an AC solution for determining circuit response to sinusoidal excitation, and (3) a transient solution to determine the time response to an arbitrary time-dependent driving function.

Six features should be incorporated into a good general network analysis program, according to Kuo.[1] First, the program should have a convenient, simple input language to describe topology of the circuit, element types, and associated values. This language should also provide a description of the circuit excitations, the analysis options desired, and the range of time or frequency over which the analysis is to be performed. The language must be so devised that a clerk with no engineering or programming knowledge will be able to prepare the input information from a given circuit schematic.

Second, the ideal program should be able to handle a wide class of models of physical devices. It should be able to change not only the parameters of a device but also its topological model.

Third, to deal effectively with active circuits, the analysis program must be able to handle nonlinearities. A linear analysis program is not suitable for obtaining quiescent solutions involving transistors with nonlinear load lines. And it is not suitable for analyzing switching phenomena. It is desirable, therefore, that an electronic circuit analysis program be able to deal with nonlinearities given either in piecewise linear (tabular) form or by means of a mathematical subroutine.

Fourth, there should be a large number of output options for such a program. The outputs can be in the form of driving-point or transfer functions, scattering parameters, poles and zeros, magnitude and phase response, or transient response. The calculated information should be presented in a compact format, with accompanying graphical output, by means of a microfilm printer, an x-y plotter, or a cathode-ray display.

Fifth, the ideal program should have an automatic parameter-modification feature. With this facility, sensitivity, tolerance, and worst case studies can be

1. F. F. Kuo, "Network Analysis by Digital Computer," *Proceedings of the Institute of Electrical and Electronic Engineers*, **54,** No. 6 (1966), 821.

performed. This feature, in many cases, can replace the breadboard stage in an experimental design.

Sixth, the program must contain error checks so the user can assess the reliability of the output.

The six features listed by Kuo summarize the requirements of a useful network analysis program. Several programs available to the designer today incorporate the majority of these features; of these, we feel ECAP is the most prominent.

1.2 WHAT IS ECAP?

The Electronic Circuit Analysis Program is an integrated system of programs developed primarily to aid the electrical engineer in the design and analysis of electronic circuits. There are four closely related programs in the system.

INPUT LANGUAGE. This program acts as the communication link between the user and the three analysis programs. The language is user-oriented and allows complex circuits to be simply described to the computer from easily developed equivalent circuits. The six basic statements used in the program completely define the topology of the circuit, the circuit element values, the type of analysis to be performed, the driving functions, and the output required. This simple scheme makes it possible to learn to use the program in a short time.

DC ANALYSIS. The DC analysis program obtains the DC or steady-state solutions of linear electrical networks and provides the worst case analysis, standard-deviation (statistical) analysis, and sensitivity coefficient calculations if requested. This program also provides an automatic parameter-modification capability.

AC ANALYSIS. The AC analysis program obtains the steady-state solution of linear electrical networks subject to sine-wave excitation at a fixed frequency. Since this program also contains the automatic parameter-modification capability, it is easy to obtain frequency- and phase-response solutions.

TRANSIENT ANALYSIS. The transient analysis program provides the time-response solution of linear or nonlinear electrical networks subject to arbitrary, user-specified driving functions. Nonlinear elements are modeled by using combinations of linear elements and switches to provide piecewise linear approximations to the nonlinear characteristics.

The details of each program are discussed in the following chapters.

ECAP is not difficult to use. It is not necessary for the user to have any knowledge of the internal mechanics of the program or any prior computer or programming experience. However, he must know the techniques of communicating with the computer via ECAP. These include (1) the means of converting the circuit schematic to a written format acceptable to the program, (2) the

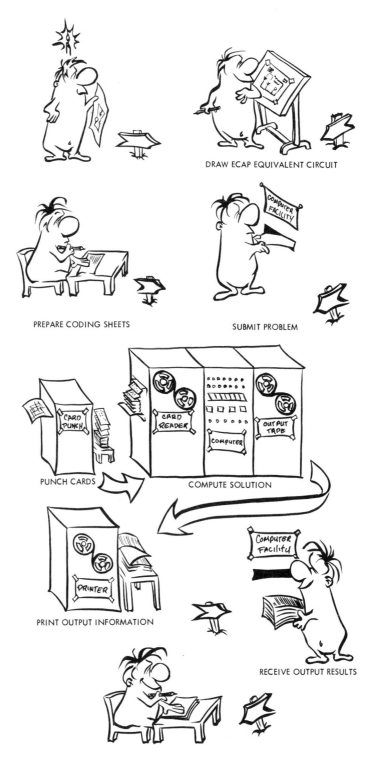

DRAW ECAP EQUIVALENT CIRCUIT

PREPARE CODING SHEETS

SUBMIT PROBLEM

PUNCH CARDS

COMPUTE SOLUTION

PRINT OUTPUT INFORMATION

RECEIVE OUTPUT RESULTS

STUDY CIRCUIT ANALYSIS

Fig. 1.2. Typical circuit analysis procedure using ECAP.

6

information required to obtain the desired analysis, and (3) the knowledge to interpret the output results. The techniques involved in using ECAP are easily learned by the electrical engineer because ECAP's input language and its output are written in electrical circuit terminology.

ECAP's capabilities are not limited to the functions of an electronic circuit simulator or to its aid in the analysis of electronic circuits. With some imagination, other capabilities become apparent. For example, heat-transfer or thermal-radiation problems can be readily solved by using an electrical-network approach and ECAP. Solutions to mechanical problems can be obtained by using electrical network analogs. These are just a few of the possible applications of ECAP. Once the reader has become familiar with the techniques of using the program and understands its limitations, he will discover other applications.

1.3 COMMENTS TO FIRST-TIME DIGITAL COMPUTER USERS

1.3.1 Computer-Installation Operation Procedures

Although computer installations usually have operating procedures tailored to their own requirements, a few general comments concerning the most common operating procedures are included for the first-time computer user. The most useful information, of course, is a brief description of a typical procedure (Fig. 1.2) that would be followed in obtaining the solution to an ECAP analysis.

1. The engineer draws an equivalent circuit from the schematic of the circuit being analyzed using elements recognized by ECAP. These elements include resistances, inductances, capacitances, voltage sources, etc.
2. The information contained in the equivalent circuit diagram is transferred, using ECAP input language, to a coding sheet, shown in Fig. 1.3, or to any sheet of paper with a similar format. The coding sheet contains at least 72 distinct columns and a convenient number of rows (usually about 25).
3. A computer card is punched for each line of information on the sheet, and results in zero to three holes in each of the card columns. A typical character set is shown in Fig. 1.4, and a typical ECAP data card is shown in Fig. 1.5.
4. The resulting deck of cards is then submitted to the computer facility along with the problem submittal forms required by that facility.
5. Inside the facility, the procedure is often as follows:
   a. The deck is processed through a card-reading device which

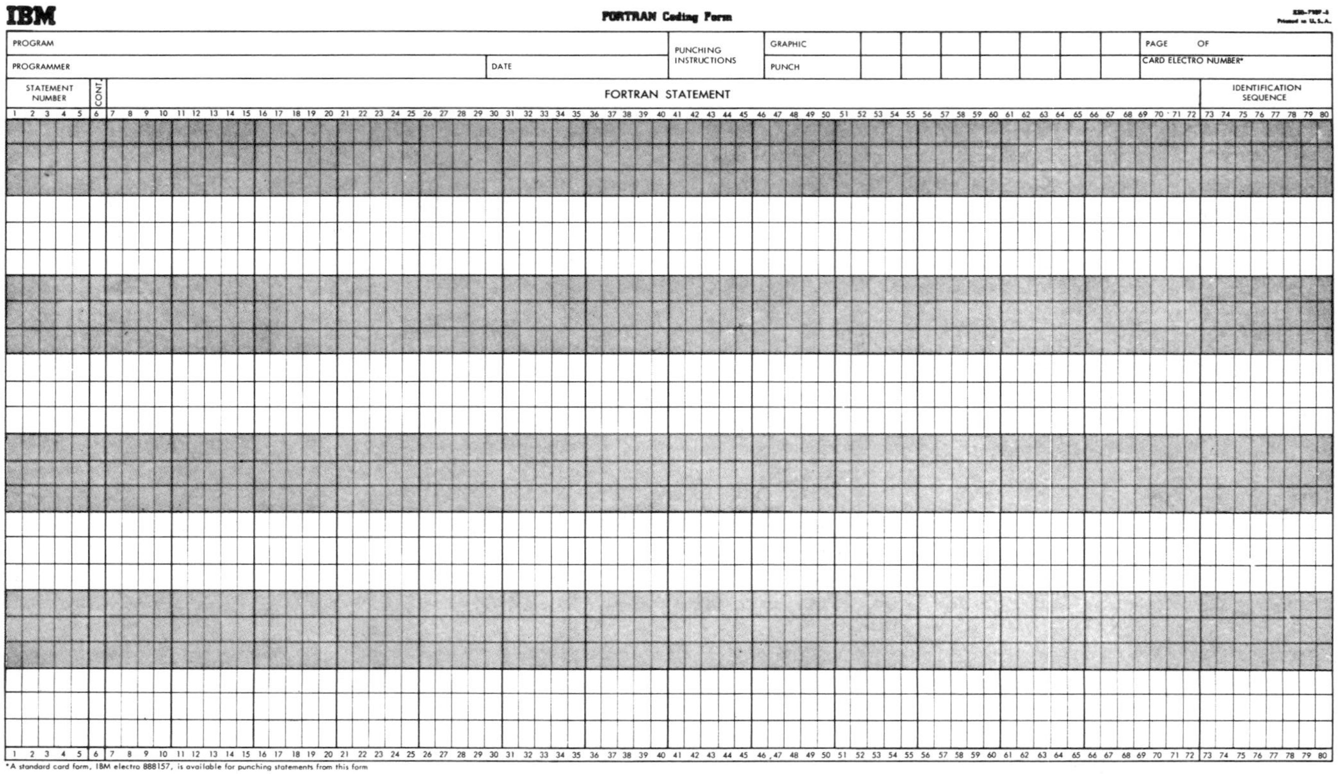

Fig. 1.3. Coding sheet format. (Courtesy of International Business Machines Corporation.)

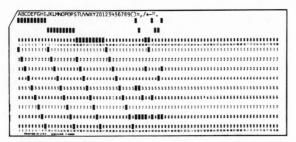

Fig. 1.4. Alphanumeric and special characters commonly used by System/360 ECAP (generated by the IBM 029 keypunch unit).

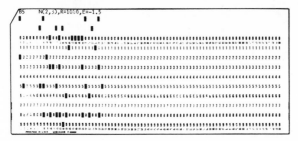

Fig. 1.5. Typical ECAP data card.

       stores the information punched in the cards on a magnetic tape (or disk) by recording groups of properly positioned magnetic "bits" corresponding to the location of the holes in the card.

    b. The tape (or disk) is read into the computer. ECAP then interprets the input statements, establishes the network topology, generates the network equations, performs the specified analysis, and produces the output information on a magnetic tape, or disk, or directly on a printer.

    c. A printer converts the information stored on the output tape into alphanumeric information which constitutes the output.

  6. The output information is returned to the user for study.

    The details of these operations (except the computer facility procedure) are discussed in the following chapters.

1.3.2 Coding-Sheet Alphanumeric Character Format

When the ECAP user prepares the coding sheets for his problem, he must take great care to ensure that the coding-sheet characters are interpreted correctly by the keypunch operators. For example, the number 1 must never be confused with the letter I, and the number 0 must be distinguished from the letter O, 2 from Z, etc., since each character has its own code. There are several other alphanumeric characters and symbols that can be easily confused if care is not taken in preparing the coding form. To remove the uncertainty in interpreting the coding, special character forms are used for writing characters unambiguously. The most commonly used character set is shown in Fig. 1.6. Distinguishing marks for some characters are used to improve clarity, such as slashing the letters $\bar{Z}$ and $\emptyset$ or capitalizing the letter I. It is also important to note the positioning and the form of punctuation marks.

A	B	C	D	E	F	G	H	I	J	K	L	M	N	Ø	P	Q	R	S	T	U	V	W	X	Y	Z
.	,	(	)	'	/	=	*	+	−	$															
1	2	3	4	5	6	7	8	9	0																

Fig. 1.6. Character format.

1.3.3 Numeric Forms

The reader will notice that the computer outputs contained in the text are written in *floating-point** form and have as their last four characters the letter E (or D) followed by a blank or minus (−) sign, which in turn is followed by two digits. This is a standard computer format (in no way special to ECAP) used to represent the power of ten by which the number (usually seven or eight places) preceding the E (or D) is to be multiplied. The letter E precedes the power of ten in the input data and in the single-precision calculation output. The letter D precedes the power of ten in the double-precision calculation output. The use of the E or D in no way implies the procedure followed in calculating the solution. Hence, the E or D does not indicate solution accuracy. According to this format

48200 would be printed 0.48200E 05 or 0.48200D 05

0.00482 would be printed 0.48200E-02 or 0.48200D-02

Fixed-point and floating-point numbers are allowed in ECAP input statements. For example, the number 482 can be written in several ways. Some of

---

*A floating-point number contains a decimal point and an exponent. In contrast, a fixed-point number may contain a decimal point (no exponent) or it may be an integer with no decimal point included. The decimal point is assumed to be immediately to the right of the rightmost digit in an integer number.

these are

$$482$$

$$48.2E\ 01$$

$$4.82E\ 02$$

$$48200E\text{-}02$$

The only restrictions imposed by ECAP are that no more than eight digits may be used before the E (i.e., the mantissa must be less than eight digits) and no more than two digits and a sign may be used following the E. The D format cannot be used in the input data. The magnitude of the numbers used must be between 2×10^{-78} and 7×10^{75}.*

1620 — 1×10^{-99} *to* $1 \times 10^{+99}$

$(1E \pm 99)$

2

DC
ANALYSIS

There are three types of analysis programs in ECAP—DC, AC, and TRANSIENT. The DC analysis program provides steady-state solutions for circuits with time-invariant sources. The techniques for using the DC program are described in this chapter.

2.1 THE SIMPLIFIED STANDARD BRANCH

Before proceeding with a step-by-step development of the first example, consider the simplified standard branch in Fig. 2.1. Its terms are defined as follows:

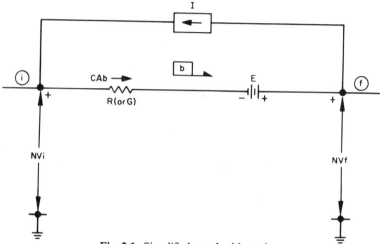

Fig. 2.1. Simplified standard branch.

- A branch b contains a nonzero resistance R or a conductance G, a voltage source E, and a current source I. E and I are assumed to be zero unless specified otherwise.
- A node voltage NV is the voltage measured at a node with respect to the reference node, which is electrical ground.
- An element current CA is the current flowing through the element in the simplified branch from the initial node i of the branch to the final node f. The choice of which node is to be the initial one is synonymous with the choice of which current direction is assumed to be positive and in most cases is arbitrary.

2.2 CODING INTRODUCTION

Suppose that the node voltages and element currents are to be calculated for the simple DC circuit shown in Fig. 2.2a. To prepare the schematic for ECAP coding all nodes and branches must be numbered. The convention of circling

14

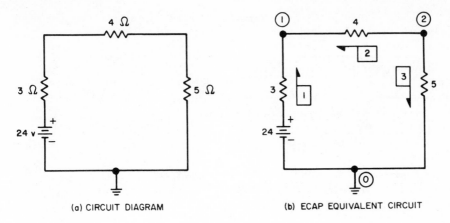

Fig. 2.2. DC circuit.

node numbers and boxing branch numbers is expedient. An arrow on each box, as in Fig. 2.2b, shows the choice of assigned positive-current direction. The ECAP input data statements can now be prepared directly from this numbered schematic and are shown in Fig. 2.3; each line is explained in Fig. 2.4.

1 2	3	4	5	6	7	8	9 10 11 12	13 14 15 16 17 18	19 20 21 22 23 24	25 26 27 28
C							C Ø D I N G	I N T R Ø D U C	T I Ø N	
					D	C				
	B	1			N	(	0 , 1) ,	R = 3 ,	E = 2 4	
	B	2			N	(	2 , 1) ,	R = 4		
	B	3			N	(	2 , 0) ,	R = 5		
					P	R	I N T ,	N V , C A		
					E	X	E C U T E			

Fig. 2.3. Coding introduction.

With an understanding of the numeric forms (see Sec. 1.3.3) the output of this example, shown in Fig. 2.5, should be virtually self-explanatory. Node voltages 1 and 2 are written on one line; the element currents of branches 1, 2, and 3 are written on another line. Note that the element current in branch 2 is negative because of the branch's assigned positive-current direction.

COMMENTS

1. In any ECAP problem:

    a. Nodes may be numbered in any order using consecutive numbers beginning with zero. ECAP always assumes that node 0 is the reference node (electrical ground). The maximum number of nodes that may be used is 50, not including node 0.

    b. Branches may be numbered in any order using consecutive

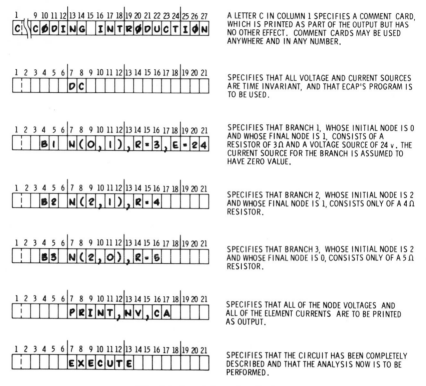

Fig. 2.4. Input data description.

Fig. 2.5. Output format introduction.

numbers beginning with one. The B cards (which contain the branch descriptions) must each contain one, and only one, passive element and must be entered sequentially on the coding sheet. The maximum number used is 200. *(1620 – 60 BRANCHES)*

2. The output is generally reliable to no more than six decimal places because of computational limitations. This reliability is more than adequate for most applications. (Note in Fig. 2.5 that NV1 is not printed as 0.18000000D 02, but is certainly close enough for most practical applications.) However, when circuit parameters range over many orders of magnitude, the solution accuracy will decrease unless care is exercised (see Sec. 2.12). The inaccuracies are due to two main factors:

   a. The conversion between the decimal input-output and the binary number system used by the digital computer is not always perfect. For example, the decimal number 0.1 cannot be represented exactly as a binary number.
   b. In general, less-than-perfect accuracy is realized from each calculation since the results are truncated after a finite number of digits (usually 8 to 16).

Again, it should be emphasized that these inaccuracies are seldom of concern in most practical applications of ECAP.

2.3 THE COMPLETE DC STANDARD BRANCH

To this point only a simplified standard branch has been described. It is now appropriate to present the complete DC standard branch, shown in Fig. 2.6, and the corresponding definitions.

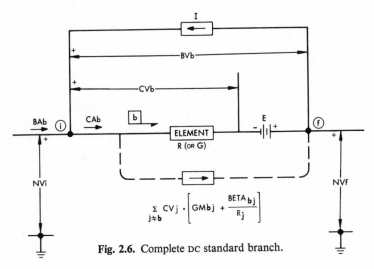

$$\sum_{j \neq b} CVj \cdot \left[GMbj + \frac{BETA_{bj}}{R_j} \right]$$

Fig. 2.6. Complete DC standard branch.

INPUT PARAMETERS (SUPPLIED BY THE USER)

- The element is a conductance (or a resistance) with a finite admittance $Y = G$(or $1/R$).
- E is an independent, time-invariant voltage source.
- I is an independent, time-invariant current source.
- BETA is the current gain and GM is the transconductance from the *from*-branch to the *to*-branch. A BETA or GM trans-term defines a dependent current source (see Sec. 2.4) in the *to*-branch. There may be more than one dependent current source in any *to*-branch.
- Any input parameter may have a negative value.

OUTPUT TERMS (CALCULATED BY ECAP)

- NV is the node voltage (the basic unknown for which the nodal equations are solved).
- BV is the branch voltage $= \text{NV}i - \text{NV}f$.
- CV is the element voltage $= \text{BV} + E$.
- CA is the element current including the effects of all trans-terms

$$\text{CA}b = \text{CV}b \cdot Yb + \sum_{j \neq b} \text{CV}j \cdot \left(\text{GM}bj + \frac{\text{BETA}bj}{R_j} \right)$$

where j is the *from*-branch and b is the *to*-branch.
- BA is the branch current $= \text{CA} - I$.
- BP is the *element* power dissipation $= \text{CV} \cdot \text{CA}$.

Note in CV, CA, and BA that C represents element and A represents current.

2.4 DEPENDENT CURRENT SOURCES

All three ECAP analysis programs permit the specification of dependent current sources across any branches, called *to*-branches. Each source is specified via

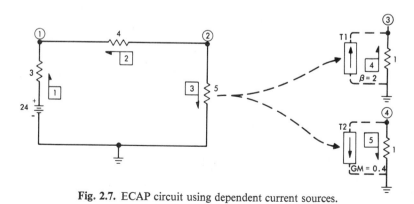

Fig. 2.7. ECAP circuit using dependent current sources.

a current gain BETA or a transconductance GM from any other branch, called a *from*-branch. Since the only terms that can be input directly to ECAP are R, G, L^*, C^*, E, I, BETA, and GM, these dependent current sources are of prime importance in modeling transistors and other physical devices. (See Chapter 5.)

Figure 2.7 shows a dependent current source across branch 4 with a value of twice the element current (i.e., a current gain BETA = 2) in branch 3; it also shows a dependent source across branch 5 whose current value is 0.4 times the element voltage (i.e., a transconductance GM = 0.4) of branch 3**. The corresponding

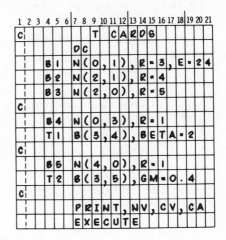

Fig. 2.8. Dependent current source introduction.

input data, introducing T cards, are shown in Fig. 2.8, and the requested solutions are in Fig. 2.9.

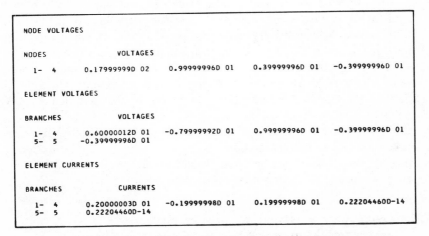

Fig. 2.9. Dependent current source introduction output.

COMMENTS

1. The numbers in the parentheses following the B on a T card specify, in order, the *from*- and *to*-branches of the dependent current source. Thus, in Fig.

---

*L and C are acceptable parameters only in the AC and transient analysis programs.

**In the figures the link between a dependent current source and the related *from*-branch will be indicated by an arrow, as in Fig. 2.7, or by a label (attached to the dependent source) of the form BETA(CA)B_{from} or GM(CV)B_{from} where B_{from} is the *from*-branch. Using the latter form, the labels in this example for T1 and T2 would read 2(CA)3 and 0.4(CV)3, respectively.

2.8, the T1 card specifies a dependent current source across *to*-branch B4, which has a value of two times the element current in *from*-branch B3; the T2 card specifies B3 and B5 as its *from*- and *to*-branches, respectively.

2. T cards may be mixed with B cards as long as the following rules are observed:

1620 Number of T
MAX Number = 10
CARDS

    a. A T card must not precede the B card defining its *from*-branch.

    b. Dependent current sources may be numbered in any order using consecutive numbers beginning with one, and must be entered sequentially on the coding sheet. Thus, in Fig. 2.8, the only restriction on the location of the T1 card is that it be after the B3 card and before the T2 card.

3. GM or BETA, the latter of which is converted by the program for its internal use to GM($=$ BETA$/R_{\text{from}}$), may be specified on any T card. In Fig. 2.7, NV3 $= -$ NV4 because

$$NV3 = \left(\frac{\text{BETA}}{R_3}\right) \cdot (CV3) \cdot R_4 = (0.4) \cdot (CV3) \cdot (1)$$

$$NV4 = -(\text{GM}) \cdot (CV3) \cdot R_5 = -(0.4) \cdot (CV3) \cdot (1)$$

Note that NV4 is negative because of the dependent current source polarity resulting from the assigned positive-current direction of branch 5.

4. A branch may be used as many times as desired as a *from*- or a *to*-branch.

5. In all ECAP analyses there must be at least one path, through the branches of the circuit, between every node and ground; that is, there can be no "floating" node. In Fig. 2.7, both branch 4 and branch 5, which were added as dummy branches, are connected to the reference node to assure that nodes 3 and 4 have paths to ground. A *dummy branch* is a branch which has no effect on the circuit performance but is added for the sake of the ECAP analysis.

6. Note that CA4 and CA5 in Fig. 2.9 differ (rather insignificantly) from zero because of system inaccuracies.

7. Although dependent voltage sources cannot be specified explicitly, they can be simulated using dependent current sources. (See prob. 2.10.)

2.5 SENSITIVITY COEFFICIENT, WORST CASE ANALYSIS, AND STANDARD DEVIATION CALCULATIONS

The ECAP DC analysis program can be used to calculate and print the sensitivity coefficients, worst case analysis solutions, and standard deviations of all node voltages.*

    • The SENSITIVITY output includes the partial derivative of every node voltage with respect to every circuit parameter as well as

*See Appendix B.1 for further explanations.

the sensitivity of every node voltage to a 1 per cent change in every circuit parameter.

- The WORST CASE analysis output includes the minimum and the maximum node voltage values for every specified node. The worst case node voltage solutions are calculated using minimum and maximum parameter values. The choices of which parameter extremes to use are individually determined by the sign of the partial derivative of the node voltage with respect to each of the parameters.

- The STANDARD DEVIATION printout includes an approximation to the standard deviation of every node voltage.

Worst case analysis and standard deviation solutions will have significance only if one or more of the circuit parameters have minimum or maximum values, or both, which differ from their nominal values. ECAP permits the minimum and maximum values of a parameter to be specified in either of two ways:

a. The algebraic minimum and maximum may be explicitly specified following the nominal value, such as $R = 3(2, 10)$ or $E = -12(-13, -9)$. The minimum value is always the least positive, or most negative, value and appears as the first number in the parentheses following the nominal value.

b. A percentage tolerance may be written following the nominal value such as $E = 24(0.25)$

```
C        SE,WO, AND ST
         DC
B1  N(0,1),R=3(2,10),E=24(.25)
B2  N(2,1),R=4
B3  N(2,0),R=5(.20)
         SENSITIVITY
         WORST CASE,2
         STANDARD DEVIATION
         PR,CA,NV,SE
         EX
```

Fig. 2.10. Sensitivity coefficient, worst case analysis, and standard deviation introduction.

to specify a 25 per cent tolerance on a 24 v nominal voltage. The program then calculates the minimum and maximum values of E as 18 and 30, respectively. If $E = -24(0.25)$ were specified, -24, -30, and -18 would be used as the nominal, minimum, and maximum values of the source, respectively.

The input data in Fig. 2.10 again uses the simple circuit of Fig. 2.2b but adds parameter tolerances and requests sensitivity coefficient, worst case analysis, and standard deviation outputs, as shown in Fig. 2.11.

COMMENTS

1. For any of the ECAP alphabetic mnemonics (like PRINT, EXECUTE, SENSITIVITY, WORST CASE, and all others yet to be introduced) only the first two letters are actually tested by the program. The remaining letters may therefore

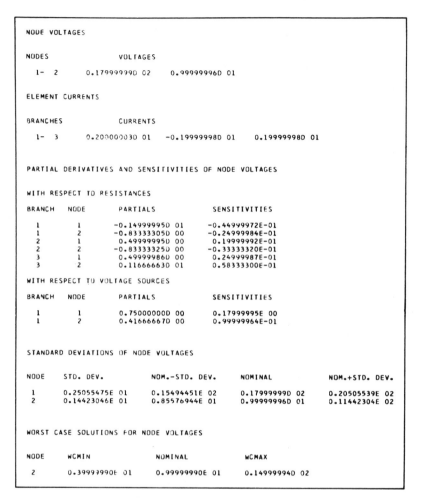

```
NODE VOLTAGES

NODES                    VOLTAGES
  1-  2       0.17999999D 02     0.99999996D 01

ELEMENT CURRENTS

BRANCHES                 CURRENTS
  1-  3       0.20000003D 01    -0.19999998D 01     0.19999998D 01


PARTIAL DERIVATIVES AND SENSITIVITIES OF NODE VOLTAGES

WITH RESPECT TO RESISTANCES

BRANCH   NODE        PARTIALS              SENSITIVITIES
   1       1       -0.14999995D 01        -0.44999972E-01
   1       2       -0.83333305D 00        -0.24999984E-01
   2       1        0.49999995D 00         0.19999992E-01
   2       2       -0.83333325D 00        -0.33333320E-01
   3       1        0.49999986D 00         0.24999987E-01
   3       2        0.11666663D 01         0.58333300E-01

WITH RESPECT TO VOLTAGE SOURCES

BRANCH   NODE        PARTIALS              SENSITIVITIES
   1       1        0.75000000D 00         0.17999995E 00
   1       2        0.41666667D 00         0.99999964E-01


STANDARD DEVIATIONS OF NODE VOLTAGES

NODE    STD. DEV.         NOM.-STD. DEV.      NOMINAL          NOM.+STD. DEV.
  1    0.25055475E 01     0.15494451E 02     0.17999999D 02    0.20505539E 02
  2    0.14423046E 01     0.85576944E 01     0.99999996D 01    0.11442304E 02


WORST CASE SOLUTIONS FOR NODE VOLTAGES

NODE    WCMIN             NOMINAL             WCMAX
  2    0.39999990E 01     0.99999990E 01     0.14999994D 02
```

Fig. 2.11. Sensitivity coefficient, worst case analysis, and standard deviation introduction output format.

be omitted (such as for the EXECUTE card in Fig. 2.10) or pertinent comments may conveniently be added, as shown in many of the upcoming examples (such as on the DC card of Fig. 2.15). The one exception to this rule is BETA, which must be written with all four letters (see Fig. 2.8).
2. The capability of calculating and printing information for fewer than all nodes of a circuit is available only in conjunction with the WO (or WORST CASE) card. If no numbers follow the letters WO, then a worst case analysis will be performed for all node voltages. However, if node numbers are written —each preceded by a comma—following WO, then the worst case node voltages will be calculated and printed only for the specified nodes. For example, in Fig. 2.10, WORST CASE,2 requests that the worst case node voltages be calculated and printed for node 2 only.

3. Worst case analysis and standard deviation solutions will be calculated and printed for any DC problem in which the appropriate WO and ST cards are included. They need not be requested on the PRINT card. The same is true for the output requested by an SE card, except when a WO or ST card or both are also included, in which case SE must be requested via the PRINT card. Thus SE is necessary on the PRINT cards in Figs. 2.10 and 2.17 but not on that in Fig. 2.20.

4. The sequence of the information blocks in the printed output is predetermined by the program and is not affected by the order in which the outputs are requested. Thus in Fig. 2.11 the node voltages are printed before the element currents, and the standard deviations are printed before the worst case solutions, even though CA precedes NV and the WORST CASE card precedes the STANDARD DEVIATION card in Fig. 2.10.

5. BETA, rather than GM, should be specified on any T card whose *from*-branch element has a nonzero tolerance. GM should not be used for these reasons:

    a. The partial derivatives are calculated with respect to BETA even when GM is specified. (See, for example, the partial derivative calculations in Fig. 2.20.)

    b. The trans-term that is entered into the nodal conductance matrix for calculating a worst case node voltage when GM is specified is $\mathrm{GM}_{\mathrm{extreme}} \cdot (R_{\mathrm{from}})_{\mathrm{nominal}}/(R_{\mathrm{from}})_{\mathrm{extreme}}$. Note that the nominal value of R_{from} is used in the numerator to convert $\mathrm{GM}_{\mathrm{extreme}}$ to $\mathrm{BETA}_{\mathrm{extreme}}$. In contrast, the trans-term when BETA is specified is simply $\mathrm{BETA}_{\mathrm{extreme}}/(R_{\mathrm{from}})_{\mathrm{extreme}}$. Note that P_{extreme} is defined as the extreme value (i.e., the minimum or maximum value) of the parameter P.

    c. R_{from} is set to its extreme values based on considerations which use BETA rather than GM as the trans-term. Thus, selecting the extreme value of R_{from} which is to be used for a particular worst case solution effectively sets $\mathrm{CA}_{\mathrm{from}}$ rather than $\mathrm{CV}_{\mathrm{from}}$ to its extreme.

(The above details can be clarified and confirmed by studying the output for MOD.C in Fig. 2.26.)

6. There are three fields in each ECAP statement. Field 1 extends from columns 1 through 5, field 2 includes only column 6, and field 3 contains columns 7 through 72 (as in Fig. 2.12). Field 2 is used for indicating a continuation of the data on the preceding card (see Sec. 4.3). Except for the letter C, which must be written in column 1 to indicate a comment card, all other information in either field 1 or field 3 may be placed anywhere within them. This is true because ECAP disregards blank columns

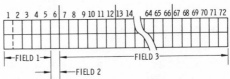

Fig. 2.12. Coding sheet fields.

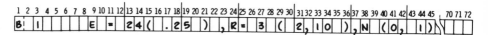

Fig. 2.13. Data format flexibility example.

within each field. For example, the B1 card of Fig. 2.10 could have been written as in Fig. 2.13. Note that the sequence of the branch data groups in columns 7 through 72 is not critical; that is, E may be specified before R, which unconventionally may be specified before the nodal information.

7. The assumptions and techniques used by ECAP in determining the worst case analysis and standard deviation solutions should be understood by users of these features. They are described in detail in Appendix B.1.

2.6   WORST CASE SOLUTIONS AND PARTIAL DERIVATIVE SIGN CHANGES

The underlying assumption (see Appendix B.1) in justifying the worst case solution technique used by ECAP is that the node voltages are monotonic functions of every nonzero-toleranced parameter in the circuit. That is, it is assumed that the signs of the partial derivatives of the node voltages with respect to all parameters do not change over the full range of parameter variations. To determine if this assumption is valid for a particular problem, the program recalculates the partial derivatives after all parameters have been set to their indicated extreme values and compares the signs of these partial derivatives with the signs of the corresponding original partial derivatives. In each case in which the signs do not agree, a message is printed as a warning that true worst case values may not have been calculated.

COMMENTS

1. Some partial derivative sign changes can be ignored, such as those in the following cases:

   a. The partial derivatives should be zero, but numerical inaccuracies in the partial derivative calculations yield small values on opposite sides of zero for the original and the sign-test derivatives.

   b. The effect on the node voltage of a particular parameter variation is very small when compared with the effect on that node voltage of other parameter variations, thereby indicating an insignificant effect due to the incorrect parameter setting.

2. Some partial derivative sign changes cannot be ignored. These cases usually have changes in current directions associated with them, as illustrated by the circuit of Fig. 2.14. A two-step solution to obtain the worst case values in such cases is desirable. For example, instead of the input data in Fig. 2.15, the original B1 card might have contained $E = 24(0, 26)$, while in a modify

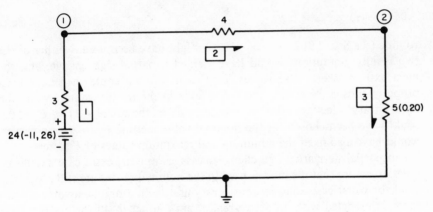

Fig. 2.14. ECAP partial derivative sign change test circuit.

```
        DC  WO AND PARTIAL SIGN CHANGES
    B1  N(0,1),R=3,E=24(-11,26)
    B2  N(1,2),R=4
    B3  N(2,0),R=5(.20)
        SE
        WO
        PR,SE,WO
        EX



PARTIAL DERIVATIVES AND SENSITIVITIES OF NODE VOLTAGES

WITH RESPECT TO RESISTANCES

BRANCH    NODE      PARTIALS            SENSITIVITIES

   1        1       -0.149999950 01     -0.44999972E-01
   1        2       -0.833333050 00     -0.24999984E-01
   2        1        0.499999950 00      0.19999992E-01
   2        2       -0.833333250 00     -0.33333320E-01
   3        1        0.499999860 00      0.24999987E-01
   3        2        0.116666630 01      0.58333300E-01

WITH RESPECT TO VOLTAGE SOURCES

BRANCH    NODE      PARTIALS            SENSITIVITIES

   1        1        0.750000000 00      0.17999995E 00
   1        2        0.416666670 00      0.99999964E-01



WORST CASE SOLUTIONS FOR NODE VOLTAGES


NODE    WCMIN               NOMINAL            WCMAX

  1  PARTIAL W.R.T.  R  1  HAS CHANGED SIGN AT MIN

  1  PARTIAL W.R.T.  R  2  HAS CHANGED SIGN AT MIN

  1  PARTIAL W.R.T.  R  3  HAS CHANGED SIGN AT MIN
  1    -0.79999990E 01     0.17999985E 02      0.200000000 02

  2  PARTIAL W.R.T.  R  1  HAS CHANGED SIGN AT MIN

  2  PARTIAL W.R.T.  R  2  HAS CHANGED SIGN AT MIN

  2  PARTIAL W.R.T.  R  3  HAS CHANGED SIGN AT MIN
  2    -0.40000000E 01     0.99999990E 01      0.120000010 02
```

Fig. 2.15. Partial derivative sign change output.

25

routine (see Sec. 2.9) $E = -5(-11, 0)$ might have been used. Neither of the two resulting solutions would have yielded a partial sign change, and the most extreme worst case values from each solution could then have been properly chosen. Note that the -5 v value in the modify routine would be an arbitrary selection. Alternatively, any E_n in the range $-11 \leq E_n < 0$ v could have been chosen for the nominal value. Setting E_n to zero, however, would have also fixed the minimum and maximum values of E to zero.

3. When partial derivative sign-change messages appear, extra care should be taken to assure that the models used in the nominal ranges are still physically valid for worst case analysis considerations. As an example, assume a transistor is modeled with an active-region BETA equal to 40 for positive base current. If parameter variations cause a reversal of base current (i.e., cause the transistor to cut off), any solution still based on the active BETA will be invalid. The user must check each solution to prevent misinterpretation of the results.

2.7 OUTPUT OPTIONS

Figures 2.16 and 2.17 present a circuit and its input data which request all directly available DC analysis output options. All terms on the PRINT card have been defined with the standard branch in Sec. 2.3 except MISC., which requests the miscellaneous output. This output includes the nodal conductance matrix

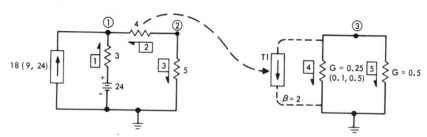

Fig. 2.16. Output options example circuit.

```
        DC  ALL OUTPUT OPTIONS
B1 N(0,1),R=3,E=24,I=-18(-24,-9)
B2 N(2,1),R=4
B3 N(2,0),R=5
B4 N(3,0),G=.25(.1,.5)
B5 N(3,0),G=.5
T1 B(2,4),BETA=2
        SE
        WO,3
        ST
        PR,NV,BV,CV,BA,CA,BP,SE,MISC.
        EX
```

Fig. 2.17. Input data for output options example.

$A_t YA$ (see Appendix B.1.1), the nodal impedance matrix $(A_t YA)^{-1}$, and the equivalent current vector $A_t(I - YE)$. These are the matrices used by ECAP to solve the nodal equations for the node voltage vector e according to

$$(A_t YA)e = A_t(I - YE)$$
$$e = (A_t YA)^{-1}[(A_t YA)e]$$
$$= (A_t YA)^{-1}A_t(I - YE)$$

NODAL CONDUCTANCE MATRIX

ROW	COLS			
1	1 - 3	0.583333310 00	-0.250000000 00	0.0
2	1 - 3	-0.250000000 00	0.444999990 00	0.0
3	1 - 3	-0.500000000 00	0.0	0.749999880 00

EQUIVALENT CURRENT VECTOR

NODE NO.	CURRENT
1	0.259999950 02
2	0.0
3	0.0

NODAL IMPEDANCE MATRIX

ROW	COLS			
1	1 - 3	0.225000010 01	0.125000010 01	0.0
2	1 - 3	0.125000010 01	0.291666680 01	0.0
3	1 - 3	0.666666790 00	-0.111111130 01	0.133333350 01

NODE VOLTAGES

NODES	VOLTAGES		
1 - 3	0.584999690 02	0.324999840 02	0.173333260 02

BRANCH VOLTAGES

BRANCHES	VOLTAGES	
1 - 4	-0.584999690 02	-0.259999850 02	0.324999840 02	0.173333260 02
5 - 5	0.173333260 02

ELEMENT VOLTAGES

BRANCHES	VOLTAGES	
1 - 4	-0.344999690 02	-0.259999850 02	0.324999840 02	0.173333260 02
5 - 5	0.173333260 02

ELEMENT CURRENTS

BRANCHES	CURRENTS	
1 - 4	-0.114999880 02	-0.649999630 01	-0.866666210 01
5 - 5	0.866666210 01

ELEMENT POWER LOSSES

BRANCHES	POWER LOSSES			
1 - 4	0.396749260 03	0.168999810 03	0.211249770 03	-0.150222080 03
5 - 5	0.150222080 03

BRANCH CURRENTS

BRANCHES	CURRENTS	
1 - 4	0.650001110 01	-0.649999630 01
5 - 5	0.866666210 01	-0.866666210 01

PARTIAL DERIVATIVES AND SENSITIVITIES OF NODE VOLTAGES

WITH RESPECT TO RESISTANCES

BRANCH	NODE	PARTIALS	SENSITIVITIES
1	1	0.862498690 01	0.258749540 00
1	2	0.479165950 01	0.143749710 00
1	3	0.255555220 01	0.766665340E-01
2	1	0.162499910 01	0.649999380E-01
2	2	-0.270833180 01	-0.108333230 00
2	3	-0.144444380 01	-0.577773660E-01
3	2	0.162499800 01	0.812498330E-01
3	2	0.379166180 01	0.189583000 00
3	3	-0.144444280 01	-0.722221140E-01
4	1	0.0	0.0
4	2	0.144444300 01	0.577777180E-01
4	3	0.0	0.0
5	1	0.0	0.0
5	2	0.0	0.0
5	3	0.577777450 01	0.115555410 00

WITH RESPECT TO BETAS

BETA	NODE	PARTIALS	SENSITIVITIES
1	1	0.0	0.0
1	2	0.0	0.0
1	3	0.866666300 01	0.173333170E 00

WITH RESPECT TO VOLTAGE SOURCES

BRANCH	NODE	PARTIALS	SENSITIVITIES
1	1	0.750000000 00	0.179999950E 00
1	2	0.416666670 00	0.999999640E-01
1	3	0.222222250 00	0.533333270E-01

WITH RESPECT TO CURRENT SOURCES

BRANCH	NODE	PARTIALS	SENSITIVITIES
1	1	-0.225000010 01	-0.404999850E 00
1	2	-0.125000010 01	-0.224999900E 00
1	3	-0.666666790 00	-0.119999950E 00

Fig. 2.18. Complete DC output data format (continued on next page).

STANDARD DEVIATIONS OF NODE VOLTAGES

NODE	STD. DEV.	NOM.-STD. DEV.	NOMINAL	NOM.&STD. DEV.
1	0.56250000E 01	0.52874954E 02	0.58499969D 02	0.64124954E 02
2	0.31250000E 01	0.29374969E 02	0.32499984D 02	0.35624969E 02
3	0.25469513E 01	0.14786374E 02	0.17333326D 02	0.19880264E 02

WORST CASE SOLUTIONS FOR NODE VOLTAGES

NODE	WCMIN	NOMINAL	WCMAX

NODAL CONDUCTANCE MATRIX

ROW	COLS			
1	1 - 3	0.58333331D 00	-0.25000000D 00	0.0
2	1 - 3	-0.25000000D 00	0.44999999D 00	0.0
3	1 - 3	-0.50000000D 00	0.50000000D 00	0.99999988D 00

EQUIVALENT CURRENT VECTOR

NODE NO.	CURRENT
1	0.16999985D 02
2	0.0
3	0.0

MATRICES FOR NV3$_{min}$

NODAL IMPEDANCE MATRIX

ROW	COLS			
1	1 - 3	0.22500001D 01	0.12500001D 01	0.0
2	1 - 3	0.12500001D 01	0.29166668D 01	0.0
3	1 - 3	0.50000007D 00	-0.83333345D 00	0.10000001D 01

NODAL CONDUCTANCE MATRIX

ROW	COLS			
1	1 - 3	0.58333331D 00	-0.25000000D 00	0.0
2	1 - 3	-0.25000000D 00	0.44999999D 00	0.0
3	1 - 3	-0.50000000D 00	0.50000000D 00	0.59999990D 00

EQUIVALENT CURRENT VECTOR

NODE NO.	CURRENT
1	0.31999985D 02
2	0.0
3	0.0

MATRICES FOR NV3$_{max}$

NODAL IMPEDANCE MATRIX

ROW	COLS			
1	1 - 3	0.22500001D 01	0.12500001D 01	0.0
2	1 - 3	0.12500001D 01	0.29166668D 01	0.0
3	1 - 3	0.83333348D 00	-0.13888891D 01	0.16666669D 01
3	0.84999933E 01	0.17333313E 02	0.26666659D 02	

NV3$_{min}$, NV3, NV3$_{max}$

Fig. 2.18 (*continued*).

Appendix B discusses ECAP solution techniques in detail. Direct usage of some of the MI output is illustrated in Sec. 2.16.

The output in Fig. 2.18 resulting from the above input data illustrates the printout formats. Any uncertainties about output information can be easily verified by hand calculation. Note that the second and third sets of the nodal conductance matrix, equivalent current vector, and nodal impedance matrix correspond to the minimum and maximum solutions, respectively, for NV3.

2.8 EMPHASIZING SIGN CONVENTIONS

The circuit of Fig. 2.19, with its input data and corresponding output of Fig. 2.20, permits the user to study any sign conventions of which he may feel unsure. Without exception the sign conventions illustrated agree with the conventions defined for the complete standard branch in Sec. 2.3.

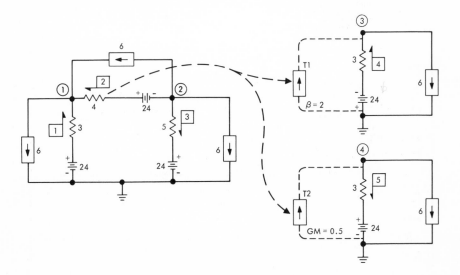

Fig. 2.19. Sign conventions example test circuit.

COMMENTS

1. Voltage sources, independent current sources, and dependent current sources should be drawn schematically with polarities consistent with the physical characteristics of the circuit. If the source polarities, as implied by the assigned positive-current direction in the branch, do not agree with those schematically specified, negative source values should be used in the input data. Thus,

```
          DC  EMPHASIZING SIGN CONVENTIONS
      B1 N(0,1),   R=3,   E= 24,   I= 6
      B2 N(2,1),   R=4,   E= 24,   I=-6
      B3 N(2,0),   R=5,   E=-24,   I=-6
      B4 N(0,3),   R=3,   E=-24,   I= 6
      B5 N(4,0),   R=3,   E=-24,   I=-6
      T1 B(2,4), BETA= 2
      T2 B(2,5), GM  =-0.5
         SE
         PR,NV,CA,MI
         EX


   NODAL CONDUCTANCE MATRIX

   ROW      COLS


      1      1 - 4    0.58333331D 00-0.25000000D 00 0.0          0.0
      2      1 - 4   -0.25000000D 00 0.44999999D 00 0.0          0.0
      3      1 - 4    0.50000000D 00-0.50000000D 00 0.33333331D 00 0.0
      4      1 - 4    0.49999994D 00-0.49999994D 00 0.0          0.33333331D 00


   EQUIVALENT CURRENT VECTOR

   NODE NO.     CURRENT


      1      0.13999999D 02
      2     -0.13200001D 02
      3     -0.19999990D 01
      4      0.13999997D 02


   NODAL IMPEDANCE MATRIX

   ROW      COLS


      1      1 - 4    0.22500001D 01 0.12500001D 01 0.0          0.0
      2      1 - 4    0.12500001D 01 0.29166668D 01 0.0          0.0
      3      1 - 4   -0.15000001D 01 0.25000002D 01 0.30000002D 01 0.0
      4      1 - 4   -0.14999999D 01 0.24999999D 01 0.0          0.30000002D 01


   NODE VOLTAGES

   NODES              VOLTAGES

    1- 4      0.14999997D 02   -0.21000004D 02   -0.60000002D 02   -0.12000004D 02


   ELEMENT CURRENTS

   BRANCHES           CURRENTS

    1- 4      0.30000007D 01   -0.30000003D 01   -0.90000002D 01    0.59999995D 01
    5- 5     -0.60000010D 01
```

Fig. 2.20. Sign conventions example.

because of the assigned positive-current direction of branch 5 in Fig. 2.19, the dependent source (T2) and the two independent sources associated with that branch must be given negative values in the input data, as shown in Fig. 2.20, to agree with their schematically indicated polarities.

2. It should be realized that the assigned positive-current direction of the *to-* branch and the sign of the trans-term determine the effective polarity of the dependent current source.

```
PARTIAL DERIVATIVES AND SENSITIVITIES OF NODE VOLTAGES

WITH RESPECT TO RESISTANCES

BRANCH    NODE      PARTIALS              SENSITIVITIES

  1        1       -0.22499994D 01       -0.67499936E-01
  1        2       -0.12499997D 01       -0.37499983E-01
  1        3        0.14999997D 01        0.44999979E-01
  1        4        0.14999995D 01        0.44999972E-01
  2        1        0.75000007D 00        0.29999994E-01
  2        2       -0.12500001D 01       -0.49999993E-01
  2        3        0.14999998D 01        0.59999976E-01
  2        4        0.14999973D 01        0.59999876E-01
  3        1       -0.22499992D 01       -0.11249989E 00
  3        2       -0.52499980D 01       -0.26249981E 00
  3        3       -0.44999985D 01       -0.22499985E 00
  3        4       -0.44999979D 01       -0.22499985E 00
  4        1        0.0                   0.0
  4        2        0.0                   0.0
  4        3       -0.11999992D 02       -0.35999966E 00
  4        4        0.0                   0.0
  5        1        0.0                   0.0
  5        2        0.0                   0.0
  5        3        0.0                   0.0
  5        4       -0.11999995D 02       -0.35999972E 00

WITH RESPECT TO BETAS

BETA      NODE      PARTIALS              SENSITIVITIES

  1        1        0.0                   0.0
  1        2        0.0                   0.0
  1        3       -0.90000005D 01       -0.17999995E 00
  1        4        0.0                   0.0
  2        1        0.0                   0.0
  2        2        0.0                   0.0
  2        3        0.0                   0.0
  2        4        0.90000005D 01        0.17999983E 00

WITH RESPECT TO VOLTAGE SOURCES

BRANCH    NODE      PARTIALS              SENSITIVITIES

  1        1        0.75000000D 00        0.17999995E 00
  1        2        0.41666667D 00        0.99999964E-01
  1        3       -0.50000001D 00       -0.11999995E 00
  1        4       -0.49999995D 00       -0.11999995E 00
  2        1        0.25000000D 00        0.59999987E-01
  2        2       -0.41666667D 00       -0.99999964E-01
  2        3        0.50000001D 00        0.11999995E 00
  2        4        0.49999995D 00        0.11999995E 00
  3        1       -0.25000000D 00       -0.59999987E-01
  3        2       -0.58333333D 00       -0.13999993E 00
  3        3       -0.50000001D 00       -0.11999995E 00
  3        4       -0.49999995D 00       -0.11999995E 00
  4        1        0.0                   0.0
  4        2        0.0                   0.0
  4        3        0.10000000D 01        0.23999989E 00
  4        4        0.0                   0.0
  5        1        0.0                   0.0
  5        2        0.0                   0.0
  5        3        0.0                   0.0
  5        4       -0.10000000D 01       -0.23999989E 00

WITH RESPECT TO CURRENT SOURCES

BRANCH    NODE      PARTIALS              SENSITIVITIES

  1        1       -0.22500001D 01       -0.13499993E 00
  1        2       -0.12500001D 01       -0.74999988E-01
  1        3        0.15000001D 01        0.89999974E-01
  1        4        0.14999999D 01        0.89999974E-01
  2        1       -0.10000000D 01       -0.59999987E-01
  2        2        0.16666667D 01        0.99999964E-01
  2        3        0.40000003D 01        0.23999995E 00
  2        4        0.39999998D 01        0.23999989E 00
  3        1        0.12500001D 01        0.74999988E-01
  3        2        0.29166668D 01        0.17499995E 00
  3        3        0.25000002D 01        0.14999998E 00
  3        4        0.24999999D 01        0.14999992E 00
  4        1        0.0                   0.0
  4        2        0.0                   0.0
  4        3       -0.30000002D 01       -0.17999995E 00
  4        4        0.0                   0.0
  5        1        0.0                   0.0
  5        2        0.0                   0.0
  5        3        0.0                   0.0
  5        4        0.30000002D 01        0.17999995E 00
```

Fig. 2.20 (*continued*).

2.9 THE MODIFY ROUTINE

A modify routine, defined as a set of ECAP statements starting with MODIFY and ending with EXECUTE, may be used to change the values of up to 50 parameters of any circuit defined by the immediately preceding routine and to request the corresponding new solutions.

The modify routine is best understood by considering the three circuits and input data in Fig. 2.21 and studying the following comments.

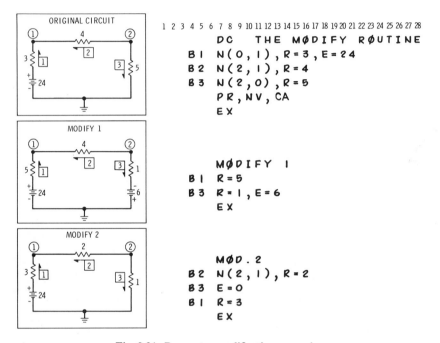

Fig. 2.21. Parameter modification example.

COMMENTS

1. All parameter values used in a modify solution remain unchanged from the previous routine unless explicitly changed, except for the two cases noted in the first and second comments of Sec. 2.11. Thus, in MODIFY1 of Fig. 2.21, $R_2 = 4$, as specified in the original circuit; in MOD.2, $R_3 = 1$, as specified in MODIFY1.

2. Voltage and current sources may be specified to have nonzero values in a modify routine even if they were previously assumed to be zero. As an example, in MODIFY1, E_3 may be set to 6 v even though it was zero in the original circuit.

3. In a modify routine the nodal information of a B card and the branch infor-

```
        MODIFY 1
    B1 R=5
    B3 R=1,E=6
       EX

NODE VOLTAGES

NODES                 VOLTAGES

  1-  2        0.899999984D 01    -0.300000003D 01

ELEMENT CURRENTS

BRANCHES              CURRENTS

  1-  3        0.300000001D 01    -0.299999997D 01    0.299999997D 01


       MOD.2
    B2 N(2,1),R=2
    B3 E=0
    B1 R=3
       EX

NODE VOLTAGES

NODES                 VOLTAGES

  1-  2        0.119999990D 02     0.399999996D 01

ELEMENT CURRENTS

BRANCHES              CURRENTS

  1-  3        0.400000001D 01    -0.399999996D 01    0.399999996D 01
```

Fig. 2.22. Parameter modification output.

mation of a T card are ignored and therefore need not be re-specified. Thus, in MOD.2, the nodal connection information on the B2 card is superfluous. This program characteristic precludes the possibility of changing the circuit topology in the modify routine.

4. The sequence of the B card entries is not critical in a modify routine. Thus, B1 need not precede B2 and B3 in MOD.2.

5. An unlimited number of modify routines can be used in a given problem. A total of only 50 parameters, regardless of type, can be modified in a single routine.

6. To illustrate the modify routine output format the output from MODIFY1 and MOD.2 is shown in Fig. 2.22.

2.10 PARAMETER ITERATION

In any modify routine any single parameter may be varied in equal increments over a specified range of values. Such a variation is called a parameter iteration. The general parameter iteration $P = n_1(n_2)n_3$ causes P to be varied from n_1 to

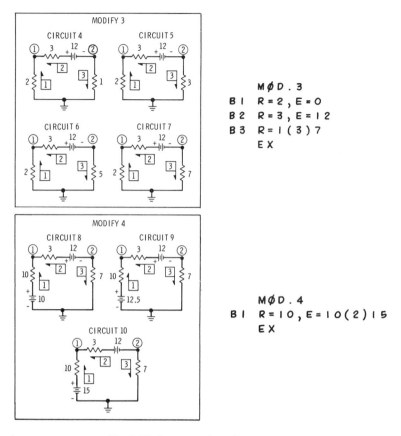

Fig. 2.23. Parameter iteration example.

n_3 in n_2 equal steps of size $(n_3 - n_1)/n_2$; that is, the general iteration P takes on values of

$$n_1, \; n_1 + (1)\left(\frac{n_3 - n_1}{n_2}\right),$$

$$n_1 + (2)\left(\frac{n_3 - n_1}{n_2}\right), \; \cdots, \; n_1 + (n_2 - 1)\left(\frac{n_3 - n_1}{n_2}\right), \; n_3$$

The circuits and corresponding input data in Fig. 2.23 illustrate this feature. (Note that Fig. 2.23 is a continuation of Fig. 2.21.)

COMMENTS

1. In MOD.4 of Fig. 2.23, E_1 successively assumes the values of 10, 12.5, and 15 in accordance with the general parameter iteration formula presented above.
2. A parameter may be iterated only in a modify routine and not in the original circuit description. No more than one parameter may be iterated in any single

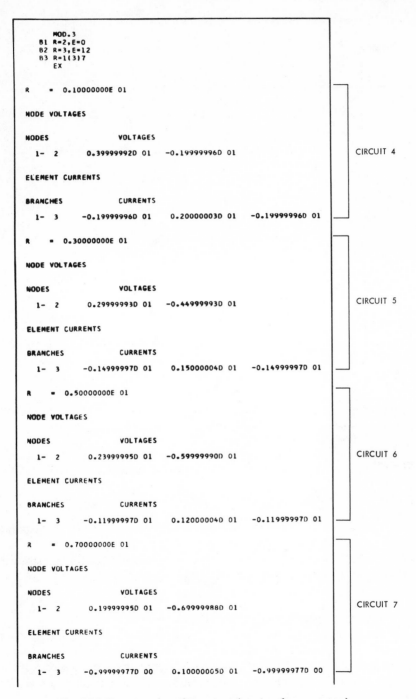

```
        MOD.3
   B1 R=2,E=0
   B2 R=3,E=12
   B3 R=1(3)7
        EX


R    =  0.10000000E 01


NODE VOLTAGES


NODES              VOLTAGES

  1-  2      0.399999920 01   -0.199999960 01


ELEMENT CURRENTS


BRANCHES             CURRENTS

  1-  3    -0.199999960 01    0.200000030 01   -0.199999960 01


R    =  0.30000000E 01


NODE VOLTAGES


NODES              VOLTAGES

  1-  2      0.299999930 01   -0.449999930 01


ELEMENT CURRENTS


BRANCHES             CURRENTS

  1-  3    -0.149999970 01    0.150000040 01   -0.149999970 01


R    =  0.50000000E 01


NODE VOLTAGES


NODES              VOLTAGES

  1-  2      0.239999950 01   -0.599999900 01


ELEMENT CURRENTS


BRANCHES             CURRENTS

  1-  3    -0.119999970 01    0.120000040 01   -0.119999970 01


R    =  0.70000000E 01


NODE VOLTAGES


NODES              VOLTAGES

  1-  2      0.199999950 01   -0.699999880 01


ELEMENT CURRENTS


BRANCHES             CURRENTS

  1-  3    -0.999999770 00    0.100000050 01   -0.999999770 00
```

CIRCUIT 4

CIRCUIT 5

CIRCUIT 6

CIRCUIT 7

Fig. 2.24. Parameter iteration output (*continued on next page*).

```
              MOD.4
        B1 R=10,E=10(2)15
             EX


        E    =  0.10000000E 02

        NODE VOLTAGES

        NODES              VOLTAGES

         1-  2        0.10999999D 02   -0.69999914D 00

        ELEMENT CURRENTS

        BRANCHES           CURRENTS

         1-  3       -0.99999872D-01    0.10000059D 00   -0.99999872D-01


        E    =  0.12500000E 02

        NODE VOLTAGES

        NODES              VOLTAGES

         1-  2        0.12249996D 02    0.17499895D 00

        ELEMENT CURRENTS

        BRANCHES           CURRENTS

         1-  3        0.25000355D-01   -0.24999133D-01    0.24999849D-01


        E    =  0.15000000E 02

        NODE VOLTAGES

        NODES              VOLTAGES

         1-  2        0.13499997D 02    0.10499991D 01

        ELEMENT CURRENTS

        BRANCHES           CURRENTS

         1-  3        0.15000028D 00   -0.14999915D 00    0.14999987D 00
```

Fig. 2.24 (*continued*).

modify routine. The iterated parameter is considered as one of the 50 parameters allowed in any modify routine.
3. The outputs of MOD.3 and MOD.4 are shown in Fig. 2.24 to illustrate the parameter iteration output format.

2.11 SPECIAL PRECAUTIONS WITH MODIFY ROUTINES

To illustrate some additional points and to note some special precautions to be taken when using a modify routine with dependent current sources or a worst case analysis request, Fig. 2.25 is presented.

1 2 3 4 5 6 7 8 9 10 11 12 13 14 15 16 17 18 19 20 21 22 23 24 25

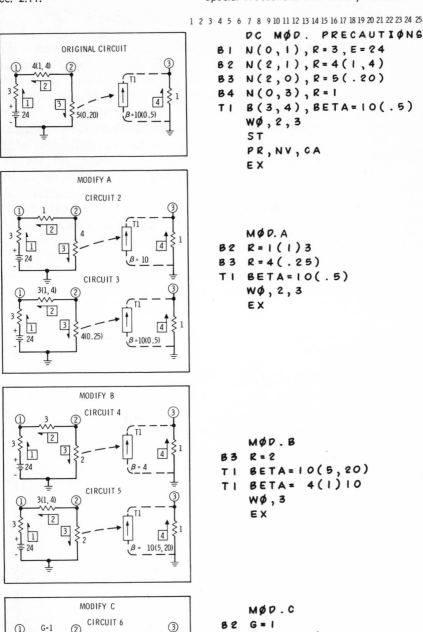

```
DC MØD. PRECAUTIØNS
B1  N(0,1),R=3,E=24
B2  N(2,1),R=4(1,4)
B3  N(2,0),R=5(.20)
B4  N(0,3),R=1
T1  B(3,4),BETA=10(.5)
    WØ,2,3
    ST
    PR,NV,CA
    EX




    MØD.A
B2  R=1(1)3
B3  R=4(.25)
T1  BETA=10(.5)
    WØ,2,3
    EX








    MØD.B
B3  R=2
T1  BETA=10(5,20)
T1  BETA= 4(1)10
    WØ,3
    EX





    MØD.C
B2  G=1
B3  R=4(2,8)
T1  GM=3(2,5)
    WØ,3
    PRINT,BP
    EX
```

Fig. 2.25. Modify routine precautions example.

COMMENTS

1. If a *from*-branch is modified and the corresponding T card specifies a BETA (rather than a GM), the T card must also be re-entered to force the program to recalculate the transconductance. Thus, in MOD.A, although its BETA was not changed, T1 was re-entered because R_3, its *from*-branch resistance, was modified.

2. When a nominal value is modified in a *non*iterated manner, its tolerance is thereafter assumed to be zero unless otherwise specified at the time of the change. In MOD.B, noniterated R_3 has its minimum, nominal, and maximum values all equal to 2 Ω even though it previously had different extreme values specified; however, in MOD.A, R_2—an iterated parameter—maintains (for the last iteration and for successive modify routines) its 1 Ω and 4 Ω extremes.

3. A PRINT card may be included in a modify routine to request additional output types for it and all succeeding modifications. In MOD.C and any other

```
      MOD.A
   B2 R=1(1)3
   B3 R=4(.25)
   T1 BETA=10(.5)
      WO,2,3
      EX


R    =  0.10000000E 01


NODE VOLTAGES


NODES              VOLTAGES
   1-  3      0.14999999D 02    0.11999999D 02    0.29999998D 02


ELEMENT CURRENTS


BRANCHES           CURRENTS
   1-  4      0.30000002D 01   -0.29999998D 01    0.29999998D 01    0.24868996D-13


R    =  0.30000000E 01


NODE VOLTAGES


NODES              VOLTAGES
   1-  3      0.16799999D 02    0.95999991D 01    0.23999998D 02


ELEMENT CURRENTS


BRANCHES           CURRENTS
   1-  4      0.24000002D 01   -0.23999998D 01    0.23999998D 01    0.21316282D-13


WORST CASE SOLUTIONS FOR NODE VOLTAGES


NODE    WCMIN              NOMINAL              WCMAX
   2    0.71999989E 01     0.95999985E 01       0.13333333D 02
   3    0.99999971E 01     0.23999985E 02       0.51428530D 02
```

Fig. 2.26. Modify routine precautions output (*continued on next page*).

```
        MOD.B
    B3  R=2
    T1  BETA=10(5,20)
    T1  BETA= 4(1)10
        WO,3
        EX
BETA =  0.40000000E 01

NODE VOLTAGES

NODES              VOLTAGES

  1-  3      0.14999999D 02    0.59999994D 01    0.11999999D 02

ELEMENT CURRENTS

BRANCHES           CURRENTS

  1-  4      0.30000002D 01   -0.29999997D 01    0.29999997D 01    0.0

BETA =  0.10000000E 02

NODE VOLTAGES

NODES              VOLTAGES

  1-  3      0.14999999D 02    0.59999994D 01    0.29999997D 02

ELEMENT CURRENTS

BRANCHES           CURRENTS

  1-  4      0.30000002D 01   -0.29999997D 01    0.29999997D 01    0.42632564D-13

WORST CASE SOLUTIONS FOR NODE VOLTAGES

NODE    WCMIN              NOMINAL          WCMAX

 3      0.13333332E 02     0.29999985E 02   0.79999993D 02

        MOD.C
    B2  G=1
    B3  R=4(2,8)
    T1  GM=3(2,5)
        WO,3
        PR,BP
        EX
NODE VOLTAGES

NODES              VOLTAGES

  1-  3      0.14999999D 02    0.11999999D 02    0.35999997D 02

ELEMENT CURRENTS

BRANCHES           CURRENTS

  1-  4      0.30000002D 01   -0.29999998D 01    0.29999998D 01    0.71054274D-14

ELEMENT POWER LOSSES

BRANCHES           POWER LOSSES

  1-  4      0.27000006D 02    0.89999985D 01    0.35999994D 02   -0.25579536D-12

WORST CASE SOLUTIONS FOR NODE VOLTAGES

NODE    WCMIN              NOMINAL          WCMAX

 3      0.15999998E 02     0.35999985E 02   0.79999993D 02
```

39

modify routines that may have followed, the element power dissipations—as well as the originally requested node voltages and element currents—will be printed.

4. All output types requested for the original circuit—except the SE, WO, and ST outputs—will be printed for all modified circuits. An SE, WO, or ST output will be printed for a modified circuit only if the appropriate card is included in that modify routine. The SE output must be requested on the PRINT card in the modify routine if a WO or an ST card or both are also included. (See comment 3 in Sec. 2.5.) In the case of an iterated parameter the SE, WO, and ST outputs will be printed only for the last circuit specified by the iteration (see Fig. 2.26). Also notice in MOD.B that the only purpose of the first T1 card is to reset the BETA limits, which are used only after the last iteration specified by the second T1 card; otherwise the limits would have been 5 and 15 as set by BETA = 10(.5) on the T1 card of MOD.A.

5. Resistances and conductances on B cards and BETA and GM values on T cards may be freely interchanged in modify routines. Thus the element in B2 properly may be specified as a G in MOD.C, although it was previously specified as an R; similarly, the trans-term of T1 may be changed from BETA to GM as shown in MOD.C.

6. For an explanation of the way the parameter extremes are selected for the worst case solution of MOD.C, see comment 5 in Sec. 2.5.

2.12   NODAL CURRENT UNBALANCES (IERROR)

Each ECAP current calculation depends on the difference between two calculated voltages. Since these voltages, in general, are themselves limited in precision (see comment 2 in Sec. 2.2), the resulting currents also are not precise. As a warning of any possible gross inaccuracies, the sum of the currents flowing into each node (which according to Kirchhoff's law should be zero) is calculated, and a check is made of the sum of the absolute values of each of these calculations. If this final sum is less than the IERROR (assumed by the program to be 0.001 amp unless explicitly specified otherwise), no special messages are written. However, if the sum is equal to or greater than the IERROR, a warning message and the current unbalance at each node are printed. A restatement of the problem, generally centered on raising the ratio of the smallest impedance at each node with a large current unbalance to the next smallest impedance at each such node, is all that is usually required to reduce the current unbalances. See Fig. 2.27.

COMMENTS

1. As a general rule, current unbalances will not be a problem if the smallest impedance at every node is not more than about six or seven orders of magnitude smaller than the next smallest impedance at that node. In addition, if

```
DC IERROR
B1 N(0,1),R=3,E=24
R2 N(1,2),R=4E-12
R3 N(2,0),R=5
IERROR=0
PR,NV,MI
EX

NODAL CONDUCTANCE MATRIX

ROW    COLS

1    1 - 2    0.24999998D 12-0.24999998D 12
2    1 - 2   -0.24999998D 12 0.24999998D 12

EQUIVALENT CURRENT VECTOR

NODE NO.    CURRENT

1    0.79999990D 01
2    0.0

NODAL IMPEDANCE MATRIX

ROW    COLS

1    1 - 2    0.18749750D 01 0.18749750D 01
2    1 - 2    0.18749750D 01 0.18749750D 01

NODE VOLTAGES

                  VOLTAGES
NODES
1- 2    0.14999798D 02    0.14999798D 02

SOLUTION NOT OBTAINED TO DESIRED TOLERANCE

              CURRENT UNBALANCES
NODES
1- 2   -0.19931930D-03    0.30708306D-03

BRANCH CURRENTS

                  CURRENTS
BRANCHES
1- 3    0.30000672D 01    0.29999594D 01
```

```
            MODIFY
B2 R=4E-3
EX

NODAL CONDUCTANCE MATRIX

ROW    COLS

1    1 - 2    0.25033352D 03-0.25000018D 03
2    1 - 2   -0.25000018D 03 0.25020018D 03

EQUIVALENT CURRENT VECTOR

NODE NO.    CURRENT

1    0.79999990D 01
2    0.0

NODAL IMPEDANCE MATRIX

ROW    COLS

1    1 - 2    0.18755623D 01 0.18740631D 01
2    1 - 2    0.18740631D 01 0.18766618D 01

NODE VOLTAGES

                  VOLTAGES
NODES
1- 2    0.15004497D 02    0.14992503D 02

SOLUTION NOT OBTAINED TO DESIRED TOLERANCE

              CURRENT UNBALANCES
NODES
1- 2    0.47683732D-06   -0.84154905D-13

BRANCH CURRENTS

                  CURRENTS
BRANCHES
1- 3    0.29985009D 01    0.29985004D 01
```

Fig. 2.27. Current unbalance example.

the branch with the smallest impedance at the node involved has its other side at node 0, there is even less likelihood of trouble.

2. The 1ERROR was specified as zero in Fig. 2.27 to guarantee a current unbalance message for illustration. Note how the current unbalances were reduced for the modified circuit when the impedance ratios R_2/R_1 and R_2/R_3 were increased (by increasing R_2).

3. Comparison of the terms of the nodal conductance matrices of the original and modified circuits of Fig. 2.27 further illustrates the reasons for the computational inaccuracies which lead to current unbalances. Note that in the original circuit the extreme impedance ratios result in a nodal conductance matrix with terms indistinguishable from one another. Certainly in such an instance, current calculations—which are dependent on this matrix—cannot be relied on.

```
            DC  ERROR MESSAGES
        B1  N(0,2),R=1, -E=10


    ERROR NO. 20
    CARD NO.=  2    APPROXIMATE COLUMN NO. IS 18
        B1 N(0,2),R=1,-E=10



        B2 N(5,4),G=4, R=6


    ERROR NO. 22
    CARD NO.=  3    APPROXIMATE COLUMN NO. IS 18
        B2 N(5,4),G=4,R=6



        B4 N(2,5),I=6
        B5 N(4,2),C=8

    ERROR NO. 33
    CARD NO.=  5    APPROXIMATE COLUMN NO. IS 14
        B5 N(4,2),C=8



        B6 N(0,5),L=3


    ERROR NO. 33
    CARD NO.=  6    APPROXIMATE COLUMN NO. IS 14
        B6 N(0,5),L=3



        B7 N(5,2),R=1
        PRNT,NO,BC


    ERROR NO. 11
    CARD NO.=  8    APPROXIMATE COLUMN NO. IS 12
        PRNT,NO,BC



        EXCTL
    INPUT ERRORS MAKE EXECUTION IMPOSSIBLE.
      5  ERROR(S) WERE DETECTED.
```

Fig. 2.28. Error messages example.

2.13 ERROR MESSAGES

Figure 2.28 presents a sample of DC analysis input data with several input errors, and illustrates ECAP's response to these errors. Note that all errors are not always detected by the program. For example, the errors of omission of branch 3 and nodes 1 and 3 in the input data of Fig. 2.28 would have been detected and appropriate messages would have been printed only if there were no other errors.

Appendix C.7 contains a complete listing of ECAP error messages.

2.14 A BUFFERING TECHNIQUE

Suppose node 2 of the simple circuit in Fig. 2.29 is to be used as a constant voltage source which maintains its terminal voltage independent of the load, and

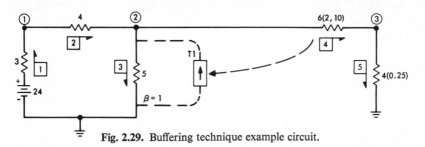

Fig. 2.29. Buffering technique example circuit.

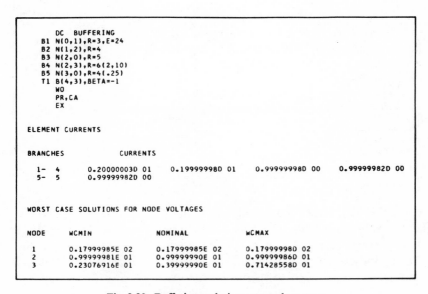

```
         DC   BUFFERING
     B1  N(0,1),R=3,E=24
     B2  N(1,2),R=4
     B3  N(2,0),R=5
     B4  N(2,3),R=6(2,10)
     B5  N(3,0),R=4(.25)
     T1  B(4,3),BETA=-1
         WO
         PR,CA
         EX


  ELEMENT CURRENTS


  BRANCHES                CURRENTS


     1-  4       0.20000003D 01     0.19999998D 01     0.99999998D 00     0.99999982D 00
     5-  5       0.99999982D 00



  WORST CASE SOLUTIONS FOR NODE VOLTAGES


  NODE     WCMIN              NOMINAL            WCMAX

     1     0.17999985E 02     0.17999985E 02     0.17999998D 02
     2     0.99999981E 01     0.99999990E 01     0.99999986D 01
     3     0.23076916E 01     0.39999990E 01     0.71428558D 01
```

Fig. 2.30. Buffering technique example output.

assume that a two-resistor network consisting of $R_4 = 6\,\Omega$ and $R_5 = 4\,\Omega$ constitutes the load at node 2. The terminal voltage of the node can be maintained regardless of load variation by feeding back into the node any current drawn out of it by the load. This can be accomplished by using a dependent current source with a BETA equal to one. Any current drawn from node 2 by branch 4 in the circuit of Fig. 2.29 is returned to the node by the dependent current source T1 across branch 3. Note that a negative BETA is required on the T card because of the assigned positive current direction of branch 3.

The output data illustrating this buffering, or unloading, technique are shown in Fig. 2.30. Note that the worst case voltages of node 2 caused by variations in branches 4 and 5 are maintained very close to the nominal value.

2.15 WORST CASE BRANCH VOLTAGES

The example in Figs. 2.31 and 2.32 illustrates a technique for determining the worst case values of a branch voltage BV2 of a circuit comprised of branches 1,

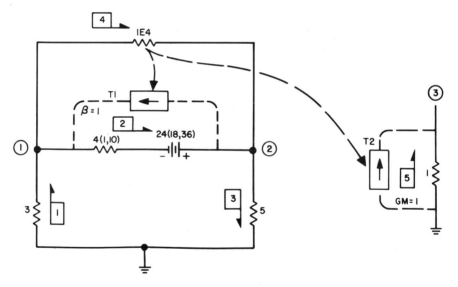

Fig. 2.31. Worst case branch voltage test circuit.

2, and 3. Note that node 3 is used only as a dummy whose node voltage is always equal to BV2. Also note that, by using the buffering technique described in Sec. 2.14, T1 prevents branch 4 from exhibiting any loading effect on the circuit.

Sensitivity coefficient and standard deviation information calculated for node 3 are also equivalent to the values for BV2.

```
          DC   WOBV
   B1 N(0,1),R=3
   B2 N(1,2),R=4(1,10),E=24(18,36)
   B3 N(2,0),R=5
C
   B4 N(1,2),R=1E4
   T1 B(4,2),BETA=-1
C
   B5 N(0,3),R=1
   T2 B(4,5),GM=1
      WO,3
      PR,BV
      EX


BRANCH VOLTAGES


BRANCHES              VOLTAGES

   1-  4      0.60000001D 01    -0.16000000D 02    0.10000000D 02    -0.16000000D 02
   5-  5      0.16000000D 02


WORST CASE SOLUTIONS FOR NODE VOLTAGES

NODE    WCMIN              NOMINAL            WCMAX

  3      -0.32000000E 02    -0.16000000E 02    -0.79999981D 01
```

Fig. 2.32. Worst case branch voltage output.

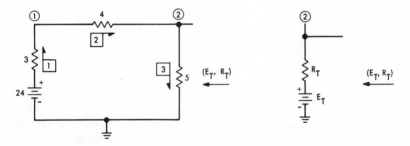

Fig. 2.33. Thévenin equivalent circuit example.

2.16 THÉVENIN EQUIVALENT CIRCUITS

Suppose the Thévenin equivalent circuit at node 2 of the circuit in Fig. 2.33 is desired. That is, E_T and R_T of the Thévenin equivalent which has the same terminal characteristics as node 2 of the example circuit are to be determined. Node voltages and the miscellaneous output provide all the necessary information, as shown in Fig. 2.34.

COMMENTS

1. The circuit impedance (Thévenin impedance) between the ith node and the reference node is equal to the ith diagonal element of the nodal impedance

```
            DC  THEVENIN
      B1 N(0,1),R=3,E=24
      B2 N(1,2),R=4
      B3 N(2,0),R=5
         PR,NV,MI
         EX


 NODAL CONDUCTANCE MATRIX

 ROW       COLS


    1      1 - 2    0.58333331D 00-0.25000000D 00
    2      1 - 2   -0.25000000D 00 0.44999999D 00


 EQUIVALENT CURRENT VECTOR

 NODE NO.     CURRENT


    1     0.79999990U 01
    2     0.0


 NODAL IMPEDANCE MATRIX

 ROW       COLS


    1      1 - 2    0.22500001D 01 0.12500001D 01
    2      1 - 2    0.12500001D 01 0.29166668D 01  R_T


 NODE VOLTAGES


 NODES            VOLTAGES

  1-  2     0.17999999D 02    0.99999996D 01  E_T
```

Fig. 2.34. Thévenin equivalent circuit data.

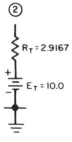

Fig. 2.35. Thévenin equivalent circuit results.

matrix, which is included with the miscellaneous output in a DC analysis. Thus, R_T for node 2 equals $Z_{22} = 2.9167 \ \Omega$.

2. The Thévenin equivalent voltage between the ith node and the reference node is equal to NVi. Thus, E_T equals NV2 = 10.0 v.

3. The desired Thévenin equivalent is shown in Fig. 2.35.

2.17 SUMMARY

This chapter has presented the information necessary for the optimal usage of ECAP's DC analysis program. The ease with which worst case analysis solutions can be obtained and circuit parameters can be modified is among the outstanding features of the program.

These points should be remembered to reduce the chance of trouble when using the DC analysis program:

1. Capacitances and inductances are not legitimate DC input parameters.
2. BP is *element* power dissipation.
3. The element current, as printed, is

$$\text{C}Ab = \text{C}vb \cdot Yb + \sum_{j \neq b} \text{C}vj \cdot \left(\text{GM}bj + \frac{\text{BETA}bj}{Rj} \right)$$

That is, the printed element current includes any effects of its being a *to*-branch.
4. BETA, rather than GM, should be specified on any T card whose *from*-branch element has a nonzero tolerance.
5. If the nominal value of a source is zero, no partial derivatives will be found with respect to it. Its extreme values therefore will also be zero, even if specified otherwise.
6. The circuit topology cannot be directly changed in a modify routine.
7. A parameter iteration may appear only in a modify routine.
▷ 8. If a *from*-branch is modified and the corresponding T card contains a BETA, the T card then must also be re-entered.
9. Previously specified extreme values of a parameter are nullified in a modify routine in which the nominal value of the parameter has been re-entered in a noniterated manner.
10. SE, WO, or ST outputs will be printed in a modify routine only if the corresponding card is included in the modify routine, and even then only for the last of the iterated solutions.

Appendix C includes tables which summarize the legitimate DC input parameters, directly available output data, assumed units, and overall program capacities.

Appendix D contains tables which summarize the allowable input statement formats for ECAP.

PROBLEMS

2.1. Using ECAP, determine the power dissipated by the 0.01 Ω internal resistance of a 24 v battery connected to a load consisting of a 47 Ω resistor in series with a coil whose DC resistance is 100 Ω.

2.2. Do the following:
a. Write the ECAP input data describing the circuits in Fig. p2.1.
b. Compare the node voltages calculated by ECAP at nodes 2, 3, and 4 of the circuits shown in Fig. p2.1 and explain the reason(s) for the difference in the magnitudes of the element currents CA3 and CA5.

2.3. Find the error(s) in the input data in Fig. p2.2.

2.4. Construct equivalent circuits and perform ECAP analyses to obtain

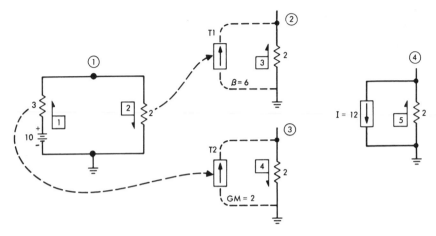

Fig. p2.1

```
    DC
B1  N(1,2),R=3.,E=2400E-02
B2  N(2,3),R=.004E3
B3  N(3,1),G=.2
    PRNT,CA,NV
    EX
```

Fig. p2.2

solutions which have *nominal* NV3 values equal to the calculated worst case minimum and maximum values for circuit 5 in Fig. 2.25. Repeat for circuit 6, using equivalent BETA's (instead of the specified GM's) to obtain the nominal NV3 values equal to the calculated worst case minimum and maximum values.

2.5. Using a minimum number of ECAP statements, determine NV2 for the original and modified circuits in Fig. p2.3.

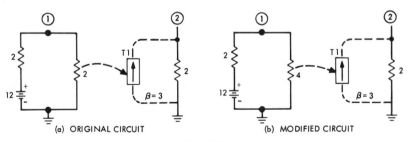

Fig. p2.3

2.6. Determine the nominal and the worst case node voltage values for the circuits shown in Fig. p2.4 using ECAP. Use a minimum number of statements.

2.7. Obtain the MI output for the circuits in Fig. p2.5 and comment on the factors (as illustrated by the nodal admittance matrices) which lead to the current unbalance problem.

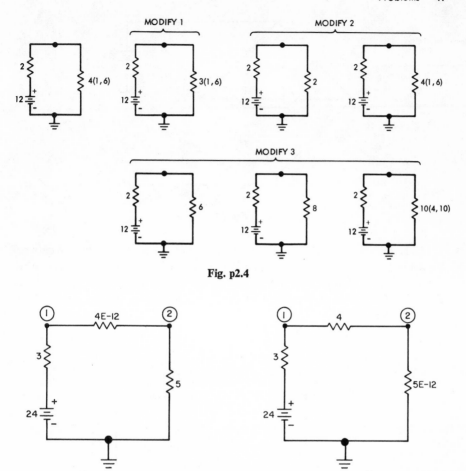

Fig. p2.4

Fig. p2.5

2.8. Using the technique described in comment 2 in Sec. 2.6, determine the extreme values of CA3 for the circuit in Fig. 2.14.

2.9. Determine the extreme values of CV2 in Fig. p2.6. Use the approach illustrated in Fig. 2.31.

2.10. What modifications must be made to the input data in Fig. 2.17 to construct a dependent voltage source, connected in series with the 0.5 ℧ conductance in branch 5, whose value is four times the voltage across branch 2 and whose negative terminal is connected to ground?

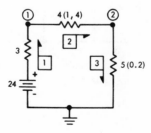

Fig. p2.6

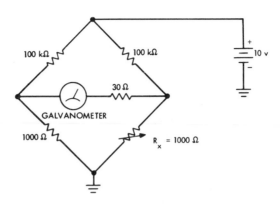

Fig. p2.7

2.11. Determine the galvanometer sensitivity (μa/ohm) in the bridge network shown in Fig. p2.7 to a small change in the value of R_x using the element values specified in the figure. Plot the galvanometer current as a function of R_x for $R_x = 100\ \Omega$ to 2000 Ω. Assume that the internal resistance of the galvanometer is negligible.

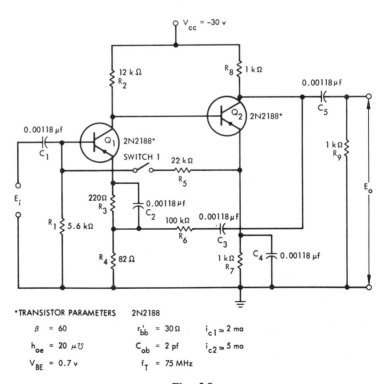

Fig. p2.8

2.12. Determine the following for the circuit shown in Fig. p2.8 with switch 1 closed:

  a. the ECAP equivalent circuit

  b. the DC operating point voltages and currents including the collector and emitter voltages for transistors Q_1 and Q_2

  c. the DC gain of the circuit from the base of transistor Q_1 to the collector of transistor Q_2

  d. the worst case collector and emitter voltages for the transistor in the circuit using the following assumptions:

    1. all component resistor tolerances $= \pm 5$ per cent

    2. transistor BETA (h_{fe}) tolerances $= \pm 30$ per cent

    3. transistor base-emitter voltage (V_{be}) tolerance $= \pm 0.1$ v

    4. transistor input resistance (h_{ie}) tolerance $= \pm 0$ per cent

    5. transistor output conductance (h_{oe}) tolerance $= \pm 5$ per cent

    6. power supply tolerance $= \pm 5$ per cent

  e. Determine which circuit parameters have the greatest effect on the collector voltage of transistor Q_2. Assuming the element voltages are linear functions of the circuit elements, re-specify the tolerances of the most sensitive elements to reduce the worst case collector voltage of transistor Q_2 to less than ± 10 per cent of its nominal value.

  f. Determine the Thévenin equivalent network for the circuit at the base of transistor Q_1 (input resistance and equivalent voltage).

A DC equivalent circuit for the transistor can be found in Appendix A.1.

2.13. Repeat parts (a) and (b) of problem 2.12 with switch 1 open. Assume the collector-emitter resistance r_{ce} of each transistor is 20 Ω and the value of h_{fe} is 0 when the transistor is saturated.

3

AC
ANALYSIS

The AC analysis program provides steady-state solutions to the analysis of linear circuits subject to sinusoidal excitation. This chapter explains how to use the program.

3.1 AC ANALYSIS CODING INTRODUCTION

The coding techniques used for analysis of AC circuits are very similar to those for DC circuit analysis. As an example, consider the simple *RLC* circuit, the corresponding AC analysis input data which request all the available AC analysis output information, and the resulting output in Figs. 3.1, 3.2, and 3.3.

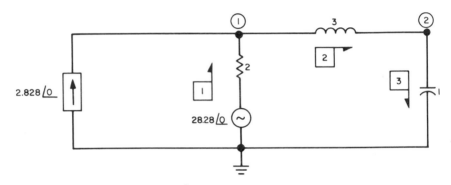

Fig. 3.1. *RLC* circuit.

			1 2 3 4 5 6	7 8 9 10 11 12	13 14 15 16 17 18	19 20 21 22 23 24	25 26 27 28 29 30	31 32 33 34 35 36	37 38 39

```
1 2 3 4 5 6| 7  8  9 10 11 12|13 14 15 16 17 18|19 20 21 22 23 24|25 26 27 28 29 30|31 32 33 34 35 36|37 38 39
         A C   C Ø D I N G     I N T R Ø D U C T I Ø N
   B 1   N ( O , 1 ) , R = 2 , E = 2 8 . 2 8 / O , I = 2 . 8 2 8 / 1 8 O
   B 2   N ( 1 , 2 ) , L = 3
   B 3   N ( 2 , O ) , C = 1
         F R E Q U E N C Y = . 1 5 9
         P R I N T , N V , B V , C V , B A , C A , B P , M I
         E X E C U T E
```

Fig. 3.2. AC coding introduction.

COMMENTS

1. The AC analysis program provides all the analysis capabilities available in the DC program except the SENSITIVITY, WORST CASE, and STANDARD DEVIATION solutions.
2. The AC analysis coding rules are the same as the DC analysis rules except for the following differences:

```
        AC CODING INTRODUCTION
   B1 N(0,1),R=2,E=28.28/0,I=2.828/180
   B2 N(1,2),L=3
   B3 N(2,0),C=1
      FREQUENCY=.159
      PRINT,NV,BV,CV,BA,CA,BP,MI
      EXECUTE


      ROW     COL
      NODE    NODE            NODAL ADMITTANCE MATRIX


REAL  1    1- 2  0.500000000 00   0.0
IMAG             -0.333658160 00   0.333658160 00
REAL  2    1- 2  0.0              0.0
IMAG             0.333658160 00   0.665367960 00


     NODES            EQUIVALENT CURRENT VECTOR


REAL  1- 2  0.169679870 02   0.0
IMAG        -0.450119840-05   0.0


FREQ =   0.15899992E 00



     NODES            NODE VOLTAGES


MAG    1-  2  0.23972946E 02   0.12021576E 02
PHA           0.45055832E 02  -0.13494415E 03


     BRANCHES           BRANCH VOLTAGES


MAG    1-  3  0.23972946E 02   0.35994522E 02   0.12021576E 02
PHA          -0.13494415E 03   0.45055832E 02  -0.13494415E 03


     BRANCHES           ELEMENT VOLTAGES


MAG    1-  3  0.20411316E 02   0.35994522E 02   0.12021576E 02
PHA          -0.56232559E 02   0.45055832E 02  -0.13494415E 03


     BRANCHES           ELEMENT CURRENTS


MAG    1-  3  0.10205664E 02   0.12009869E 02   0.12009869E 02
PHA          -0.56232559E 02  -0.44944153E 02  -0.44944153E 02


     BRANCHES           BRANCH POWER

MAG    1-  3  0.20831117E 03   0.0                0.0


     BRANCHES           BRANCH CURRENTS


MAG    1-  3  0.12009876E 02   0.12009869E 02   0.12009869E 02
PHA          -0.44944127E 02  -0.44944153E 02  -0.44944153E 02
```

Fig. 3.3. AC analysis output introduction.

a. A FREQUENCY card, or FR card (see comment 1 of Sec. 2.5), must be included with every AC analysis problem and must specify a nonzero frequency in Hertz (cps). In Fig. 3.2 the choice of 0.159 Hz corresponds to an angular frequency ω of approximately 1 radian per sec, which simplifies any calculations that the reader may make for verification.

b. As for DC input data, a branch must contain one, and only one, passive element, but of course inductances and capacitances are permitted AC elements. Thus, in Fig. 3.2, the inductance value (which we refer to as L_2) in branch 2 is specified as 3 henries, and the capacitance value (which we refer to as C_3) in branch 3 is specified as 1 farad. (See Sec. 3.3 for magnetic coupling techniques.)

c. If an L or a C is the *from*-branch specified on a T card, that T card must contain a GM (not a BETA). An L or a C may be used in any *to*-branch.

d. Every source must have a phase angle specified following the source magnitude and a slash (/). The program assumes that all angles are specified in degrees relative to a 0° reference. Note that in Fig. 3.2 the angle of E in B1 is specified even though it is zero, and the current source in B1 could have been written alternatively as $I = -2.828/0$, or as $I = 2.828/-540$.

3. The program prints all voltage and current phases in degrees between ±180°. For example, $-315°$ and $+225°$ angles will be printed as $+45°$ and $-135°$ angles, respectively.

4. The complete AC analysis standard branch is very similar to the DC analysis standard branch described in Sec. 2.3. The exceptions are:

a. For an AC analysis, the passive element may be a resistance, a capacitance, or an inductance with a finite admittance magnitude. That is,

$$Y = \frac{1}{R} \quad \text{or} \quad 2\pi fC \quad \text{or} \quad \frac{1}{2\pi fL}$$

where $f = $ frequency, as specified by the FREQUENCY card. Negative element values may be specified.

b. The independent voltage and current sources, for which the magnitude and phase must be specified, are sinusoidal.

c. All blocks of output information are printed in magnitude and phase except for the element power dissipations BP, which are scalar quantities, and the MISC output, for which the real and imaginary components of the complex nodal admittance matrix and equivalent current vector are printed. (The nodal impedance matrix is not used in the AC analysis program and is not available in the analysis output.)

d. Except where element power dissipations are requested, either rms or peak values may be assumed for the independent source magnitudes. The element power dissipation is calculated from

$$\text{BP} = \text{Real} \, [\text{CV} \cdot \overline{\text{CA}}]$$

where $\overline{\text{CA}}$ = complex conjugate of CA; the calculation is valid only when rms values are used. In all cases the output voltages and currents have the units assumed for the inputs.

3.2 PARAMETER MODIFICATION AND FREQUENCY RESPONSE CALCULATIONS

Modify routines may be used in an AC analysis just as they are used in a DC analysis. The parameter iteration feature also makes it possible to perform frequency "sweeps" in an AC analysis. The input data of Fig. 3.4 illustrates the use of the AC analysis parameter modification capability. Portions of the resulting output from both element and frequency iterations are shown in Fig. 3.5.

COMMENTS

The rules for making AC analysis modifications are the same as those for making DC modifications with the following rules added:

```
AC FREQ. ITERATIONS
R1  N(0,1),R=2,F=28.28/0,I=2.828/180
R2  N(1,2),L=3
R3  N(2,0),C=1
    FREQ=.159
    PR,NV
    FX
    MOD.1
R1  I=0/0,F=14.14(2)28.28/-180(2)0
    FX
    MOD.2
    FREQ=.01(1.1).08
    FX
    MOD.3
    FREQ=.085(+15).10
    FX
```

Fig. 3.4. AC parameter modification example.

a. Source magnitude iterations must always be accompanied by source phase iterations with the same number of steps. For example, consider the parameter iteration in MOD.1 in Fig. 3.4. Since the magnitude of the source is specified to go in two steps from the lower limit 14.14 v to the upper limit 28.28 v, the associated phase angle must also go in two steps between the limits $-180°$ and $0°$. Even if the phase angle remains unchanged, this rule must be observed. Thus, if the source in MOD.1 is to maintain a $0°$ phase angle for all magnitudes, it should be written $E = 14.14(2)28.28/ 0(2)0$. Since a negative sign before an independent source magni-

MOD.3
FREQ=.085(+15).IC
EX

FREQ = 0.84999919E-01

NODES   NODE VOLTAGES

MAG   1- 2   0.47327862E 01   0.32796906E 02
PHA   0.76939993E 01   0.76939993E 01

FREQ = 0.85999906E-01

NODES   NODE VOLTAGES

MAG   1- 2   0.40313225E 01   0.32496643E 02
PHA   0.65482006E 01   0.65482006E 01

FREQ = 0.86999893E-01

NODES   NODE VOLTAGES

MAG   1- 2   0.33337326E 01   0.32189957E 02
PHA   0.54113293E 01   0.54113293E 01

MOD.2
FREQ=.01(1.1).08
EX

FREQ = 0.99999942E-02

NODES   NODE VOLTAGES

MAG   1- 2   0.35067551E 02   0.35487854E 02
PHA   0.82752487E 02   0.82752487E 02

FREQ = 0.10999981E-01

NODES   NODE VOLTAGES

MAG   1- 2   0.35007401E 02   0.35516373E 02
PHA   0.82016800E 02   0.82016800E 02

FREQ = 0.12099974E-01

NODES   NODE VOLTAGES

MAG   1- 2   0.34934219E 02   0.35550674E 02
PHA   0.81203949E 02   0.81203949E 02

MOD.1
B1 1=0/0,E=14.14(2)28.28/-180(2)10
EX

E   = 0.14139999E 02 / -0.18000000E 03

FREQ = 0.15899992E 00

NODES   NODE VOLTAGES

MAG   1- 2   0.99887381E 01   0.50089931E 01
PHA   -0.13494405E 03   0.45055939E 02

E   = 0.21209991E 02 / -0.90000000E 02

FREQ = 0.15899992E 00

NODES   NODE VOLTAGES

MAG   1- 2   0.14983102E 02   0.75134878E 01
PHA   -0.44944061E 02   0.13505589E 03

F   = 0.28279984E 02 / 0.0

FREQ = 0.15899992E 00

NODES   NODE VOLTAGES

MAG   .1- 2   0.19977663E 02   0.10017982E 02
PHA   0.45055847E 02 -0.13494409E 03

E   = 0.35349976E 02 / 0.90000000E 02

FREQ = 0.15899992E 00

NODES   NODE VOLTAGES

MAG   1- 2   0.24971617E 02   0.12522476E 02
PHA   0.13505573E 03 -0.44944214E 02

Fig. 3.5. AC parameter modification output.

tude causes an effective 180° shift in phase, the voltage source E in B1 of MOD.1 could have been written alternatively as $E = -28.28(2) - 14.14/180(2)0$.

b. Either a multiplicative or an additive frequency iteration may be included in any modify routine which does not include any other parameter iteration. The multiplicative frequency step used for logarithmic frequency response calculations is requested by using a FREQUENCY card of the general form FREQ $= L(M)U$ where L is the lower frequency limit, U is the upper frequency limit, and M is a multiplication factor. Such a FREQ card requests analyses at frequencies L, ML, M^2L, ..., T where $T \geq U$. For example, the solutions in MOD.2 of Fig. 3.5 are obtained at frequencies 0.0100, 0.0110, 0.0121, ...

The additive frequency step used for linear frequency response calculations is requested by using a FREQUENCY card of the general form FREQ $= L(+M)U$ where $+M$ is the number of frequency steps required excluding the lower frequency limit L. In this case the FREQ card requests analyses at frequencies L, $L + [(U - L)/M]$, $L + 2[(U - L)/M]$, ..., U. For example, the solutions in MOD.3 of Fig. 3.5 are obtained at frequencies 0.085, 0.086, 0.087, ..., 0.099, 0.100.

The presence or absence of a plus ($+$) sign before the variable M in parentheses on the FREQUENCY card determines the type of frequency iteration to be used.

3.3 INDUCTIVELY COUPLED CIRCUITS (M CARDS)

A mutual inductance existing between inductively coupled circuits may be specified directly on an M card. For example,

$$\text{M1 B(4, 7), L} = 0.5\text{E}{-}3$$

indicates that the inductances in branches 4 and 7 have a mutual inductance of 0.5 mh. When the inductance value L on the M card is positive, the initial node of each of the two coupled inductances is considered to have the *dot**; when the mutual inductance L is specified with a negative sign, the dot is assumed to be at the final node of one of the two inductances (which effectively reverses the secondary-current direction).

As an example of the use of the M card, consider the magnetic circuit in Fig. 3.6a. The circuit is represented by the electrical equivalent network in Fig. 3.6b, including the dot relationships which account for flux linkage and induced-

---

*The transformer *dot* convention states that current flowing into the dotted end of the primary winding produces a voltage rise at the dotted end of the secondary winding.

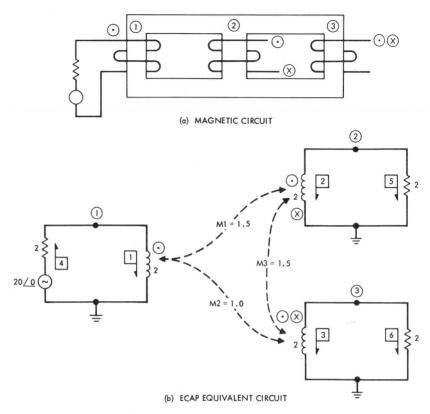

(a) MAGNETIC CIRCUIT

(b) ECAP EQUIVALENT CIRCUIT

Fig. 3.6. Inductively coupled circuit.

current directions. The input data for the electrical equivalent of the magnetic circuit are shown in Fig. 3.7.

```
AC   MUT. INDUCTANCES
B1  N(1,0),L=2
B2  N(2,0),L=2
B3  N(3,0),L=2
B4  N(0,1),R=2,E=20/0
B5  N(2,0),R=2
B6  N(3,0),R=2
M1  B(1,2),L=1.5
M2  B(1,3),L=1.0
M3  B(2,3),L=-1.5
    FR=.159
    PR,NV
    EX
```

Fig. 3.7. Inductively coupled circuit input data.

COMMENTS

1. The order of the inductively coupled branches specified on an M card is of no consequence.
2. The combination of the assigned branch current directions and the mutual-inductance sign determine the relative positioning of the dots for each pair of inductances in a coupled circuit. Thus an alternative set of input data, completely

consistent with the data of Fig. 3.7, contains only the following changes:

$$\text{B3 N}(0, 3), L = 2$$
$$\text{M2 B}(1, 3), L = -1.0$$
$$\text{M3 B}(3, 2), L = 1.5$$

A negative sign is necessary in this circuit because the physical construction of the magnetic circuit requires branch 2 to have a dot at its initial node with respect to branch 1 but at its final node with respect to branch 3.

3. The coupling coefficient k implied by an M card must be less than 0.999995 to avoid difficulties due to the precision limitations of the program. That is, the inductance value L specified on an M card should always be less than $0.999995 \sqrt{L_p L_s}$, where L_p and L_s are the coupled branch inductances. If a coupling coefficient k greater than 0.999995 is required for an analysis, an ideal transformer model must be used. The ideal transformer model ($k = 1.0$) is developed in Sec. 5.3.2.

4. M card numbers must be assigned sequentially beginning with 1.

5. If one inductance is coupled to a second inductance (via an M card), which in turn is coupled to a third inductance (via a second M card), then the first and third inductances must also have an M card assigned to indicate their coupling. In general, if there are n mutually coupled inductances there must be $C(n, 2) = n!/[2!(n - 2)!]$ M cards, which allows for one M card per pair of inductances. Thus in the input data of Fig. 3.7, since there are mutual inductances coupling B1 with B2 and with B3, it follows that B2 and B3 also require a mutual inductance specified by an M card; hence there are $C(3, 2) = 3!/[2!(3 - 2)!] = 3$ M cards required.

6. The mutual inductance specified on an M card may be modified and iterated just as any other inductance. For example, the M3 card in Fig. 3.7 could be re-entered in a modify routine as

$$\text{M3 } L = -1.6(2)-1.4$$

to vary the mutual inductance from -1.6 to -1.5 to -1.4. Note that the branch information need not be re-entered.

3.4 NODAL CURRENT UNBALANCES (1ERROR)

Nodal current unbalances arise and are checked in AC analyses for the same reasons described in Sec. 2.12 for DC analyses. The only difference is that for the AC analysis program the real and imaginary branch current components BA_r and BA_i are summed separately at each node m; the real and imaginary branch current unbalances at each node are then summed over the entire n node circuit as

$$(BA_r)_{\text{total}} = \sum_n \left| \sum_m BA_r \right| \quad \text{and} \quad (BA_i)_{\text{total}} = \sum_n \left| \sum_m BA_i \right|$$

```
AC IERR,R
B1 N(0,1),R=2,E=28.28/0
B2 N(1,2),L=3E-12
B3 N(2,3),C=1
FR=159
IERR=0
PR,NV,MI
EX

ROW  COL
NODE NODE           NODAL ADMITTANCE MATRIX

REAL   1   1- 2   0.50000000 00    0.0
IMAG             -0.33365819D 12   0.33365819D 12
REAL   2   1- 2   0.0              0.0
IMAG              0.33365819D 12  -0.33365819D 12

NODES        EQUIVALENT CURRENT VECTOR

REAL  1- 2   0.14139993D 02   0.0
IMAG         0.0              0.0

FREQ = 0.15899992E 00

NODES           NODE VOLTAGES

MAG   1- 2   0.12657860E 02,  0.12657860E 02
PHA          0.63410751E 02  -0.63410751E 02

SOLUTION NOT OBTAINED TO DESIRED TOLERANCE

NODES           CURRENT UNBALANCES

REAL  1- 2  -0.89135743E-03  -0.32398648E-05
IMAG        -0.37703826E-03  -0.70917100E-04

BRANCHES        BRANCH CURRENTS

MAG   1- 3   0.12645633E 02   0.12645498E 02   0.12645533E 02
PHA          0.26589188E 02   0.26588913E 02   0.26589188E 02
```

PRINTED AS RESULT OF CURRENT UNBALANCES EXCEEDING IERROR

```
B2 MO B2 IMPED. MADE REASONABLE
   L=3F-3
FX

ROW  COL
NODE NODE           NODAL ADMITTANCE MATRIX

REAL   1   1- 2   0.50000000D 00   0.0
IMAG             -0.33365796D 03   0.33365796D 03
REAL   2   1- 2   0.0              0.0
IMAG              0.33365796D 03  -0.332658893D 03

NODES        EQUIVALENT CURRENT VECTOR

REAL  1- 2   0.14139999D 02   0.0
IMAG         0.0              0.0

FREQ = 0.15899992E 00

NODES           NODE VOLTAGES

MAG   1- 2   0.12026726E 02   0.12664646E 02
PHA         -0.63481308E 02  -0.63481308E 02

SOLUTION NOT OBTAINED TO DESIRED TOLERANCE

NODES           CURRENT UNBALANCES

REAL  1- 2  -0.17047402E-07  -0.17047963E-07
IMAG        -0.81051951E-06  -0.81051979E-06

BRANCHES        BRANCH CURRENTS

MAG   1- 3   0.12652312E 02   0.12652312E 02   0.12652312E 02
PHA          0.26518677E 02   0.26518677E 02   0.26518677E 02
```

PRINTED AS RESULT OF CURRENT UNBALANCES EXCEEDING IERROR

Fig. 3.8. Nodal current unbalances.

A current-unbalance message will be printed unless each of these final two sums is less than the 1ERROR.

COMMENTS

1. Current unbalances generally will not be a problem if the smallest impedance at every node is not more than about six or seven orders of magnitude smaller than the next smallest impedance at that node. This rule must be considered throughout the frequency range used in the analysis.
2. Since 1ERROR equals 0.001 unless explicitly specified otherwise, the 1ERROR in Fig. 3.8 was specified as zero to assure (for illustration) the printing of the current-unbalance messages, as shown in the bracketed portions of the figure. Note the current-unbalance reduction in the modified circuit when the impedance in branch 2 is increased (i.e., when the ratios of the smallest impedance to the next smallest impedance at the nodes are increased). As can be seen from the results for the original circuit in Fig. 3.8, $(\mathrm{BA}_r)_{\text{total}}$ and $(\mathrm{BA}_i)_{\text{total}}$—as defined by the above equations—are:

$$(\mathrm{BA}_r)_{\text{total}} \simeq 0.89135743\mathrm{E}{-}03 + 0.32398648\mathrm{E}{-}05$$

$$(\mathrm{BA}_i)_{\text{total}} \simeq 0.37703826\mathrm{E}{-}03 + 0.70917100\mathrm{E}{-}04$$

Since both of these sums are less than 0.001, the current-unbalance message for this solution would not have been printed if 1ERROR had not been explicitly specified.
3. Comparing the terms of the nodal admittance matrices of the original and modified circuits in Fig. 3.8 should clarify the computer precision limitations which can lead to nodal current unbalances.

3.5 THÉVENIN EQUIVALENT CIRCUITS

Suppose the Thévenin equivalent circuit in Fig. 3.9b is required for node 2 of the circuit in Fig. 3.9a. The Thévenin voltage source $E_T \underline{/\theta_T}$ is equal to NV2 as calculated in a normal solution for the *RLC* circuit in Fig. 3.9a. The Thévenin

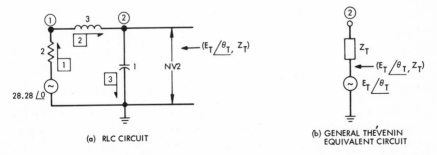

(a) RLC CIRCUIT   (b) GENERAL THÉVENIN
                                      EQUIVALENT CIRCUIT

Fig. 3.9. Thévenin equivalent circuit example.

```
              AC THEVENIN EQUIVALENT
          B1 N(0,1),R=2,E=28.28/0
          B2 N(1,2),L=3
          B3 N(2,0),C=1
          B4 N(2,0),R=1E5
             FR=.159
             PR,NV,CA
             EX


 FREQ =   0.15899992E 00



        NODES          NODE VOLTAGES                ┌──────┐
                                                    │ E_T/θ_T │
                                                    └──────┘
 MAG     1-  2  0.19977417E 02  ┌0.10017962E 02 ┐
 PHA            0.45055695E 02  └-0.13494339E 03┘



      BRANCHES        ELEMENT CURRENTS


 MAG     1-  4  0.10008205E 02  0.10008205E 02  0.10008205E 02  0.10017962E-03
 PHA           -0.44944000E 02 -0.44944000E 02 -0.44943420E 02 -0.13494339E 03



        MODIFY SHORTING NODE 2
      B4 R=1E-3
         EX

 FREQ =   0.15899992E 00



        NODES          NODE VOLTAGES


 MAG     1-  2  0.23519699E 02  0.78475364E-02
 PHA            0.33709930E 02 -0.56328171E 02



      BRANCHES        ELEMENT CURRENTS                         ┌──────────┐
                                                               │CA4|R=1E-3│
                                                               └──────────┘
 MAG     1-  4  0.78475418E 01  0.78475418E 01  0.78398921E-02 ┌0.78475380E 01┐
 PHA           -0.56270935E 02 -0.56270935E 02  0.33671814E 02 └-0.56328171E 02┘
```

Fig. 3.10. Thévenin equivalent circuit data.

impedance Z_T can be calculated by dividing NV2 by the current flowing through a short circuit (i.e., a small resistance) placed between node 2 and ground. Figure 3.10 shows the ECAP solutions required to obtain the values of

$$E_T \underline{/\theta_T} = 10 \underline{/-135}$$

and

$$Z_T = \frac{\text{NV2}}{\text{CA4}}\Big|_{R=1E-3} = \frac{10 \underline{/-135}}{7.84 \underline{/-56.3}} = 1.28 \underline{/-78.7}$$

COMMENTS

1. In effect, two solutions are necessary for calculating an AC Thévenin equivalent circuit, in contrast to the single solution needed to calculate a DC Thévenin equivalent circuit, because a nodal impedance matrix is not calculated in the AC analysis program.

2. Note that the addition of a large resistance in branch 4 of the original circuit description made it possible to use the modify routine to conveniently short node 2 to ground.

3.6 COMPLEX BETA SIMULATION AT A FIXED FREQUENCY

Only the *magnitude* of BETA or GM may be specified on a T card. This section illustrates a method for the realization of a *complex* trans-term at a specified frequency. The basic approach is to simulate the complex trans-term with a simple phase-shifting network completely isolated from the circuit being analyzed.

Suppose a complex beta, denoted as $\bar{\beta}$, is desired for the circuit shown in Fig. 3.11a. The complex beta is replaced by an arbitrary system consisting of two scalar beta trans-terms and the simple RL phase-shifting network in Fig. 3.11b. The values of the elements in this arbitrary system are calculated as follows: Defining CA3′ as the portion of CA3 due to the trans-term and defining $\bar{\beta}_{3,2}$ as the complex current gain from branch 2 to branch 3, in Fig. 3.11a we have

$$\text{CA3}' = \bar{\beta}_{3,2} \cdot (\text{CA2})$$
$$= \text{M} \underline{/\theta} \cdot (\text{CA2})$$

(3.1)

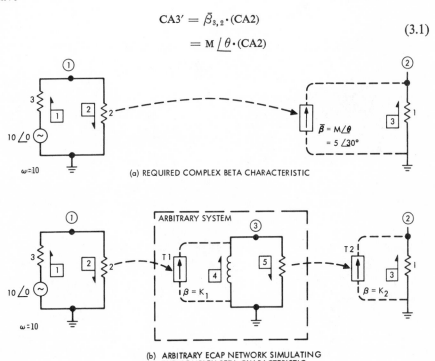

(a) REQUIRED COMPLEX BETA CHARACTERISTIC

(b) ARBITRARY ECAP NETWORK SIMULATING THE COMPLEX BETA CHARACTERISTIC

Fig. 3.11. Complex BETA simulation.

From Fig. 3.11b

$$CA3' = \beta_{3,5} \cdot (NV3)\left(\frac{1}{R_5}\right)$$

$$= \beta_{3,5} \cdot (\beta_{4,2}) \cdot (CA2) \cdot \left(\frac{j\omega R_5 L_4}{(R_5 + j\omega L_4)}\right) \cdot \frac{1}{R_5}$$

Arbitrarily letting $\beta_{3,5} = 1$ and $R_5 = 1$, we have

$$CA3' = \beta_{4,2} \cdot (CA2) \cdot \left(\frac{j\omega L_4}{1 + j\omega L_4}\right) \tag{3.2}$$

Combining Eqs. (3.1) and (3.2) results in

$$M \underline{/\theta} = \beta_{4,2} \cdot \left(\frac{j\omega L_4}{1 + j\omega L_4}\right)$$

Solving for L_4 and $\beta_{4,2}$ yields

$$L_4 = \frac{1}{\omega \tan \theta} \tag{3.3}$$

$$\beta_{4,2} = M\sqrt{1 + (1/\omega L_4)^2} \tag{3.4}$$

$$\beta_{3,5} = 1$$

$$R_5 = 1$$

```
        AC  COMPLEX BETA
  C
     B1  N(0,1),R=3,E=10/0
     B2  N(1,0),R=2
     B3  N(0,2),R=1
  C
     B4  N(0,3),L=.1732
     B5  N(3,0),R=1
     T1  B(2,4),BETA=5.775
     T2  B(5,3),BETA=1
  C
         FREQ=1.59
         PR,NV,CA
         EX

  FREQ =  0.15899982E 01


         NODES            NODE VOLTAGES

    MAG   1-  3  0.39999990E 01  0.10000080E 02  0.10000080E 02
    PHA          0.0             0.30024902E 02  0.30024902E 02


         BRANCHES         ELEMENT CURRENTS

    MAG   1-  4  0.19999990E 01  0.19999990E 01  0.0      0.10000080E 02
    PHA          0.0             0.0             0.0      0.30024902E 02

    MAG   5-  5  0.10000080E 02
    PHA          0.30024902E 02
```

Fig. 3.12. Complex BETA simulation example.

With $\text{M}\underline{/\theta} = 5\underline{/30}$ at $\omega = 10$, $L_4 = 0.1732$ and $\beta_{4,2} = 5.775$. This completes the specification of the arbitrary system in Fig. 3.11b used to produce the complex beta $= \text{M}\underline{/\theta}$ at an angular frequency ω. The input data for the circuit can now be prepared from the network in Fig. 3.11b; they result in the output in Fig. 3.12.

COMMENTS

1. The realization of the desired complex gain $\beta_{3,2} = 5\underline{/30}$ can be verified by noting, from the output in Fig. 3.12, the relation between NV2 and CA2

$$\bar{\beta}_{3,2} = \frac{\text{NV2}/R_3}{\text{CA2}}$$

$$= \frac{(10\underline{/30})/1}{2\underline{/0}}$$

$$= 5\underline{/30}$$

2. Element current CA3 in the output is not the same as CA3′ used in the development of this section, but rather CA3 $=$ CA3′ $+$ (CV3)$\cdot Y_3$ in accordance with the standard-branch terminology defined in Sec. 2.3. Since branch 3 is a dummy branch, CV3$\cdot Y_3 = -$CA3′ and therefore CA3 $= 0$.
3. If the desired complex trans-term has a negative phase, the negative value for L_4 resulting from Eqs. (3.3) and (3.4) can be used, or an RC network may be used instead of the RL network.
4. Complex transfer functions with specified frequency characteristics may be realized by using the isolated-network approach described in this section. (See prob. 3.11.) Each desired function must be treated individually. This concept is used extensively in control system analysis, as described in Chapter 7.

3.7 SUMMARY

This chapter has presented the details necessary for the optimum usage of the ECAP AC analysis program. The frequency iteration capability, used for calculating the frequency response characteristics of a circuit, stands out as one of the program's most useful features.

Points to be emphasized to reduce the likelihood of trouble in AC analysis program usage include these:

1. A nonzero frequency must be specified for every AC analysis problem.
2. An L or a C may be used as the *from*-branch of a T card with a GM only (i.e., the *from*-branch for a BETA must be a resistance).
3. Every source must have a specified phase angle.

4. As in the DC analysis program, if a *from*-branch is modified and the corresponding T card contains a BETA trans-term, the T card must also be re-entered.
5. A parameter iteration (including a frequency iteration) may not be included in the original circuit description. It may be included only in a modify routine.
6. Source magnitude iterations must always be accompanied by phase iterations with the same number of steps.
7. The effective coupling coefficient implied by an M card should be less than 0.999995.
8. Nodal impedance unbalances must be considered throughout the frequency range of analysis.

Appendix C includes tables which summarize the legitimate AC analysis input parameters, directly available output data, assumed units, and overall program capacities.

PROBLEMS

3.1. Determine the power dissipated in the 0.25 Ω resistance in series with the 1 h inductance in the circuit shown in Fig. p3.1a. The unloaded source terminal voltage E is as shown in Fig. p3.1b.

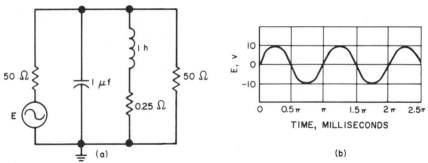

Fig. p3.1

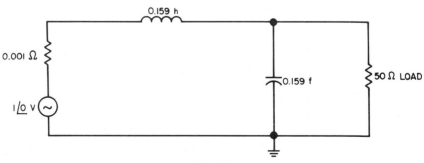

Fig. p3.2

3.2. Plot the resulting voltage (magnitude and phase) across the 50 Ω load in the circuit shown in Fig. p3.2 when the voltage source frequency is varied, by using the multiplicative iteration, over the frequency range 0.1 Hz to 10 Hz. By using the additive iteration, obtain additional points in the 0.7 to 1.3 Hz range. Calculate only as many points as are necessary to draw a smooth curve.

3.3. Derive an equation to compute the multiplication factor M for any required number of solutions n from L to U, in which the terms of the equation are as defined for the relationship FREQ $= L(M)U$ of Sec. 3.2.

3.4. Determine the power dissipated at 1 radian/sec in a 4 Ω resistive load connected across a transformer which has a primary inductance of 5 h, a secondary inductance of 20 h, and a coupling coefficient k of 0.8. The transformer primary winding is driven by a 1 vrms source with a 1 Ω source resistance.

3.5. Find the six errors in the input data shown in Fig. p3.3 which would either prevent a solution or lead to an incorrect solution.

```
       AC
    B1  N(0,1),C=4
    B2  N(1,2),L=8
    B3  N(2,3),M=2
    B3  N(1,3),G=4
    B4  N(3,0),L=2
    B5  N(0,1),R=2(4)10,E=10/0
    M1  B(2,4)
    T1  B(1,5),BETA=10
        FREQ=0
        PR,MI,NV
        EX
        MOD
    B5  R=5,L=10(5)20/0(2)9C
        PR,CA
        EX
```

Fig. p3.3

3.6. ECAP can be used to solve general equations by utilizing electrical analogs. For the following set of equations, draw an electrical analog network and use ECAP to determine the value of x which satisfies the equations:

$$5x - 3y = 20$$

$$-3x + (3 + j4)y = 0$$

3.7. For the following set of equations

$$2(1 + j)t - j3u = w$$

$$-j3t + 8(1 + j)u = 0$$

draw an analogous electrical network which includes mutual inductances, and use ECAP to determine the value of u which satisfies the equations for the conditions: (a) $w = 10$, (b) $w = 2(3 + j)$. Repeat the analyses using an electrical analog that does not contain mutual-inductance terms.

3.8. Using ECAP, analyze the circuit in Fig. p3.4 at frequencies of 10^{-10}, 10^{-5}, 1, 10^5, and 10^{10} Hz. Compare the calculated values for CA1 and CA2 at the analysis frequencies. Explain why CA1 is not always equal to CA2. What practical significance does this have on ECAP AC analyses in general?

3.9. Find the AC impedance at $\omega = 1$ radian/sec across terminals A and B of the circuit in Fig. p3.5.

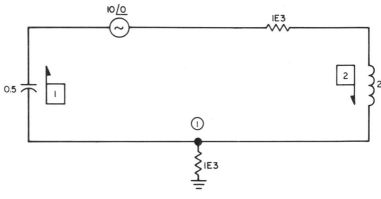

Fig. p3.4

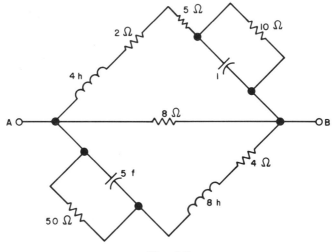

Fig. p3.5

3.10. For the circuit in Fig. p3.6, determine the value of the inductance (between 5 and 30 h) in series with the 10 Ω resistance which produces a current of 1 ma or less in the 0.01 Ω resistance at a frequency of 0.159 Hz.

3.11. Derive an electrical network to simulate a complex transfer function which results in NV3 = [2/(3 + j4)](NV2) for the circuit in Fig. p3.7. Obtain an ECAP solution to verify the realization of the required characteristic.

3.12. Determine the frequency response of the double-tuned coupled circuit in Fig. p3.8 for coupling coefficient values of $k = 0.02, 0.05, 0.08$. Plot $G(s) = E_o(s)/E_i(s)$ in decibels.

3.13. Repeat prob. 3.12 with $R_2 = 1$ Ω.

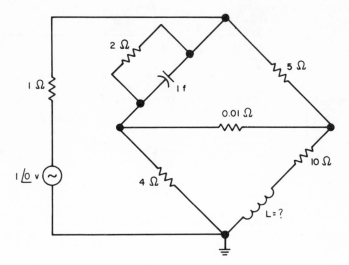

Fig. p3.6

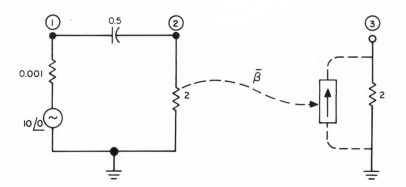

Fig. p3.7

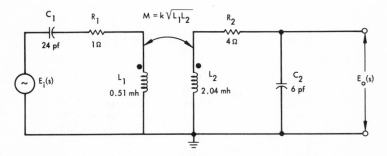

Fig. p3.8

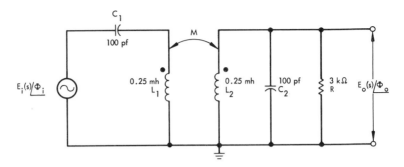

Fig. p3.9

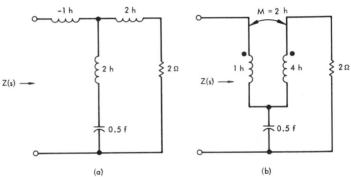

Fig. p3.10

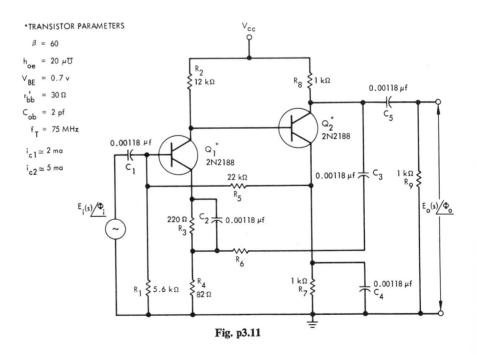

Fig. p3.11

3.14. Determine the frequency response of the double-tuned circuit in Fig. p3.9 for coupling coefficient values of $k = 0.3$, 0.5, and 0.8. Plot the gain $G(s) = E_o(s)/E_i(s)$ in decibels, and the phase shift $(\Phi_o - \Phi_i)$ in degrees.

3.15. Verify that the networks in Fig. p3.10 are Brune network realizations of the positive real function

$$Z(s) = \frac{s^2 + s + 2}{2s^2 + s + 1}$$

The *T*-equivalent circuit of a transformer shown in this problem can be used in circuits requiring a unity coupling coefficient and a finite value of primary inductance which differs from the infinite primary inductance present in the ideal transformer.

3.16. Determine the frequency response of the circuit in Fig. p3.11 with $R_6 = 100 \, \Omega$ over the frequency range 100 Hz to 100 MHz. Plot the output gain $G(s) = E_o(s)/E_i(s)$ in decibels, and the phase shift $(\Phi_o - \Phi_i)$ in degrees. Repeat the analysis using values of $R_6 = 1000 \, \Omega$ and 10 kΩ. (The equivalent circuit developed in Appendix A.1 can be used for the transistors.)

3.17. Repeat prob. 3.16 with C_1, C_2, C_3, C_4, and C_5 equal to 1.0 μf.

4

TRANSIENT

ANALYSIS

The transient analysis program, using the capability of piecewise linear modeling of the circuit nonlinearities, provides time-dependent solutions to the analyses of linear and nonlinear circuits. The techniques required for optimum use of this program are described in this chapter.

4.1 TRANSIENT ANALYSIS CODING
INTRODUCTION

Figures 4.1 and 4.2 introduce the coding techniques necessary to perform a transient analysis. The ECAP statements, represented by their two-letter codes, first introduced in Fig. 4.2 are defined as follows:

TR   specifies that a transient analysis is to be performed by ECAP.

$TI = 0.1$ specifies that the transient analysis program is to calculate each

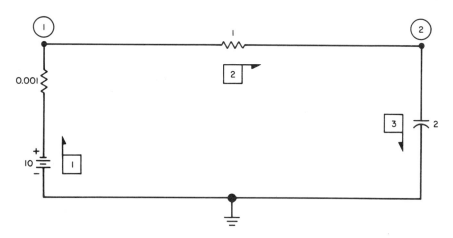

Fig. 4.1. Transient analysis circuit.

1 2 3 4 5 6	7 8 9 10 11 12	13 14 15 16 17 18	19 20 21 22 23 24	25 26 27 28 29 30	31 32
	T R A N S I E N T	A N A L Y S I S	I N T R Ø .		
B 1	N (0 , 1) , R = . 0 0 1 , E = 1 0				
B 2	N (1 , 2) , R = 1				
B 3	N (2 , 0) , C = 2				
	T I M E S T E P = 0 . 1				
	Ø U T P U T I N T E R V A L = 5				
	F I N A L T I M E = 4				
	P R I N T , N V , C A , C V , B A , B V , B P , M I				
	E X E C U T E				

Fig. 4.2. Transient analysis input data.

76

new solution using 0.1 sec time steps. The TIME STEP defines the interval to be used by the program in solving the algebraic approximations to the general integral-differential equations to produce a transient analysis. (See Appendix B.3 for a description of the program usage of TI and Sec. 4.6 for a discussion on time-step selection.)

OU = 5   specifies that solutions will be printed at intervals of five time steps; that is, only every fifth solution will be printed. (If OU is not specified, OU = 1 will be used.)

FI = 4   specifies that the final solution will be calculated at a time equal to or greater than 4 sec.

Figure 4.3 contains the output from this example.

COMMENTS

1. The transient analysis program provides all the features available in the DC analysis program except the MODIFY capability and SENSITIVITY coefficient, WORST CASE analysis, and STANDARD DEVIATION calculations.
2. Transient analysis coding follows the same basic rules as DC analysis with the following additions:
   a. A TIME STEP and a FINAL TIME must be specified by the user for every transient analysis except an EQUILIBRIUM solution. (See Sec. 4.9.)
   b. Inductances and capacitances are permitted input parameters for a transient analysis.
   c. If an L or a C is the *from*-branch specified on a T card, that T card must contain a BETA (i.e., not a GM) trans-term. This is different from the format allowed in the AC analysis program. An L or a C may be used as any *to*-branch.
   d. No branch should be specified as both a *from*- and a *to*-branch. A dummy *to*-branch shunting the *from*-branch can be added to the circuit to achieve the same effect.
3. The complete standard branch used in the transient analysis program is similar to the one used in the DC analysis program (described in Sec. 2.3) except for the following points:
   a. The passive element in the branch may be a resistance, a capacitance, or an inductance. If the element is reactive, the equivalent conductance is a function of the selected time step and is equal to $Y = C/\Delta t$ or $\Delta t/2L$. The equivalent conductances and their relationship to the numerical integration techniques used in calculating the circuit response are explained in detail in Appendix B.3.
   b. Time-independent voltage and current sources on a B card are

```
TIME STEP       =  0.99999964E-01
OUTPUT INTERVAL =  5
1ERROR          =  0.99999993E-03
2ERROR          =  0.99999993E-03
3ERROR          =  0.10000000E 01
SHORT           =  0.99999979E-02
OPEN            =  0.10000000E 08


NODAL CONDUCTANCE MATRIX

ROW     COLS
 1      1 - 2   -0.10010000D 04  0.10000000D 01
 2      1 - 2    0.10000000D 01 -0.10100002D 03

EQUIVALENT CURRENT VECTOR

NODE NO.   CURRENT
 1        -0.10000000D 05
 2         0.0

T =  0.0

NODES            NODE VOLTAGES
1 - 2     0.99901085E 01   0.98911941E-01

BRANCHES         BRANCH VOLTAGES
1 - 3    -0.31316000E-06   0.98911963E 01   0.98911941E-01

BRANCHES         ELEMENT VOLTAGES
1 - 3     0.99999990E 01   0.98911963E 01   0.98911941E-01

BRANCHES         ELEMENT CURRENTS
1 - 3     0.98911968D 01   0.98911968D 01   0.98911968D 01

BRANCHES         INSTANTANEOUS ELEMENT POWER
1 - 3     0.98911957E 02   0.97835754E 02   0.97835743E 00

BRANCHES         BRANCH CURRENTS
1 - 3     0.98911963E 01   0.98911963E 01   0.98911963E 01


T =  0.4999998E 00

NODES            NODE VOLTAGES
1 - 2     0.99922476E 01   0.22403927E 01

BRANCHES         BRANCH VOLTAGES
1 - 3    -0.56330799E-06   0.77518539E 01   0.22403927E 01

BRANCHES         ELEMENT VOLTAGES
1 - 3     0.99999990E 01   0.77518539E 01   0.22403927E 01

BRANCHES         ELEMENT CURRENTS
1 - 3     0.77518552D 01   0.77518552D 01   0.77518552D 01

BRANCHES         INSTANIANEOUS ELEMENT POWER
1 - 3     0.77518533E 02   0.60091248E 02   0.17367187E 02

BRANCHES         BRANCH CURRENTS
1 - 3     0.77518549E 01   0.77518549E 01   0.77518549E 01


T =  0.4099982E 01

NODES            NODE VOLTAGES
1 - 2     0.99986591E 01   0.86579494E 01

BRANCHES         BRANCH VOLTAGES
1 - 3    -0.15665722E-06   0.13407087E 01   0.86579494E 01

BRANCHES         ELEMENT VOLTAGES
1 - 3     0.99999990E 01   0.13407087E 01   0.86579494E 01

BRANCHES         ELEMENT CURRENTS
1 - 3     0.13407094D 01   0.13407094D 01   0.13407094D 01

BRANCHES         INSTANTANEOUS ELEMENT POWER
1 - 3     0.13407092E 02   0.17975006E 01   0.11607794E 02

BRANCHES         BRANCH CURRENTS
1 - 3     0.13407087E 01   0.13407087E 01   0.13407087E 01
```

Fig. 4.3. Transient analysis output.

        set to their specified E and I values at the time of the initial solution. That is, E specifies a function $Eu(t - \text{INITIAL TIME})$ where $u(t)$ is the unit step-function and INITIAL TIME is the time of the first solution (see Sec. 4.8).

    c. Time-dependent source specification is described in Sec. 4.2.

    d. All output is printed at every OU·TI interval, except the MISC. output, which is printed only for the initial solution. A nodal impedance matrix is not printed with the MI output since it is not used in the transient analysis program.

    e. The instantaneous element power dissipation BP is calculated at each time t_k according to $\text{BP}(t_k) = \text{CV}(t_k) \cdot \text{CA}(t_k)$.

4. Nodal current unbalances arise and are checked in a transient analysis for the same reasons described in Sec. 2.12 for the DC analysis program. In the transient analysis program the equivalent resistances $2L/\Delta t$ and $\Delta t/C$, which are described in Appendix B.3, must be considered. The program automatically terminates any transient analysis which obtains 21 solutions violating the 1ERROR criterion. If nodal current-unbalance messages are written by the program, the nodal impedance ratios of the nodes with large current unbalances must be improved—usually by increasing the smallest impedances.

5. M cards may be used in a transient analysis to specify the coupling between inductances, as described in Sec. 3.3 for the AC analysis program.

4.2 TIME-DEPENDENT SOURCES

Three types of time-dependent sources are directly available in the transient analysis program. The three source types—nonperiodic, periodic, and sinusoidal —are described as follows:

    a. *Nonperiodic*
    The input data for a circuit might include (anywhere after the B27 card):

E27 (4),15,20,0,10

    This statement defines a time-dependent voltage source in branch 27 with the characteristic shown in Fig. 4.4. The general form of the nonperiodic time-dependent source is

Wxx (m), $p_0, p_1, p_2, \ldots, p_n$

    where W is the source type (E for a voltage source; I for a current source)
        xx is the assigned branch of the source
        m is a positive integer defining the number of time steps between specified source values
        $p_0, p_1, p_2, \ldots, p_n$ are the values of the source at $m \cdot$ TI intervals

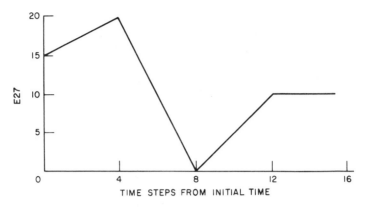

Fig. 4.4. Time-dependent nonperiodic voltage source.

Note that p_0 is the source value at $t =$ INITIAL TIME (see Sec. 4.8), and that the source value is p_n for $t \geq n \cdot m \cdot$ TI. Also note that the list of source values must not contain more than 126 entries (i.e., $n \leq 125$).

b. *Periodic*

The input data for a circuit might include (anywhere after the B46 card):

$$\text{I46 P(10),} -2,4,4,4,-2,-2,-2$$

This defines a time-dependent current source in branch 46 with the characteristic shown in Fig. 4.5. The general form of the periodic source is

$$\text{W} xx \text{ P}(m), p_0, p_1, p_2, \ldots, p_n$$

where w, xx, m, and $p_0, p_1, p_2, \ldots, p_n$ are as defined for the non-periodic source

P indicates to ECAP that the source is periodic

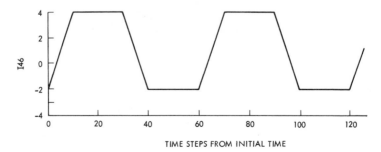

Fig. 4.5. Time-dependent periodic current source.

Since the source value entries $p_0, p_1, p_2, \ldots, p_n$ describe one full period, p_n must have the same value as p_0.

1620 Note c. *Sinusoidal*

The input data for a circuit might include (anywhere after the B9 card):

E9 SIN(.01),8,6,.0025

This defines a time-dependent voltage source in branch 9 with the characteristic shown in Fig. 4.6. The general form of the sinu-

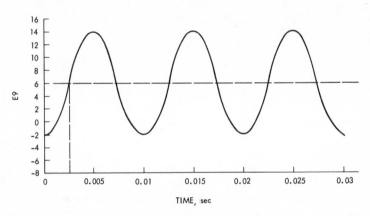

Fig. 4.6. Sinusoidal voltage source.

soidal source is

Wxx SIN (t_p), w_1, w_0, t_0

where w and *xx* define the source type and the source branch number

 SIN indicates to ECAP that the source is sinusoidal

 t_p defines the period of the source in seconds (the period must be in the range $3 \cdot \text{TI} \le t_p \le 126 \cdot \text{TI}$)

 w_1 defines the amplitude of the sinusoidal output in volts (or amperes) from the average to the peak

 w_0 specifies the DC offset in volts (or amperes)

 t_0 defines the time offset in seconds

The values for the entries w_1, w_0, and t_0 may be positive or negative. The general equation (presented graphically in Fig. 4.7) which describes the output of the sinusoidal source is

$$W xx = w_0 + w_1 \text{ SIN} \left[\frac{2\pi(t - t_0)}{t_p} \right]$$

Figure 4.8 further illustrates the time-dependent source input data format.

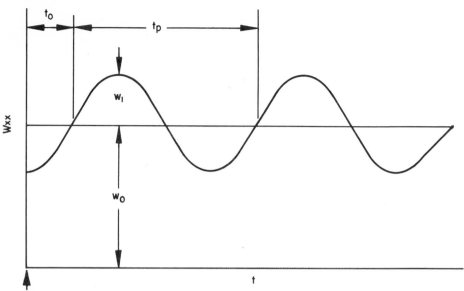

Fig. 4.7. General sinusoidal source.

COMMENTS

1. As many as five time-dependent voltage and five time-dependent current sources may be used in any one problem.
2. The specification of a time-dependent voltage (or current) source for a branch replaces any fixed voltage (or current) source also specified for that branch. Thus, in Fig. 4.8, the fixed current source $I = -10$ in branch 3 is used, but the fixed voltage source $E = 30$ is replaced by the time-dependent source E3.

```
     TR TIME-DEPENDENT SOURCES
  B1  N(0,1),R=2
  E1  SIN(4),2,5,-1
  B2  N(0,2),R=2
  E2  P(2),-1,1,-1
  I2  (3),0,4
  B3  N(0,3),R=1,E=30,I=-10
  E3  P(1),5,4,-2,-1,5
      TI=1
      FI=6
      PR,NV
      EX
```

Fig. 4.8. Time-dependent source input data.

3. The period t_p of any sinusoidal source must be specified within the range $3 \cdot \mathrm{TI} \leq t_p \leq 126 \cdot \mathrm{TI}$. If t_p is specified outside this range, the program will re-specify it and print a message accordingly. If a sinusoidal source must have a period greater than $126 \cdot \mathrm{TI}$, the source can be approximated using the periodic time-dependent source. Alternatively, a special function-generating network, using the techniques described in chapters 5 and 7, could be constructed to generate a sine wave which is independent of the selected time step.
4. It should be noted that all time-dependent sources use straight-line interpolations to determine values at times other than those explicitly specified.
5. Infinite-sloped pulses or step changes in source values cannot be generated by time-dependent sources. If such pulses are required, special-function generators can be built such as shown in Sec. 4.11.

(a) TIME-DEPENDENT SOURCE

(b) BRANCH CARD

Fig. 4.9. Continuation cards.

4.3 THE CONTINUATION CARD

When more than one card is needed to describe a time-dependent source or a B card, a continuation card—specified by an asterisk (*) in column 6—is used to include the additional input data. As many continuation cards as are needed to completely define the time-dependent source or branch input data may be used, as shown in Fig. 4.9a. Note that the card prior to a continuation card must end with a comma, and, if it is a B card, the comma must separate data subgroups, as shown in Fig. 4.9b.

Continuation cards may be used in any ECAP analysis. However, they may be used only with B cards and, in transient analyses, with time-dependent sources.

4.4 THE SWITCH: A CURRENT-DIRECTION SENSING DEVICE

Elements with nonlinear characteristics can be represented in ECAP analyses using piecewise linear models. The key tool in modeling nonlinear characteristics is a switch. A switch is a branch whose element current direction determines which of two possible specified values is in effect for each parameter in each of the branches controlled by the switch. Switches are specified by s cards. The general format of an s card is

$$Sxx \quad B = a_s, (a_i, a_j, \ldots, a_n), \left\{ {ON \atop OFF} \right\}$$

where xx is the serial number of the s card

a_s is the branch number of the branch assigned as the switch (i.e., the branch whose current direction is being monitored)

$a_i, a_j, \ldots, a_n$ are the branch numbers of the branches controlled by the switch

ON (or OFF) specifies that the first of the two values for all controlled parameters

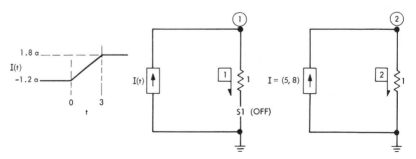

Fig. 4.10. Switch introduction circuit.

84

```
    TR  THE SWITCH
B1  N(1,0),R=1
I1  (3),-1.2,+1.8
B2  N(2,0),R=1,I=(5,8)
S1  B=1,(2),OFF
    TI=1
    FI=3
    PR,NV,CA
    EX


    T =   0.0


NODES                   NODE VOLTAGES

  1- 2        -0.11999998E 01      0.50000000E 01

BRANCHES              ELEMENT CURRENTS

  1- 2        -0.11999998D 01      0.50000000D 01


    T =   0.1000000E 01

NODES                   NODE VOLTAGES

  1- 2        -0.20000017E 00      0.50000000E 01

BRANCHES              ELEMENT CURRENTS

  1- 2        -0.20000017D 00      0.50000000D 01


    T =   0.1200195E 01

NODES                   NODE VOLTAGES

  1- 2         0.19512163E-03      0.50000000E 01

BRANCHES              ELEMENT CURRENTS

  1- 2         0.19512163D-03      0.50000000D 01


SWITCH  1 IS ON
```

```
    T =   0.1200195E 01


NODES                   NODE VOLTAGES

  1- 2         0.19512163E-03      0.80000000E 01

BRANCHES              ELEMENT CURRENTS

  1- 2         0.19512163D-03      0.80000000D 01


    T =   0.2000000E 01

NODES                   NODE VOLTAGES

  1- 2         0.79999948E 00      0.80000000E 01

BRANCHES              ELEMENT CURRENTS

  1- 2         0.79999948D 00      0.80000000D 01


    T =   0.3000000E 01

NODES                   NODE VOLTAGES

  1- 2         0.17999983E 01      0.80000000E 01

BRANCHES              ELEMENT CURRENTS

  1- 2         0.17999983D 01      0.80000000D 01
```

Fig. 4.11. Switch introduction output.

is in effect when the current in the switch is positive (or negative), and the second value is in effect when the current is negative (or positive). See comment 6 of this section for exceptions to this rule.

Figures 4.10 and 4.11 illustrate the use of the s card.

COMMENTS

1. Any branch may be specified as a switch branch or a controlled branch. A switch can control the branch containing it, such as

$$S4 \ B = 8, (5, 8, 11, 12), ON$$

2. s cards may be intermixed with other cards, but they must be numbered consecutively beginning with 1. For example, the following is a legitimate input data arrangement:

```
B6  N(3,8),L=(.1,5),E=(10,-20),I=5
S1  B=3,(5,3),ON
T2  B(2,5),GM=(3E-6,8E-6)
T3  B(4,1),BETA=(200,400)
S2  B=4,(4,8),OFF
B7  N(4,0),R=2,E=(5,15),I=(-10,4)
S3  B=7,(2,8,11,19,27),ON
B8  N(1,3),R=(24,100)
```

3. A parameter of a controlled branch will be affected by switches controlling the branch only if it has two specified values. Thus, in Fig. 4.11, I_2 is changed and R_2 is unchanged by a switch actuation.

4. All standard branch input parameters, other than time-dependent sources, can be switch-controlled.

5. Values of BETA and GM are controlled by specifying the *from*-branches as controlled branches. For example, in the input data in comment 2 of this section, the current direction in branch 7 controls the GM value of T2, and the current direction in branch 4 controls the BETA value of T3.

R_8	SWITCH STATE			CURRENT DIRECTION		
	S1	S2	S3	B3	B4	B7
24	ON	OFF	ON	+	–	+
24	ON	ON	OFF	+	+	–
24	OFF	ON	ON	–	+	+
24	OFF	OFF	OFF	–	–	–
100	ALL OTHER COMBINATIONS					

Fig. 4.12. Controlled parameter values.

6. A parameter may be controlled by any number of switches. The following rule can be used to determine which of the two possible parameter values is in effect at any given time: The first value of the ordered pair of values of a parameter controlled by any number of switches is in effect when the currents in all but an even number of the controlling switches (zero is considered an even number) are as specified by the ON and OFF labels assigned to the switches; the second value is in effect otherwise. Thus in the input data of comment 2, assuming that no s card other than 1, 2, and 3 as shown controls branch 8, the value of R_8 is a function of the current directions in branches 3, 4, and 7, as shown in Fig. 4.12.

7. The ON or OFF label on an s card should be specified to agree with the initial current direction in the switch branch.

8. Section 4.5 describes the 2ERROR feature which specifies the accuracy to which the printed switching time approximates the actual time of change of

the switch current direction. Appendix B.3.3 discusses the technique used by the program in determining the switching time.

4.5   *SWITCH ACTUATION TIME (2ERROR)*

A switch is actuated whenever its element current changes direction, as described in Sec. 4.4. The accuracy to which the printed (i.e., program-determined) switching time approximates the actual switching time is a function of the 2ERROR,

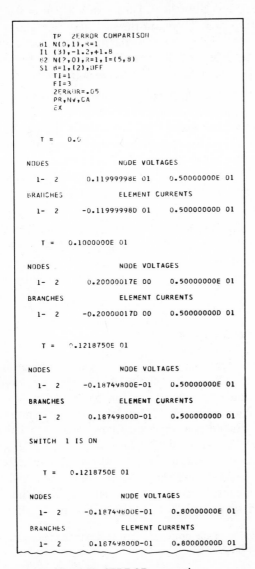

Fig. 4.13. 2ERROR comparison.

which can be written 2E. The printed switching time will be within $2E \cdot TI$ seconds of the actual switching time.* For example, compare the output in Fig. 4.13 with the output for the same circuit in Fig. 4.11. The 2ERROR specified in Fig. 4.13 is 0.05 rather than the 0.001 value assumed by the program when the 2ERROR was not specified in Fig. 4.11.

COMMENTS

1. The number of solutions n calculated (but not printed) from the time of the last solution prior to a switch actuation to the computed switching time is given by $n \geq 3.32 \log (2ERROR)^{-1}$. Thus with 2ERROR = 0.001, $n = 10$; with 2ERROR = 0.05, $n = 5$. Therefore the switching time shown in Fig. 4.11 is closer to the true switching time than the time calculated in Fig. 4.13.
2. Increasing the 2ERROR of 0.001 normally assumed by the program may be desired to do both or either of the following:

    a. Minimize problem execution time when many switch actuations are expected.

    b. Improve the nodal impedance ratios during the switching time calculations to decrease the nodal current unbalances as described in comment 4 of Sec. 4.1. The impedance ratios at nodes that connect branches containing inductances or capacitances will be affected because Δt may be as small as TI/n when switch actuations occur (see Appendix B.3.3).

However, it is recommended that the 2ERROR not be specified greater than 0.2. Also, it should be realized that decreasing the 2ERROR to less than 0.001 in an attempt to determine switch actuation times more accurately may be undesirable for the same reasons that increasing it may be desirable.

4.6 TIME STEP SELECTION

The selection of an appropriate TIME STEP for each transient analysis problem is a task left to the ECAP user. The choice of TI usually represents a compromise between good solution accuracy and a reasonable execution time. The numerical integration technique used by ECAP requires the specification of a time increment over which straight-line approximations to smooth curves will supply sufficient accuracy. (See Appendix B.3 for a description of ECAP's numerical solution techniques.) If the time increment is too large, the results may be grossly inaccurate; if too small, either the required computer calculation time may become excessive or a loss in accuracy may result because of the effects of limited computer precision, or possibly both results will occur. Significant inaccuracies

*A detailed description of the procedure used to determine the switch actuation time is presented in Appendix B.3.3.

are usually indicated by a current-unbalance message, generated by the 1ERROR feature.

Proper time step selection depends on the circuit's time response at the nodes of interest. The user has to determine whether the time response at the nodes is primarily exponential (RC or L/R) or oscillatory ($2\pi\sqrt{LC}$). This can be accomplished by using Table 4.1, in which R, L, and C are defined as the parallel combinations of each element type connected at the node. It should be emphasized that the table gives only suggested values for TI since other factors must also be considered. These include the use of time steps compatible with the requirements of time-dependent source specifications (see Sec. 4.2) and the use of time steps which assure acceptable current unbalances as mentioned in comment 4 of Sec. 4.1.

When time steps different by about three or four orders of magnitude are required for two different nodes, two separate analyses should be prepared, each with the appropriate time step, to study the response of the individual nodes. If the required time steps differ by greater than about four orders of magnitude, separate equivalent circuits should be analyzed, each omitting the parameters that are unimportant over the time range being examined.

Table 4.1.   Time step selection.

Elements at node of interest	Suggested TIME STEP	
R, C only	$0.01RC$	
R, L only	$0.01L/R$	
L, C only	$0.001\,(2\pi\sqrt{LC})$	
$R, L,$ and C	$0.01RC$	when $\dfrac{2RC}{\sqrt{LC}} \le 1$
$R, L,$ and C	$0.001\,(2\pi\sqrt{LC})$	when $\dfrac{2RC}{\sqrt{LC}} > 1$

Probably the surest way of guaranteeing a good time step selection is by a trial-and-error procedure, in which the results of interest obtained for an analysis with one TI value are compared with those for a TI value which is an order of magnitude different. If the results from the two analyses agree, then a time step at least as large as the larger of the two time steps can confidently be used. If the two analyses show significant variations, a smaller TI value must be tried. To minimize execution time, the final time FI should be reduced for these trials to a value for which only enough output is obtained to permit an adequate comparison of time step effects.

The transient analysis examples in the text further illustrate proper time step selection.

4.7 INITIAL CONDITION SOLUTIONS AND THE SHORT AND OPEN STATEMENTS

The first solution in any transient analysis problem, as well as all solutions immediately following switch actuations, are called initial condition solutions. In these solutions a small resistance is substituted for each capacitance and a large resistance is used for each inductance. (See Appendix B.3 for a discussion of the transient analysis capacitance and inductance models.) The use of these resistance values in initial condition calculations is consistent with the fact that

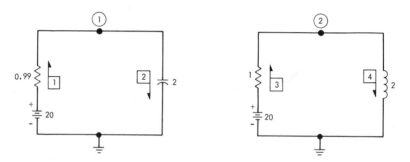

Fig. 4.14. Initial condition solution example.

```
       TR INIT. COND. SOLUTIONS
    B1 N(0,1),R=.99,E=20
    B2 N(1,0),C=2
  C
    B3 N(0,2),R=1,E=20
    B4 N(2,0),L=2
       OPEN=99
       TI=1
       FI=2
       PR,NV,CA
       EX


    T =    0.0

  NODES              NODE VOLTAGES

   1-  2      0.19999981E 00      0.19799988E 02
  BRANCHES           ELEMENT CURRENTS

   1-  4      0.19999987D 02      0.19999987D 02      0.19999998D 00      0.19999998D 00


    T =   0.1000000E 01

  NODES              NODE VOLTAGES

   1-  2      0.68442917E 01      0.11879999E 02
  BRANCHES           ELEMENT CURRENTS

   1-  4      0.13288585D 02      0.13288585D 02      0.81200000D 01      0.81200000D 01
```

Fig. 4.15. Initial condition solution output illustrating SHORT and OPEN effects.

neither the voltage across a capacitance nor the current through an inductance can change instantaneously. The magnitudes of the small and large resistances can be specified by using the SHORT and OPEN statements, which, if not included in the input data, are assumed by the program to be 0.01 Ω and 1E7 Ω, respectively.

The effects of the SHORT and OPEN values used in the $t = 0$ solution of the circuits in Fig. 4.14 should be obvious by inspection of Fig. 4.15. That is, at $t = 0.0$ the capacitance is replaced by the implicit SHORT value of 0.01, and the inductance is replaced by the explicitly specified OPEN value of 99.

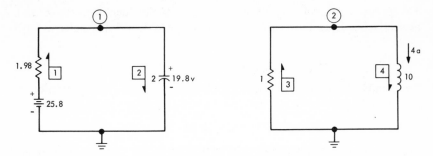

Fig. 4.16. Initial conditions example.

```
        TR   E0,I0
     B1  N(0,1),R=1.98,E=25.8
     B2  N(1,0),C=2,E0=-19.8
     B3  N(0,2),R=1
     B4  N(2,0),L=10,I0=4
         OPEN=99
         SHORT=.02
         INITIAL TIME=8.0
         TI=.01
         OU=50
         FI=14
         PR,NV,CA
         EX


     T =    0.7999999E 01


 NODES                  NODE VOLTAGES

    1-  2       0.19859985E 02    -0.39599991E 01

 BRANCHES               ELEMENT CURRENTS

    1-  4       0.30000018D 01     0.30000018D 01     0.39600000D 01     0.39600000D 01


     T =    0.8499963E 01


 NODES                  NODE VOLTAGES

    1-  2       0.20563721E 02    -0.37668686E 01

 BRANCHES               ELEMENT CURRENTS

    1-  4       0.26445724D 01     0.26445724D 01     0.37668687D 01     0.37668687D 01
```

Fig. 4.17. Initial conditions and INITIAL TIME output.

4.8 INITIAL CONDITIONS AND INITIAL TIME

The initial voltages across capacitances and currents through inductances may be specified for any transient analysis problem. EO in a capacitance branch and IO in an inductance branch are used to specify the initial conditions.

The first solution of any transient analysis can be labeled with a time other than zero by use of the INITIAL TIME card.

These features are illustrated in Figs. 4.16 and 4.17.

COMMENTS

1. The program properly treats EO and IO virtually identically to E and I, respectively. Thus B2 in Fig. 4.17 would yield identical results if written instead as

$$\text{B2 } N(1, 0), C = 2, E = -19.8$$

2. The number 0 can be substituted for the letter O in the initial condition voltage and current specifications. Thus B4 in Fig. 4.17 could be described as

$$\text{B4 } N(2, 0), L = 10, I0 = 4$$

3. Initial condition calculations yield initial values which are not identical to

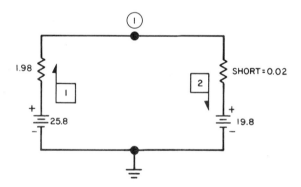

Fig. 4.18. Equivalent circuit for initial condition solution.

the values inserted as initial conditions via the EO and IO specifications. For example, in Fig. 4.17, NV1 does not exactly equal the $EO = 19.8$ originally specified across the capacitance. The reason for this is clarified by studying the circuit (Fig. 4.18) used by the program for the initial condition solution. If the SHORT value is decreased, the error in the initial condition value will be lessened.

4. The value on an INITIAL TIME card only specifies that the first solution will be labeled with that time value and that the following solutions will be successively labeled; the initial time specification does not affect the results of

any calculation. Thus, if the INITIAL TIME card in Fig. 4.17 were not present, the first solution would be labeled 0.00, the second solution would be labeled 0.50, etc., until the final time of $t = 14$ were reached.

4.9 EQUILIBRIUM SOLUTIONS

If an EQUILIBRIUM (or EQ) card is included with the input data for a transient analysis problem, only one solution is calculated. This is called a steady-state solution, calculated by substituting the OPEN value for all capacitances and the SHORT value for all inductances to represent the behavior of those elements after the circuit transients have died out. It corresponds to a solution at $t \to \infty$. The circuits and output in Figs. 4.19 and 4.20 illustrate the equilibrium condition solution.

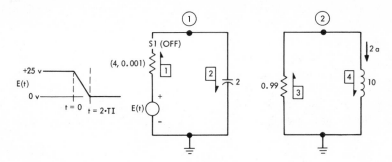

Fig. 4.19. EQUILIBRIUM solution example.

```
       TR EQUILIBRIUM
    B1 N(0,1),R=(4,.001)
    B2 N(1,0),C=2
    E1 (2),25,0
    S1 B=1,(1),OFF
C
    B3 N(0,2),R=.99
    B4 N(2,0),L=10,I0=2
       SH=.1
       OP=96
       EQUILIBRIUM
       PR,NV,CA
       EX



STEADY STATE SOLUTION


NODES              NODE VOLTAGES

  1-  2     0.24000000E 02    -0.18165129E 00

BRANCHES           ELEMENT CURRENTS

  1-  4     0.249999994D 00    0.249999994D 00    0.18348608D 00    0.18348608D 00
```

Fig. 4.20. EQUILIBRIUM solution output.

COMMENTS

1. TIME STEP and FINAL TIME may be omitted from a transient analysis problem only when an EQ card is present.
2. If no inductances or capacitances are present and if there are no time-dependent sources specified, then only an equilibrium solution will be calculated.
3. The first values of all ordered pairs of switch-controlled parameters are used in an equilibrium solution; all time-dependent sources assume their $t =$ INITIAL TIME values. Thus, in Fig. 4.20, $R_1 = 4$ even though S1 changes state in a normal solution; and $E_1 = 25$.
4. Initial condition values included in the input data are used during an equilibrium solution. For example, the initial condition $IO = 2$ for the inductance in branch 4 is used in the equilibrium solution calculations.
5. Equilibrium solutions are useful in determining the initial condition values (discussed in Sec. 4.8) before the application of a transient signal.

4.10 A NEGATIVE ELEMENT AND
A NONLINEAR MODEL

Although modeling techniques will be discussed in detail in Chapter 5, a resistance with the nonlinear characteristic of Fig. 4.21 is modeled here to introduce

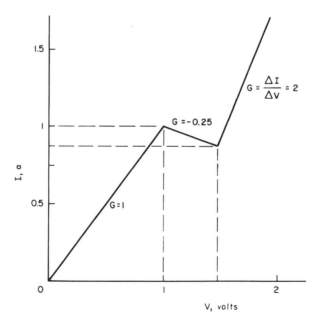

Fig. 4.21. Nonlinear element characteristic.

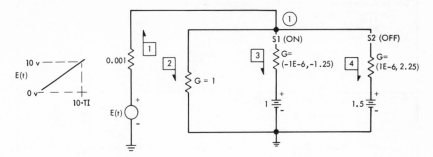

Fig. 4.22. Equivalent circuit using negative element.

the negative element and a nonlinear model. Branches 2, 3, and 4 in Fig. 4.22 represent a nonlinear element with the required characteristic. The time-dependent source in branch 1 is used to sweep the voltage across the element over the desired range of test values. The output data in Fig. 4.23 present the ECAP analysis of the nonlinear model.

COMMENTS

1. A negative element produces a voltage rise when positive current flows through it. Thus, in Fig. 4.22 when NV1 < 1 v, the element current CA3 flows from node 1 to "ground" (that is, in the assumed positive current direction) and S1 is ON.
2. Both element values of a controlled pair should have the same sign. For example, the sign of the arbitrarily small conductance of 1E$-6\mho$ in branch 3 is specified as negative because the sign of the alternate value of the pair (-1.25) is negative.
3. Note that NV1 is equal to the voltage across the nonlinear resistance and that CA1 is the current flowing through it. The realization of the nonlinear conductance specified in Fig. 4.21 can best be confirmed by checking ΔCA1/ΔNV1 in each region.

**Table 4.2. Summary of the nonlinear model
characteristic.**

NV1	S1	S2	G_2	G_3	G_4	Approximate nonlinear resistance
NV1 ≤ 1	ON	OFF	1	$-$1E-6	1E-6	$\frac{1}{1}$
$1 <$ NV1 ≤ 1.5	OFF	OFF	1	-1.25	1E-6	$\frac{1}{-0.25}$
$1.5 <$ NV1	OFF	ON	1	-1.25	2.25	$\frac{1}{2}$

```
          TR A NEGATIVE ELEMENT
       B1  N(0,1),R=.001
       E1  (10),0,10
       B2  N(1,0),G=1
       B3  N(1,0),G=(-1E-6,-1.25),E=-1
       B4  N(1,0),G=(1E-6,2.25),E=-1.5
       S1  B=3,(3),ON
       S2  B=4,(4),OFF
          TI=1
          FI=5
          PR,NV,CA
          EX


   T =   0.0

NODES              NODE VOLTAGES

  1-  1      0.49949911E-09

BRANCHES           ELEMENT CURRENTS

  1-  4     -0.49949926D-06     0.49949926D-09     0.99999943D-06     -0.14999982D-05


   T =   0.1000000E 01

NODES              NODE VOLTAGES

  1-  1      0.99900097E 00

BRANCHES           ELEMENT CURRENTS

  1-  4      0.99900050D 00     0.99900100D 00     0.99889999D-09     -0.50099776D-06

SWITCH  1 IS OFF


   T =   0.1000000E 01

NODES              NODE VOLTAGES

  1-  1      0.99997652E 00

BRANCHES           ELEMENT CURRENTS

  1-  4      0.10000054D 01     0.99997656D 00     0.29303553D-04     -0.50002220D-06


   T =   0.1499023E 01

NODES              NODE VOLTAGES

  1-  1      0.14991245E 01

BRANCHES           ELEMENT CURRENTS

  1-  4      0.87521928D 00     0.14991248D 01    -0.62390550D 00     -0.87426511D-09

SWITCH  2 IS ON


   T =   0.1499023E 01

NODES              NODE VOLTAGES

  1-  1      0.14991264E 01

BRANCHES           ELEMENT CURRENTS

  1-  4      0.87325611D 00     0.14991267D 01    -0.62390795D 00     -0.19626796D-02


   T =   0.2000000E 01

NODES              NODE VOLTAGES

  1-  1      0.19991026E 01

BRANCHES           ELEMENT CURRENTS

  1-  4      0.18732093D 01     0.19991034D 01    -0.12488782D 01      0.11229842D 01
```

Fig. 4.23. Negative element example output.

```
     T =    0.3000000E 01

   NODES                  NODE VOLTAGES
     1-   1       0.29971066E 01
   BRANCHES               ELEMENT CURRENTS
     1-   4       0.38692173D 01    0.29971073D 01    -0.24963823D 01     0.33684922D 01


     T =    0.4000000E 01

   NODES                  NODE VOLTAGES
     1-   1       0.39951105E 01
   BRANCHES               ELEMENT CURRENTS
     1-   4       0.58652253D 01    0.39951113D 01    -0.37438863D 01     0.56140003D 01


     T =    0.5000000E 01

   NODES                  NODE VOLTAGES
     1-   1       0.49931145E 01
   BRANCHES               ELEMENT CURRENTS
     1-   4       0.78612333D 01    0.49931153D 01    -0.49913904D 01     0.78595083D 01
```

Fig. 4.23 (*continued*).

4. The nonlinear element in the example has a resistance of $1/(G_2 + G_3 + G_4)$. Table 4.2 demonstrates how parallel conductances are combined to yield the required element characteristic. The approximate nonlinear resistance in each region is also shown in the table.
5. Conductances (rather than resistances) have been used in the branches which constitute the nonlinear element simply because summing parallel conductances permits easy user determination of component values.

4.11 *FUNCTION GENERATION*

Although time-dependent sources provide a reasonable capability for generating arbitrary signal waveforms, it is occasionally necessary to construct a separate

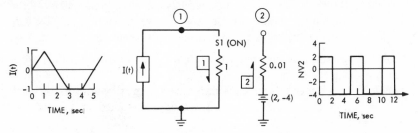

Fig. 4.24. Function generator circuit.

```
      TR   FUNCTION GENERATOR
   B1 N(1,0),R=1
   I1 P(1),0,1,0,-1,-1,0
   S1 B=1,(1),ON
   B2 N(0,2),R=.01,E=(2,-4)
C  B3 IS THE LOAD ON B2
   B3 N(2,0),R=1E3
      TI=1
      FI=20
      PR,NV
      EX
```

Fig. 4.25. Function generation input data.

ECAP network to generate a special signal for a specific application. Such networks can be designed to generate almost any physically realizable waveform. Of course, the complexity of the signal-generating network is usually directly proportional to the complexity of the required waveform.

Because the time-dependent source is limited to finite-sloped pulses, the technique for generating a pulse train with zero rise-and-fall times is a simple practical example of an ECAP signal-generating network. A signal-generating network and its associated discontinuous periodic signal are shown in Fig. 4.24. The input data describing the network are shown in Fig. 4.25. A buffering technique such as described in Sec. 2.14 can be used to keep the generated waveform at node 2 invariant with loading.

Further discussion of the generation of arbitrary functions using ECAP networks is presented in chapters 5 and 7.

4.12 SUMMARY

This chapter presented the details necessary for the use of ECAP's transient analysis program. The piecewise linear modeling capability is one of the most important features of the program. This feature allows the dynamic modification of model parameters as a function of the circuit variables and is important in the flexibility of ECAP.

These points should be re-emphasized to reduce the possibility of trouble in using the transient analysis program:

1. MODIFY routines are not permitted in a transient analysis.
2. A TIME STEP and a FINAL TIME must be specified for every problem unless an equilibrium solution is requested. Both values must be chosen judiciously.
3. An L or a C may be used as the *from*-branch of a T card with a BETA only (i.e., the *from*-branch for a GM must be a resistance).
4. No branch can be simultaneously used as a *from*- and a *to*-branch.
5. Switch conditions must be specified realizing that positive current flowing through a negative element produces a voltage rise.

Table 4.3 summarizes the solution-control input statements which are used only in a transient analysis.

Appendix C includes tables that summarize the legitimate transient analysis input parameters, directly available output data, assumed units, and overall program capacities.

Table 4.3.   Transient analysis solution-control input statements.

Input statement		Definition of x	Assumed value if not specified	Reference section
INITIAL TIME	$= x$	Indicated time of first solution	0.0 sec	4.8
TIME STEP	$= x$	Effective solution interval	Must be specified	4.1, 4.6
OUTPUT INTERVAL	$= x$	Only every xth solution calculated is to be printed	1	4.1
FINAL TIME	$= x$	Time of last requested solution	Must be specified	4.1
SHORT	$= x$	Short-circuit resistance	$0.01\ \Omega$	4.7, 4.9
OPEN	$= x$	Open-circuit resistance	$1E7\ \Omega$	4.7, 4.9
1ERROR	$= x$	Acceptable total current unbalance	0.001 a	4.1
2ERROR	$= x$	Determines switching-time accuracy	0.001	4.5
EQUILIBRIUM		Requests steady-state solution	—	4.9

PROBLEMS

4.1. Complete the input data shown in Fig. p4.1a to obtain, at node 1, the voltage waveform illustrated in Fig. p4.1b.

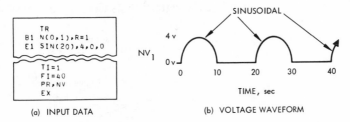

(a) INPUT DATA   (b) VOLTAGE WAVEFORM

Fig. p4.1

4.2. Write the input data statement that specifies (in branch 143) an approximately sinusoidal current source with a period of $80 \cdot TI$, a peak-to-peak amplitude of 20 amp, and an average value of 0. Use a periodic time-dependent source and at least 17 points to specify the source waveform.

4.3. Write the simplest ECAP input data which represent these two-valued sources:
  a. The source has a value of 24 v from $t = 0$ sec to $t = 1.25$ sec, and a value of 50 v thereafter.
  b. The source has a value of 12 v from $t = 0$ sec to $t = 10$ sec, $t = 15$ sec to $t = 25$ sec, $t = 30$ sec to $t = 40$ sec, etc., and a value of 18 v for the intermediate 5 sec intervals.

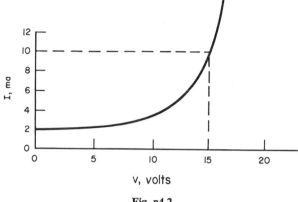

Fig. p4.2

4.4. Write the ECAP input data which simulate the voltage-current characteristic shown in Fig. p4.2. Use at least four line segments to approximate this characteristic. Test the simulation by varying the voltage across it and observing the corresponding currents.

4.5. Determine the switch response as a function of time for each of the following conditions:
  a. The switch is initially in the ON state with its element current as shown in Fig. p4.3a.
  b. The switch is initially in the OFF state with its element current as shown in Fig. p4.3b.
  c. The switch is initially in the ON state with its element current as shown in Fig. p4.3b.

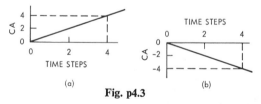

Fig. p4.3

```
        TR
B1  N(0,1),C=4,EO=2
B2  N(2,1),L=20
B3  N(2,0),R=4,E=5
T1  B(2,3),BETA=2
T2  B(1,2),GM=.4
    IN=5
    FI=15
    PR,NV
    EX
```

Fig. p4.4

4.6. Find the error(s) in the input data shown in Fig. p4.4. Which error(s) will elicit an ECAP error message?

4.7. By considering ECAP's technique for determining switch actuation times (see Appendix B3.3), derive the equation $n \geq 3.32 \log (2\text{ERROR})^{-1}$ of comment 1 in Sec. 4.5.

4.8. Write the algebraic nodal equations at $t = 3, 5.375, 5.375+$, and 6 sec which are solved by the ECAP transient analysis program when executing the input data in Fig. B.17 in Appendix B. Write the equations in the format shown below for $t = 5$ sec.

$$t = 5 \left\{ \begin{aligned} \left(\frac{1}{R_1} + \frac{1}{\Delta t/C} \right) e_1(5) &= \frac{E_1}{R_1} + \frac{e_1(4)}{\Delta t/C} \\ \Delta t = 1 \left\{ \left(\frac{1}{R_3} + \frac{1}{2L/\Delta t} \right) e_2(5) &= \frac{E_3}{R_3} - \left[J_L(4) + \frac{\Delta t}{2L} \cdot e_2(4) \right] \right. \end{aligned} \right.$$

4.9. What will be the transient response computed by ECAP to the input data in Fig. p4.5? (Note that if S1 had been specified as OFF initially, the problem would have run indefinitely because of *switch lockup*.) What effect does branch 3 have on the problem solution?

```
            TR
  B1  N(0,1),R=3,E=(10,-2)
  B2  N(1,0),R=2
  S1  B=2,(1),ON
  B3  N(0,2),C=1
      TI=1
      FI=5
      PR,NV,CA
      EX
```

Fig. p4.5

```
          TR
  B1  N(0,1),R=2
  E1  (3),-1.4,1.6
  B2  N(1,0),R=(2,3)
  S1  B=2,(2),OFF
      IN=5
      TI=1
      FI=10
      2ERROR=.1
      PR,NV,CA
      EX
```

Fig. p4.6

4.10. What is the smallest value of Δt used in the transient analysis to determine the switch actuation time in the circuit described by the input data in Fig. p4.6? Determine the printed switching time.

4.11. Plot NV1(t) for Circuit B in Fig. p4.7 when Circuit A is used to control the 10 v source in branch 2 in such a way that the source has a 10 sec period and 20 per cent duty cycle.

4.12. Repeat prob. 4.11 with an initial voltage of $+5$ v at node 1.

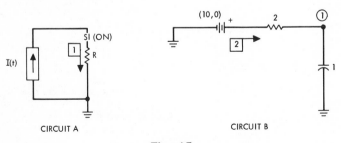

Fig. p4.7

```
       TR
B1 N(0,1),R=1
E1 SIN(8),1,0,0

TI=.1
OU=5
FI=25
PR,NV,CA
EX
```

Fig. p4.8

4.13. Complete the input data of Fig. p4.8 so that there will be a branch across branch 1 whose element current is proportional to the derivative of the unloaded terminal voltage of branch 1. Plot the waveforms of NV1 and the element current of the new branch shunting branch 1 to verify that the relationship between the sinusoidal voltage source E1 and its derivative has been realized.

4.14. Determine the transient response of the double-tuned circuit in Fig. p4.9 for the 5 μsec period after the switch is closed at $t = 0$ sec. Calculate the response for $k = M(L_1 L_2)^{-1/2} = 0.3$. Repeat for $k = 0.5$, and $k = 0.8$. Correlate the results from these analyses with the AC analyses of prob. 3.14 and explain the transient analysis results obtained. (The IBM 7094 and 1620 versions of ECAP require the use of the nonideal transformer model described in Sec. 5.3.3.)

4.15. Calculate the transient response of the circuit in Fig. p3.11 with C_1, C_2, C_3, C_4, C_5 equal to 1.0 μf and $R_6 = 10$kΩ to the input signal shown in Fig. p4.10. Repeat with $R_6 = 100\Omega$. An equivalent circuit for the transistors can be found in Appendix A.1.

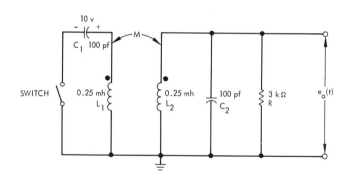

Fig. p4.9

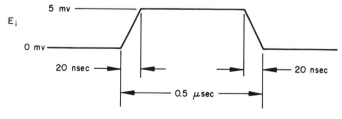

Fig. p4.10

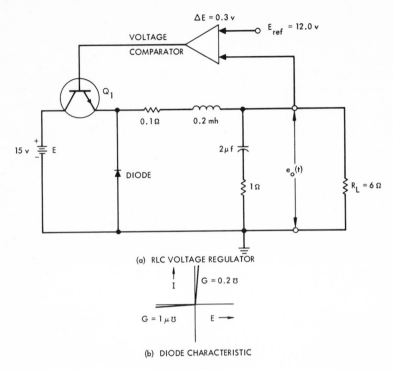

(a) RLC VOLTAGE REGULATOR

(b) DIODE CHARACTERISTIC

Fig. p4.11

4.16. Calculate the transient response of the *RLC* voltage regulator in Fig. p4.11. The output voltage is regulated by alternately applying current from a power source (transistor Q_1 and voltage source E) into a closed *RLC* circuit. The transistor is switched on and off by the voltage comparator, which maintains the output voltage between 11.7 v and 12.0 v. (When $e_o(t)$ is less than $+11.7$ v, transistor Q_1 is saturated; when $e_o(t)$ is greater than $+12$ v, Q_1 is cut off.) *Note:* In this analysis the transistor can be modeled as a simple controlled branch with a collector saturation resistance value of 0.1 Ω. The voltage comparator can also be modeled as a simple controlled branch.

5

MODELING
TECHNIQUES

5.1 INTRODUCTION

In the application of the Electronic Circuit Analysis Program, the need often arises for a circuit element that is not included in the list of standard ECAP elements, which consists of resistances (or conductances), capacitances, inductances, switches, dependent current sources, independent voltage sources, and independent current sources. In all the programs, applications may require the use of linear but nonstandard elements—such as small-signal transistors or ideal transformers (coupling coefficient $k_{ij} = |M_{ij}|/\sqrt{L_i L_j} = 1$). In the transient analysis program, there are many requirements for nonlinear elements, such as a transistor to operate in a large-signal application, a diode or nonlinear resistance, or possibly a tunnel diode, which not only is nonlinear but also exhibits a negative resistance characteristic in one region of operation. In the case of linear nonstandard elements, general equivalent circuits have often been developed by others; they can be applied to ECAP. For example, the hybrid-π transistor equivalent circuit, whose characteristics are summarized in Appendix A.1, is ideal for most small-signal transistor applications in ECAP.

This chapter will present four basic approaches to developing equivalent circuits for elements that are not standard in ECAP. These approaches are (1) the piecewise linear approximation, (2) the mathematical function simulation, (3) the use of physical analogies to develop a piecewise linear approximation, and (4) the use of physical analogies to simulate a mathematical function. These approaches can be used individually or can be combined to form more complicated element models. Using the techniques in this chapter, one should be able to construct almost any required nonstandard element.

Each approach to modeling is first discussed conceptually to establish the necessary groundwork. Then one or more equivalent circuits are developed in detail.

5.2 PIECEWISE LINEAR MODELING

5.2.1 General Concepts

The transient response of a circuit containing nonlinear elements can be analyzed by replacing all the nonlinear components with equivalent circuits, constructed with the linear elements recognized by the transient analysis program. All standard elements in the program are linear, except when they are used in conjunction with a switch. The switch is the key element here because it provides the capability to formulate piecewise linear approximations to nonlinear characteristics.

The operation of the switch is described in detail in Sec. 4.4. Briefly, as a review, a change in the state of a switch can cause the elements in its own branch or in other branches or in both to change from an initial to a second value.

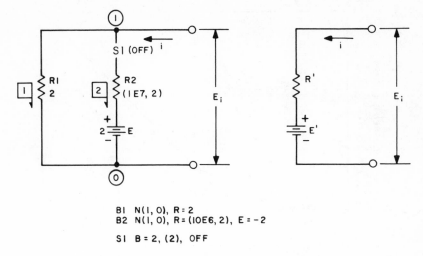

BI N(I, 0), R = 2
B2 N(I, 0), R = (IOE6, 2), E = -2

SI B = 2, (2), OFF

Fig. 5.1. Approximation of nonlinear *E-I* function.

Each parameter can have two values; each value can be controlled by one or more switches.

The circuit in Fig. 5.1 is an excellent example of the piecewise linear approximation that can be obtained by using the switch in combination with other linear elements. When node voltage 1 is less than $+2.0$ v, S1 is in its assumed state, and the resistance value in branch 2 is 10 MΩ. Under this condition the Thévenin equivalent circuit of the network is

$$R_T = \frac{R_1 R_2}{R_1 + R_2} = \frac{(2)(10^7)}{2 + 10^7} \simeq 2 \ \Omega$$

$$E_T = \frac{ER_1}{R_1 + R_2} = \frac{(2)(2)}{2 + 10^7} \simeq 0 \ \text{v}$$

When node voltage 1 is equal to or greater than $+2.0$ v, S1 changes states, and the resistance value in branch 2 is changed to 2 Ω. The Thévenin equivalent circuit now becomes

$$R_T = \frac{(2)(2)}{2 + 2} = 1 \ \Omega$$

$$E_T = \frac{(2)(2)}{2 + 2} = 1.0 \ \text{v}$$

Graphically, the voltage-current characteristic of this network is illustrated in Fig. 5.2. The reader should note that the resistance value assumed in branch 2 when S1 was in the OFF state was selected large enough so the equivalent network resistance was equal for all practical purposes to the resistance in branch 1. Also, the Thévenin equivalent voltage of the network was essentially equal to zero. Selecting a smaller value of resistance in branch 2 when the switch was

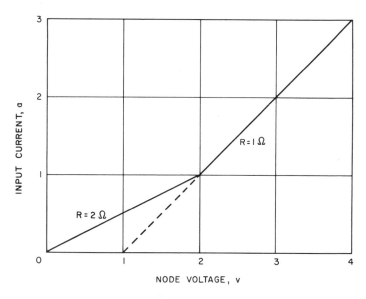

Fig. 5.2. Response of nonlinear *E-I* function approximation.

in its assumed state would have caused a shift in the resistance characteristic and the intercept voltage on the abscissa, which was not desired in this case.

Generally, only three ECAP elements are used to simulate nonlinear characteristics. These are the resistance, the voltage source, and the switch. Four possible combinations of a resistance and a voltage source ($\pm R$, $\pm E$) can be used to form linear segments. The resistance value determines the slope of the segment (i.e., a positive resistance produces a positive slope and a negative resistance produces a negative slope). The voltage source determines the point at which the segment intercepts the abscissa.

When building an equivalent circuit for a nonlinear component, the first step is always the definition of the characteristic to be modeled. The characteristic is often a function of the voltage across the element, or a function of the current through the element. For example, the performance of a nonlinear resistance can be defined by relating the current that flows through it to the voltage impressed across it.

Similarly, the charge on a nonlinear capacitance is a nonlinear function of the voltage applied to it, and the behavior of a nonlinear inductance can be described by relating the magnetic flux to the current.

The technique used in constructing the latter two nonlinear models is discussed in Sec. 5.4, in which physical analogies become necessary to derive equivalent circuits. In this section, however, only the techniques used to formulate a piecewise linear approximation to a nonlinear voltage-current characteristic are discussed. The principles of piecewise linear network synthesis for a general nonlinear network using the switch element (or ideal diode) are discussed in great detail by Stern[1] and others. If additional information on the subject is desired, Stern's report is recommended.

Two equivalent circuits will be constructed to demonstrate the application of the piecewise linear approximation concept. The first example develops a model of the commonly used 1N3600 diode. The second example develops a model of the General Electric TD-254A tunnel diode, to demonstrate the modeling of a negative resistance characteristic.

5.2.2 Diode Model

The voltage-current relationship of the 1N3600 diode is defined in the specifica-

[1] T. E. Stern, "Piecewise-Linear Network Theory," *Technical Report No. 315* (Cambridge, Mass.: Research Laboratory for Electronics, Massachusetts Institute of Technology, 1956).

DATA SHEET AND PROCUREMENT SPECIFICATIONS • FEBRUARY 1964 ■

HIGH CONDUCTANCE

1N3600

HIGH CONDUCTANCE ULTRA FAST EPITAXIAL PLANAR DIODE

REGISTERED SPECIFICATIONS

ELECTRICAL SPECIFICATIONS (25°C unless otherwise noted)

Symbol	Characteristic	Min.	Max.	Units	Test Conditions
V_{F_1}	Forward Voltage	.87	1.00	Vdc	$I_F = 200$ mA
V_{F_2}	Forward Voltage	.82	.92	Vdc	$I_F = 100$ mA
V_{F_3}	Forward Voltage	.76	.86	Vdc	$I_F = 50$ mA
V_{F_4}	Forward Voltage	.66	.74	Vdc	$I_F = 10$ mA
V_{F_5}	Forward Voltage	.54	.62	Vdc	$I_F = 1$ mA
I_{R_1}	Reverse Current		0.1	μA	$V_R = 50$ V
I_{R_2}	Reverse Current (150°C)		100	μA	$V_R = 50$ V
BV	Breakdown Voltage	75			$I_R = 5 \mu$A
t_{rr_1}	Reverse Recovery Time		4.0	nSec	See Table III
t_{rr_2}	Reverse Recovery Time		6.0	nSec	See Table III
C_O	Capacitance [Note 2]		2.5	pf	$V_R = 0$ V, $f = 1$ mc

Copyright 1964 by Fairchild Semiconductor, a Division of Fairchild Camera and Instrument Corporation

NOTES:
(1) The maximum ratings are limiting values above which life or satisfactory performance may be impaired.

(2) Capacitance as measured on Boonton Electronic Corporation Model No. 75-AS8 Capacitance Bridge or equivalent.

(3) Leads are tinned. Gold plate with nickel strike may be obtained when specified.

FAIRCHILD

SEMICONDUCTOR
A DIVISION OF FAIRCHILD CAMERA AND INSTRUMENT CORPORATION

MANUFACTURED UNDER ONE OR MORE OF THE FOLLOWING U. S. PATENTS: 2959681, 2971139, 2981877, 3013955, 3015048, 3025589, 3034106. OTHER PATENTS PENDING.

Fig. 5.3. Diode specification sheet.

tion sheet* in Fig. 5.3. The nominal voltage-current characteristic will be modeled in this example.

The first step in the development of an equivalent circuit is to draw a piecewise linear approximation for the nonlinear characteristic. Figure 5.4a shows a piecewise linear approximation consisting of five connected straight-line segments superimposed on the nominal voltage-current curve of the 1N3600 diode.

The second step in the construction of the model is to reduce the characteristic of Fig. 5.4a to an equivalent set of functions whose sum is the characteristic 1 → 6 . The individual elements of the set are combined as discussed in Sec. 5.2.1 to form the piecewise linear approximation shown in Fig. 5.4a.

Next, the slopes of the segments of the piecewise linear approximation are calculated and listed in Table 5.1.

The slopes of the line elements A through E are also given in Table

*Courtesy of Fairchild Semiconductor, a division of Fairchild Camera and Instrument Corp., Mountain View, Calif.

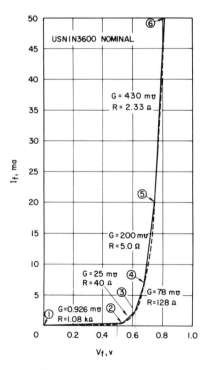

(a) PIECEWISE LINEAR APPROXIMATION

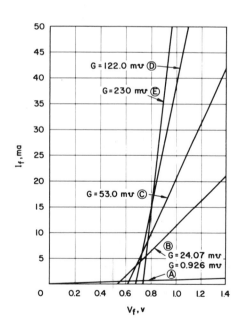

(b) ELEMENTS FORMING IN3600 NOMINAL PIECEWISE LINEAR APPROXIMATION EQUIVALENT CIRCUIT

Fig. 5.4. 1N3600 nominal diode characteristic.

Table 5.1.   Slopes of the piecewise linear voltage-current characteristic of the 1N3600 diode.

Line	Slope, m℧	Line	Slope, m℧
1 → 2	0.926	A	0.926
2 → 3	25	B	24.07
3 → 4	78	C	53
4 → 5	200	D	122
5 → 6	430	E	230

5.1. The slope of A is the same as the slope of 1 → 2 since these are essentially the same line segment. The slope of B is the difference between the slopes of 2 → 3 and 1 → 2 . That is, the slope of 2 → 3 is 25 m℧ and the slope

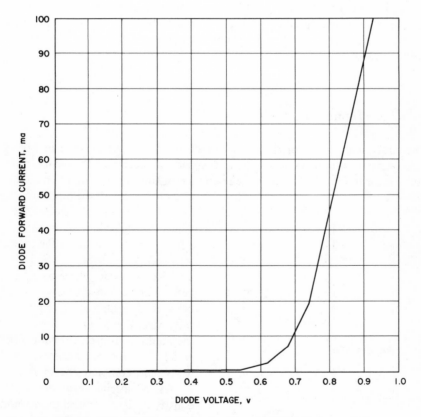

Fig. 5.5. Piecewise linear approximation of 1N3600 diode nominal forward characteristic.

```
C    1N3600 DIODE MODEL
     TR
B1  N(1,0),G=(.926E-3,.4E-9),E=.001
B2  N(1,0),G=(0.4E-9,24.074E-3),E=-.540
B3  N(1,0),G=(0.4E-9,.053),E=-.620
B4  N(1,0),G=(0.4E-9,.122),F=-.680
B5  N(1,0),G=(0.4E-9,.230),E=-.740
B6  N(1,0),R=1E7
I6  (100),0,0.1
S1  B=1,(1),ON
S2  B=2,(2),OFF
S3  B=3,(3),OFF
S4  B=4,(4),OFF
S5  B=5,(5),OFF
    TI=1
    OU=1
    FI=100
    PRINT,NV,CA
    EXECUTE
```

Fig. 5.6. Typical input data including the five-segment nominal 1N3600 equivalent circuit.

of A is 0.926 m℧; so then the slope of B must be approximately 24.07 m℧. Similarly, the slope of C is the difference between the slopes of $3 \rightarrow 4$ and $2 \rightarrow 3$, etc.

The resistance of the diode characteristic, if the terminal voltage is less than zero, is the sum of the OFF-state conductances of the equivalent circuit branches. In this example, the value of the reverse conductance is $I_R/V_R = 5(0.4 \times 10^{-9})$ ℧.

The values of the voltage sources in branches 1 through 5 of the model are the voltages at the breakpoints of the piecewise linear approximation. That is, the value of the voltage source specified in a branch is the lower voltage limit at which the conductance of that branch should be added to the characteristic.

The five-segment piecewise linear approximation to the 1N3600 diode nominal characteristic is shown in Fig. 5.5. The input data describing this approximation are illustrated in Fig. 5.6.

5.2.3   Tunnel Diode Model

The equivalent circuit of the tunnel diode is of the form shown in Fig. 5.7. It consists of an inductance L_s due to the lead inductance of the device, a resistance R_s primarily due to the bulk resistance of the diode, a shunt capacitance C associated with the diode junction, and a resistance characteristic R whose magnitude and polarity are a function of the value of the forward bias voltage.

The parasitic elements (L_s, R_s, and C) are assumed constant so that no special treatment is required. A piecewise linear approximation of the nonlinear resistance characteristic which is voltage sensitive can be devised to complete the model.

As a specific example of the construction of a tunnel diode model let us develop a model for the General Electric TD-254A tunnel diode. The

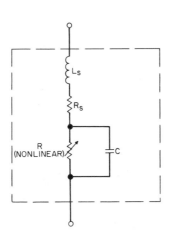

Fig. 5.7. Tunnel diode equivalent circuit.

ELECTRICAL CHARACTERISTICS: (25°C)		TD–254A		
		MIN	TYP	MAX
PEAK POINT CURRENT, I_p	(ma)	20	22	24
VALLEY POINT CURRENT, I_v	(ma)		2.7	3.8
PEAK POINT VOLTAGE, V_p	(mv)		100	120
VALLEY POINT VOLTAGE, V_v	(mv)		425	
FORWARD VOLTAGE				
$(I_f = I_p)\ V_{fp}$	(mv)	550	600	650
$(I_f = 0.25\ I_p)\ V_{fs}$	(mv)	460	540	
TOTAL SERIES INDUCTANCE, L_s	(nh)		1.5	
TOTAL SERIES RESISTANCE, R_s	Ω		2.0	
VALLEY POINT TERMINAL CAPACITANCE, C	(pf)		2.5	4.0
RISE TIME, t_r	(psec)		64	

Fig. 5.8. GE TD-254A specification.

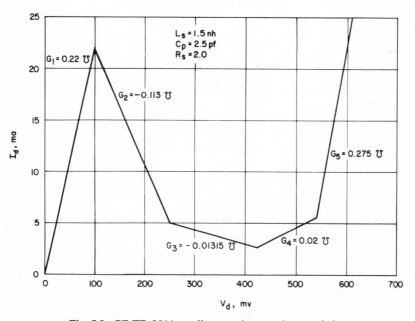

Fig. 5.9. GE TD-254A nonlinear resistance characteristic.

specifications of this device* are listed in Fig. 5.8. The values for L_s, R_s, and C are available directly from this specification.

The peak point, valley point, and slope above the valley point ($V_d > V_v$) of the nonlinear resistance characteristic are specified. Using this information, an approximation to the characteristic can be formed. The specification does not provide the value of negative resistance between the peak point and valley point. Therefore, the slope of the negative resistance region must be either measured or arbitrarily defined. For this example, the assumed resistance characteristic is shown in Fig. 5.9. The resistance characteristic must now be constructed using the standard ECAP elements.

The conductance of the first segment ($V_d < 100$ mv) is 0.220 ℧. The conductance G_2 of the second segment (100 mv $< V_d < 250$ mv) is negative and assumed equal to -0.113 ℧, and can be obtained by adding line element B to element A of Fig. 5.10. The third segment (250 mv $< V_d < 425$ mv) contains conductance G_3, which is less negative than G_2, and is assumed equal to -0.01315 ℧. This portion of the characteristic is formed by adding line element C of Fig. 5.10 to the combination of elements A and B. The conductance G_4 of the fourth segment (425 mv $< V_d < 540$ mv), which is the combination of the third segment and line element D, is positive and equal to 0.02 ℧. The fifth segment representing $G_5 = 2.275$ ℧ is the combination of functions A through E of Fig. 5.10. Table 5.2 contains the values of conductance for G_1 through G_5 for line elements A through E.

The resistance of the tunnel diode characteristic for $V_d < 0$ is the sum of the OFF-state conductances of the equivalent circuit branches.

Table 5.2.   Slopes of the piecewise linear voltage-current characteristics of the TD-254A tunnel diode.

Conductance	Slope, mhos	Line	Slope, mhos
G_1	+0.220	A	+0.220
G_2	−0.113	B	−0.333
G_3	−0.01315	C	+0.09985
G_4	+0.02	D	+0.03315
G_5	+0.275	E	+0.255

The breakpoints of the piecewise linear approximation are the values of the voltage sources in the branches representing elements A through E. That is, the value of the voltage source specified in a branch is the lower voltage limit at which the conductance of that branch should be changed to the second listed value.

*Courtesy of General Electric Company, Syracuse, N. Y.

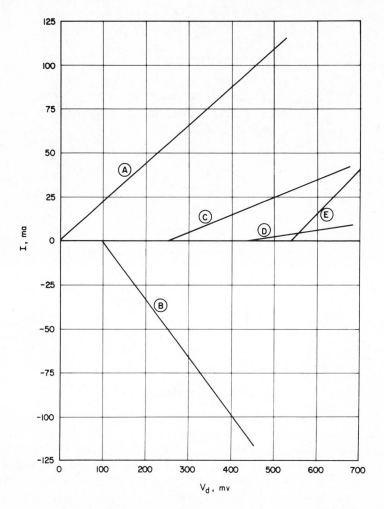

Fig. 5.10. Elements forming GE TD-254A piecewise linear approximation resistance characteristic.

The shunt capacitance, which is present in the tunnel diode, is always required in the tunnel diode model. It is also required with any nonlinear resistance in which the resistance assumes both positive and negative values as a function of voltage, and the component model operates across a region not defined by a load line. The reason for this can best be explained in the following manner.

In Fig. 5.11a, as i is increased from zero to the peak point i_p, the operating point of the tunnel diode follows line segment G_1. When the current reaches i_p, the switch actuation occurring at peak voltage E_p changes the resistance characteristic to a new slope G_2. As the current is further increased, the operating point will still follow the characteristic since a node voltage solution cannot be

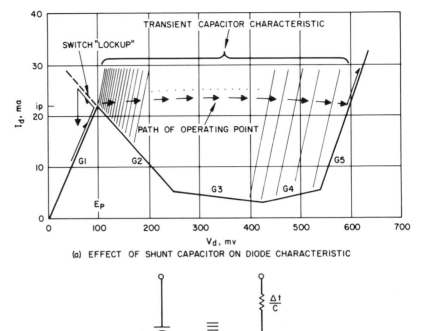

(a) EFFECT OF SHUNT CAPACITOR ON DIODE CHARACTERISTIC

(b) CAPACITOR EQUIVALENT CIRCUIT

Fig. 5.11. Tunnel diode model operation during switching time.

obtained at any point not on the characteristic. The line segments forming the characteristic extend beyond it through the current and voltage axes. Thus, when the current is increased above i_p, the operating point will attempt to follow the characteristic toward the current axis, causing the diode voltage to drop below E_p and the slope of the characteristic to return to G_1. This cycle, referred to as *switch lockup*, will repeat until the program is manually stopped.

The problem is solved by shunting a capacitance across the nonlinear resistance. The transient analysis capacitance equivalent circuit, shown in Fig. 5.11b, is a conductance $C/\Delta t$ in series with a voltage source $E(t)$, whose value equals the branch voltage of the previous time step. If the value of the capacitance is selected so that

$$C/\Delta t > |G_2| \tag{5.1}$$

the characteristic formed by combining the capacitance equivalent circuit and the nonlinear resistance will always have a positive slope.

The effect of the capacitance is very important after the drive current i exceeds i_p. Since the slope of the combined characteristic is always positive, the

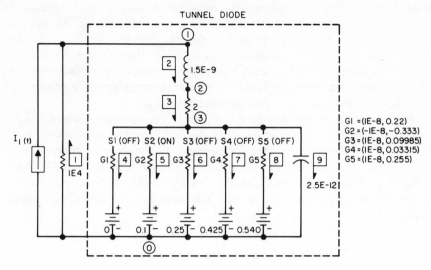

Fig. 5.12. Piecewise linear approximation to GE TD-254A tunnel diode.

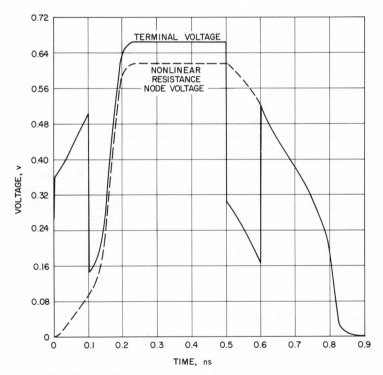

Fig. 5.13. Tunnel diode model response to a 26-ma current pulse with $T_r = 0.1$ nanosecond $= T_f$.

switch lockup experienced without the capacitance will not occur. Instead, the operating point will tend to follow the new positive characteristic. At each time step, the voltage source $E(t)$ will increase in value since the capacitance is being charged by the Thévenin equivalent voltage E_T through the conductance $(-G_2)$. The increase in $E(t)$ at each time step will shift the contribution due to the capacitance characteristic. This shift with each time step will continue until the operating point comes to rest on conductance G_5.

The transition of the operating point from conductance G_5 to G_1 across the forbidden region occurs in a similar manner. The complete model for the TD-254A tunnel diode, including a five-segment piecewise linear approximation to the nonlinear resistance characteristic and the parasitic elements, is shown in Fig. 5.12. The response of the model to a 26 ma current pulse with 0.1 nsec rise and fall times is plotted in Fig. 5.13. The input data for this example appear in Fig. 5.14.

```
C       TUNNEL DIODE MODEL
        TR
B1  N(0,1),R=1E4
B2  N(1,2),L=1.5E-9
B3  N(2,3),R=2
B4  N(3,0),G=(1E-8,0.22)
B5  N(3,0),G=(-1E-8,-0.333),E=-0.1
B6  N(3,0),G=(1E-8,0.09985),E=-0.25
B7  N(3,0),G=(1E-8,0.03315),E=-0.425
B8  N(3,0),G=(1E-8,0.255),  E=-0.540
B9  N(3,0),C=2.5E-12
S1  B=4,(4),OFF
S2  B=5,(5),ON
S3  B=6,(6),OFF
S4  B=7,(7),OFF
S5  B=8,(8),OFF
T1  (1000),0,-26E-3,-26E-3,-26E-3,-26E-3,-26E-3,0
    TI=0.1E-12
    OU=20
    FI=1E-9
    PRINT,NV,CA
    EX
```

Fig. 5.14. Input data to compute tunnel diode model response to a 26-ma current pulse with $T_r = 0.1$ nanosecond $= T_f$.

RESTRICTIONS. There are only two major restrictions to be considered in the use of the tunnel diode model. First, the value of the capacitance shunting the nonlinear resistance characteristic must be greater than $\Delta t \, |1/(-R)|$. If C is less than $\Delta t \, |1/(-R)|$, the model will not function properly.

Second, the time constants of the parasitic elements must be considered when making the time-step selection for the analysis. The parasitic elements are very important when considering the transient response of tunnel diode models.

5.3   MATHEMATICAL MODELING

5.3.1   General Concepts

Linear circuit elements not standard to ECAP often can be modeled using only the mathematical relationships which describe their terminal characteristics.

Reasonably accurate equivalent circuits for many linear elements are available and in general use. A good example of a circuit element for which an equivalent circuit has not been established is the ideal transformer. This particular element is very commonly used in analysis but has always been treated as a fundamental element or as one which is in its simplest form and which serves as its own equivalent circuit. (Other examples of fundamental elements are the standard ECAP elements.) However, the ideal transformer is not available as a standard ECAP element, which indicates a definite need for an equivalent circuit that can be formulated using ECAP elements. The first mathematical modeling example (Sec. 5.3.2) develops the ideal transformer model.

The procedure to follow in developing equivalent circuits using mathematical modeling techniques is quite straightforward. In most cases, the modeling problem can be reduced to the synthesis of a simple *RLC* network sometimes requiring the use of dependent current sources. In the other cases, a little inventiveness is useful. The model of the ideal transformer developed in the following section requires that a small series resistance be inserted in the drive winding so that the primary and secondary currents can be directly related to the primary and secondary voltages. In reality, a small resistance usually does exist in the primary winding. If the resistance is negligibly small, however, a resistance that is small in relation to that of the primary circuit can be added without adversely affecting the validity of the model.

The equations describing the component must be linear since only linear elements are used in the development of the model. The availability of R, L, and C elements in the construction of the equivalent circuit permits the modeling of linear differential equations with constant coefficients.

Sections 5.3.2 and 5.3.3 demonstrate the application of the mathematical modeling technique to the development of the ideal transformer and nonideal transformer models. The examples illustrate the use of linear algebraic equations and linear differential equations in the construction of component models.

5.3.2 Ideal Transformer[2]

In the AC analysis and transient analysis programs, the capability to form transformers is provided via the use of the M card with coefficients of coupling < 0.999995. If a transformer is required with a coupling coefficient $k \geq 0.999995$, an equivalent circuit must be defined with the necessary properties. It is often necessary to analyze circuits containing transformers on a DC basis assuming a DC condition exists after the transients in the circuit have subsided. This transformer equivalent circuit must be constructed using only resistance elements and dependent current sources. The object of this example is to develop this model.

An ideal transformer (Fig. 5.15) is a device that consists of a group of two

[2]*The 1620 Electronic Circuit Analysis Program (ECAP) (1620-EE-02X) User's Manual* (White Plains, N.Y.: International Business Machines Corp., 1965), p. 86.

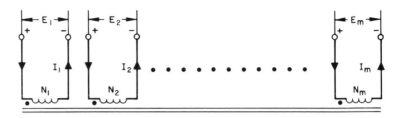

Fig. 5.15. Ideal transformer.

or more inductors which are perfectly coupled, i.e., $k_{ij} = 1$ (unity coefficient of coupling). The following voltage and current relationships define the ideal transformer:

$$\sum_{i=1}^{m} N_i \cdot I_i = 0 \tag{5.2}$$

$$\frac{E_1}{N_1} = \frac{E_2}{N_2} = \cdots \frac{E_m}{N_m} \tag{5.3}$$

where N_i = number of turns of the ith winding
  I_i = current flowing in the ith winding
  E_i = voltage developed across the terminals of the ith winding

For simplicity, the ideal transformer model development will be limited to the two-winding transformer. In this case, the basic Eqs. (5.2) and (5.3) reduce to

$$N_1 I_1 + N_2 I_2 = 0 \tag{5.4}$$

$$\frac{E_1}{N_1} = \frac{E_2}{N_2} \tag{5.5}$$

These equations provide the information necessary to construct a model of an ideal two-winding transformer.

Consider the transformer in Fig. 5.16. Resistance R_1 in series with winding N_1 is necessary to relate the currents I_1 and I_2 to voltages E_1 and E_2. The need for this resistance will become more apparent as the development progresses.

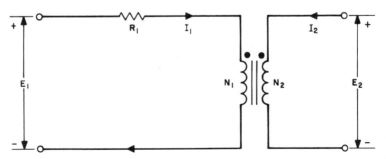

Fig. 5.16. Modified ideal transformer.

The loop equations for the circuit (Fig. 5.16) are:

$$E_1 = I_1 R_1 + E_2(N_1/N_2) \tag{5.6}$$

$$I_2 = -(N_1/N_2)I_1 \tag{5.7}$$

If R_1 is kept small, the equations are a good approximation of the ideal transformer.

By rearranging Eqs. (5.6) and (5.7) and solving for currents I_1 and I_2, the expressions become

$$I_1 = \frac{E_1}{R_1} - \left[\frac{E_2}{R_1(N_2/N_1)^2} \right] \cdot \frac{N_2}{N_1} \tag{5.8}$$

and

$$I_2 = -\frac{E_1}{R_1}\left(\frac{N_1}{N_2}\right) + \frac{E_2}{R_1(N_2/N_1)^2} \tag{5.9}$$

The equations are written in this form to isolate the terms involving the primary and secondary voltages and to establish the relationships between the terms of the primary and secondary currents.

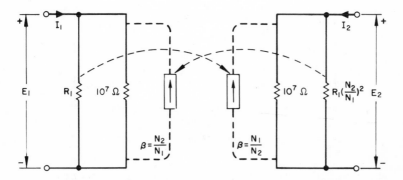

Fig. 5.17. Ideal transformer equivalent circuit (two windings).

The equations are obviously of the forms

$$I_1 = A - (1/K)B \tag{5.10}$$

$$I_2 = -KA + B \tag{5.11}$$

where $A = E_1/R_1$
$B = E_2/R_1(N_2/N_1)^2$
$K = N_1/N_2$

Utilizing the relationship established in Eqs. (5.8) and (5.9), the equivalent circuit (Fig. 5.17) can be constructed. The circuit shown possesses the same "black-box" characteristics as the transformer in Fig. 5.16. The 10 MΩ resistors are used as *to*-branches for the dependent current sources since a branch cannot be used as both a *to*- and a *from*-branch in the transient analysis program.

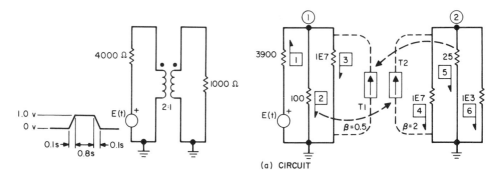

(a) CIRCUIT

```
      TR
   B1 N(0,1),R=3900
   E1 (1),0,1,1,1,1,1,1,1,1,1,1,0
   B2 N(1,0),R=100
   B3 N(1,0),R=1F7
   B4 N(2,0),R=1F7
   B5 N(2,0),R=25
   B6 N(2,0),R=1000
   T1 B(5,3),BETA=-0.5
   T2 B(2,4),BETA=-2.0
      TI=0.1
      OU=1
      FI=1.2
      PRINT,NV,CA
      EX
```

(b) INPUT DATA

Fig. 5.18. Ideal transformer example.

The application of the two-winding ideal transformer can be illustrated by the following example.

> Suppose the response of the ideal 2:1 transformer, driven by a 1 v trapezoidal signal as shown in Fig. 5.18a, is to be determined.

The input data for this example are shown in Fig. 5.18b; the input and output voltages are plotted in Fig. 5.19.

The concept applied here can be simply extended to ideal transformers of three or more windings. Transformers of this type must have a small series resistance in each drive winding to relate currents I_1 through I_m to voltages E_1 through E_m, respectively.

Restrictions on the use of this model are primarily on the selection of the series resistance R_i value. To maintain a balance of currents at the transformer nodes, it is necessary that the current in the elements constituting the primary side of the transformer be less than approximately 10^5 times the terminal current in the primary. For example, if the secondary load impedance R_L is very large, the primary current is small, and the primary voltage is relatively large, the element current E_i/R_i will be large, as will the current due to the dependent current source in the primary. The primary current I_i is equal to the difference between the element currents. As this difference approaches zero, the primary current becomes less accurate because of the limited computer precision. This inaccuracy

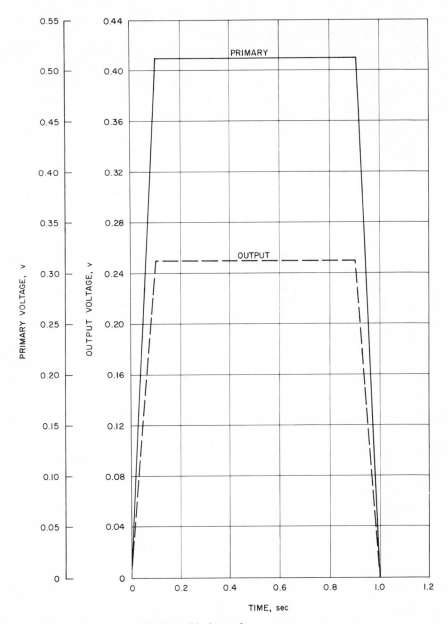

Fig. 5.19. Ideal transformer response.

can generally be avoided by increasing the value of R_i to a value which will improve the nodal current balance without degrading the model validity.

Since this model contains only resistance elements and dependent current sources, it can be used in the DC analysis (DC is defined as any time after transients in the circuit have subsided), the AC analysis, and the transient analysis programs.

5.3.3 Nonideal Transformer

The physical characteristics of a component often cannot be described with simple linear algebraic equations. In many of these instances, the mathematical relationships which define a component consist of linear differential equations with constant coefficients. This does not present any serious problems in the formulation of equivalent circuits.

Development of the nonideal transformer is not required in the IBM 360 version of ECAP since the M card is available in its transient analysis program, as well as in the AC analysis program. However, the model is very useful in the 1620 and 7094 versions of ECAP because the M card feature is not available in their transient analysis programs. This model makes it possible to perform a transient analysis on a circuit containing a multiple-winding nonideal transformer with coupling coefficients < 0.999995.

Unlike the ideal transformer model (Sec. 5.3.2), there is no requirement to use only resistances and dependent current sources since the nonideal transformer has no physical meaning in the DC analysis program. The object of this section is to develop the two-winding nonideal transformer model.

The nonideal transformer is a device consisting of a group of two or more windings which are not perfectly coupled, i.e.,

$$k_{ij} = |M_{ij}|/\sqrt{L_i L_j} < 0.999995$$

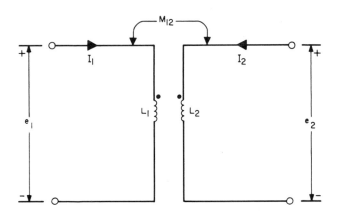

Fig. 5.20. Nonideal transformer.

The two-winding transformer illustrated in Fig. 5.20 can be described by the following equations:

$$e_1 = L_1 \frac{di_1}{dt} + M_{12} \frac{di_2}{dt} \tag{5.12}$$

$$e_2 = M_{12} \frac{di_1}{dt} + L_2 \frac{di_2}{dt} \tag{5.13}$$

Integrating Eqs. (5.12) and (5.13) and solving for i_1 and i_2

$$i_1 \left(1 - \frac{M_{12}^2}{L_1 L_2}\right) = \frac{1}{L_1} \int e_1 \, dt - \frac{M_{12}}{L_1} \cdot \frac{1}{L_2} \int e_2 \, dt \tag{5.14}$$

$$i_2 \left(1 - \frac{M_{12}^2}{L_1 L_2}\right) = -\frac{M_{12}}{L_2} \cdot \frac{1}{L_1} \int e_1 \, dt + \frac{1}{L_2} \int e_2 \, dt \tag{5.15}$$

since

$$k_{12}^2 = M_{12}^2 / L_1 L_2 \tag{5.16}$$

$$i_1 = \frac{1}{L_1(1 - k_{12}^2)} \int e_1 \, dt - \frac{M_{12}}{L_1} \cdot \frac{1}{L_2(1 - k_{12}^2)} \int e_2 \, dt \tag{5.17}$$

$$i_2 = -\frac{M_{12}}{L_2} \cdot \frac{1}{L_1(1 - k_{12}^2)} \int e_1 \, dt + \frac{1}{L_2(1 - k_{12}^2)} \int e_2 \, dt \tag{5.18}$$

Using the expression relating current and voltage in an inductance, we have

$$i = \frac{1}{L} \int e \, dt \tag{5.19}$$

which is of the form required to fit the equations.

Then, using the inductance L as a *from*-branch for a dependent current source (BETA only), the individual terms in the current equations can be generated.

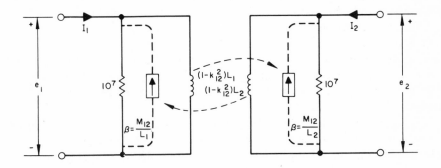

Fig. 5.21. Nonideal transformer equivalent circuit (two windings).

Using Eq. (5.19), the equivalent circuit for the two-winding nonideal transformer can be constructed as shown in Fig. 5.21. The transformer equivalent circuit possesses the same black-box characteristics as the transformer in Fig. 5.20. Application of the nonideal transformer is illustrated by the following example.

Suppose the response of the nonideal transformer driven by a 1 v trapezoidal signal, as shown in Fig. 5.22, is to be determined.

The first step is the calculation of the equivalent circuit for the primary inductance, turns ratio, and the coefficient of coupling:

$$L_2 = (1/n^2)L_1 = 0.25(1) = 0.25 \text{ h}$$

$$M_{12} = k_{12}\sqrt{L_1 L_2} = 0.75\sqrt{(1)(0.25)} = 0.37 \text{ h}$$

The inductance values for the equivalent circuit are

$$L_1' = (1 - k_{12}^2)L_1 = 0.4375 \text{ h}$$

$$L_2' = (1 - k_{12}^2)L_2 = 0.109375 \text{ h}$$

The dependent current sources T1 and T2 required in the equivalent circuit have values

$$\text{BETA}_{\text{T1}} = M_{12}/L_1 = 0.375$$

$$\text{BETA}_{\text{T2}} = M_{12}/L_2 = 1.5$$

The complete equivalent circuit and the input data for the analysis are shown in Fig. 5.23. The maximum time step for the solution is determined by the L/R time constant of the circuit. In this case, the time step should be approximately

$$\text{TI} = 0.01 \left(\frac{(1 - k_{12}^2)L_1}{R_p} \right) = 0.0625 \times 10^{-5} \simeq 1 \times 10^{-6} \text{ sec}$$

The results of the analysis are plotted in Fig. 5.24.

The techniques used in the development of the nonideal two-winding transformer can easily be extended to transformers of three or more windings.

The restrictions on the use of this model are primarily on the coefficient of coupling k_{ij}. As k_{ij} approaches unity, the inductance values used in the transformer model approach zero as a limit. In problems where a large load impedance exists and the coefficient of coupling approaches unity, the input terminal current will be very small in comparison with the element currents that combine to form the primary current.

The larger the value of k_{ij}, the greater the difference between the element currents and the terminal current. Since there is a limit to the precision of the calculations, the upper limit to the value of k_{ij} in the 360 version of the program is 0.999995; it is 0.9995 in the 7094 and 1620 versions.

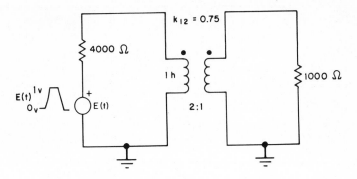

Fig. 5.22. Nonideal transformer example.

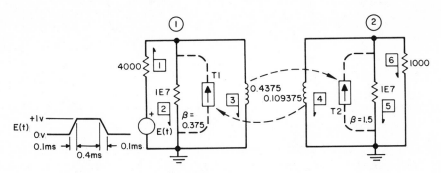

(a)   ECAP EQUIVALENT CIRCUIT

```
C
C        NONIDEAL TRANSFORMER EXAMPLE
C        TWO WINDINGS
         TR
    B1   N(0,1),R=4000
    E1   (100),0,1,1,1,1,1,1,0
    B2   N(1,0),R=1E7
    B3   N(1,0),L=0.4375
    B4   N(2,0),L=0.109375
    B5   N(2,0),R=1E7
    B6   N(2,0),R=1000
    T1   B(4,2),BETA=-0.375
    T2   B(3,5),BETA=-1.50
         TI=1E-6
         OU=10
         FI=1.2E-3
         PRINT,NV,CA
         EX
```

(b) INPUT DATA

Fig. 5.23. Equivalent circuit and input data for nonideal transformer example.

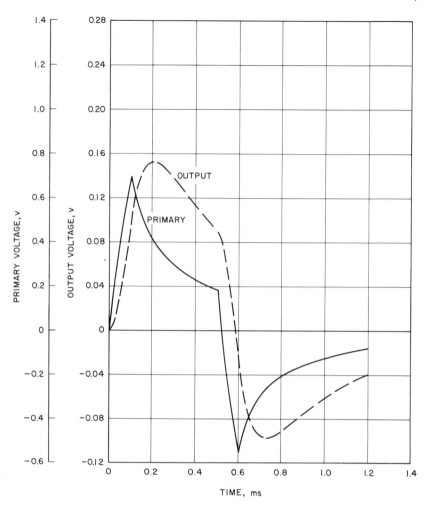

Fig. 5.24. Primary and output voltage waveforms—nonideal transformer example.

5.4 *PHYSICAL ANALOGY:*
PIECEWISE LINEAR MODELING

5.4.1 General Concepts

The techniques for modeling a nonlinear element with an algebraic voltage-current relationship, such as a nonlinear resistance, were discussed in detail in Sec. 5.2. The techniques for modeling a nonlinear reactive element with a non-algebraic voltage-current relationship, such as an inductance or a capacitance, will be described in this section.

Consider the nonlinear characteristics of these elements. The current flowing in a capacitance can be described by the expression

$$i = \frac{dq}{dt} = \frac{d(Ce)}{dt} \tag{5.20}$$

where C is the value of the capacitance and e is the voltage across the capacitance. The product $Ce = \int i\, dt$ is defined as the charge q on the capacitance. If C is constant, then q will be a linear function of e; if C is not constant but a function of e, the product $Ce = q$ must be modeled as a function of e. The characteristic of the capacitance is then defined by a nonlinear voltage-charge relationship.

The inductance characteristics can be expressed as a nonlinear current magnetic-flux relationship. The voltage developed across an inductance is described by

$$e = \frac{d\Phi}{dt} = \frac{d(Li)}{dt} \tag{5.21}$$

where L is the value of the inductance and i is the current flowing in it. The product Li is the magnetic flux Φ associated with the inductance. If L is constant, the product Li will be a linear function of i. If L is not constant, the magnetic flux Φ must be modeled as a function of the current flowing in the inductance.

Since neither of these elements possesses the simple voltage-current relationship discussed in Sec. 5.2, it is necessary to expand the piecewise linear approximation technique to model these components. The nonlinear resistance can be modeled in voltage-current terms readily accepted by ECAP. Nonlinear inductance and capacitance elements are algebraic functions of magnetic flux Φ and charge q, respectively, which are not recognized by the program.

An obvious solution to the modeling problem is to find a physical analogy which will allow magnetic flux or charge to be treated in units recognized by ECAP. In a sense, the physical analogy acts as a black box which converts the physical relationship into terms which are acceptable to the program as illustrated in Fig. 5.25.

Elements such as nonlinear inductances and capacitances can be modeled if an analogy between the characteristic to be modeled and a voltage-current relationship acceptable to ECAP can be established. The best way to describe the techniques used in this approach is by the use of a simple example. The fol-

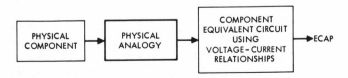

Fig. 5.25. Physical analogy concept use in physical component modeling.

lowing example, which develops a model for a nonlinear capacitance, outlines
a general approach in the development of models using physical analogies in
conjunction with the piecewise linear modeling techniques described earlier.

5.4.2   Nonlinear Capacitance Model[3]

The current flowing in a capacitance as defined earlier in Eq. (5.20) is

$$i = \frac{d(Ce)}{dt}$$

If C is a function of the voltage across the capacitance, it is possible to plot the
function Ce(or q) as a function of e. The resulting curve Ce versus e can be used
in the derivation of the capacitance model.

The first step is to model the voltage-charge characteristic of the nonlinear
capacitance. This is the portion of the problem that requires the use of a physical
analogy since the unit of charge is not a standard parameter in ECAP. It is
possible, using the physical analogy, to relate the voltage-charge characteristic
to a voltage-current characteristic in which the charge level is directly proportional to the current flowing in an element. Hence, the curve relating the charge
and voltage appears similar to the nonlinear resistance characteristic and is
modeled as described in Sec. 5.2.2.

The second step in the capacitance-model construction incorporates the
physical analogy network (Ce versus e) into the remainder of the model so that
the terminal characteristics of the model and of the nonlinear capacitance are
identical. In the Ce versus e network derived, the current flow is proportional
to the charge on the capacitance. The current is converted to a voltage of equal
magnitude by transferring it to a 1 Ω resistance through a dependent current
source. Applying this voltage, equal in magnitude to the charge, to a fixed
linear capacitance yields a current in the capacitance equal to the time rate
of change of the charge and hence also equal to the current flowing in the nonlinear capacitance. The current is then transferred to the network containing
the nonlinear capacitance via a dependent current source so that the terminal
characteristics of the nonlinear capacitance are realistically simulated.

For example, suppose a model for a nonlinear capacitance is defined by the
expression

$$C = C_0 e^{-\alpha e} \qquad (5.22)$$

where $C_0 = 1000$ pf, $\alpha = 0.1$, and $e < 10$ v.

The voltage-charge characteristic for this capacitance is shown in Fig. 5.26.
Once the definition of the voltage-charge characteristic is established, the next
step in the derivation is to construct a piecewise linear approximation to the
characteristic using a current-charge analogy. Figure 5.27a shows the piecewise

[3]*The 1620 Electronic Circuit Analysis Program (ECAP) (1620-EE-02X) User's Manual*
(White Plains, N.Y.: International Business Machines Corp., 1965), pp. 133–37.

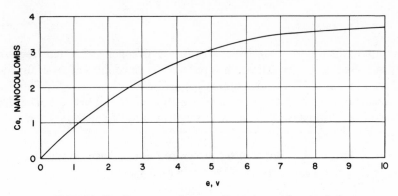

Fig. 5.26. Nonlinear capacitance voltage-charge characteristic.

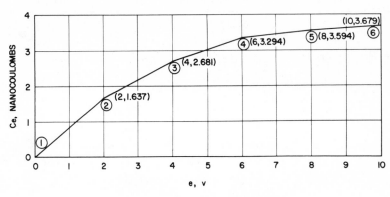

(a) PIECEWISE LINEAR APPROXIMATION

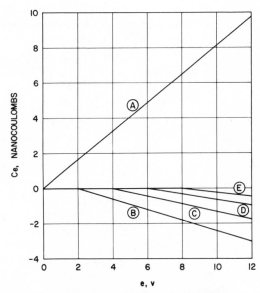

(b) ELEMENTS FORMING PIECEWISE LINEAR
APPROXIMATION AT BOTTOM OF FIGURE

Fig. 5.27. Nonlinear capacitance characteristic.

131

linear approximation $1 \rightarrow 6$ consisting of five connected line segments. The piecewise linear approximation is reduced to an equivalent set of functions (as shown in Fig. 5.27b) whose sum is the characteristic $1 \rightarrow 6$. This set of functions is obtained in the same manner as the models of Sec. 5.2. The slopes of the characteristic and the individual segments are tabulated in Table 5.3.

Table 5.3.   Slopes of the piecewise linear voltage-charge characteristic of the nonlinear capacitance.

Line	Slope, nanofarads	Line	Slope, nanofarads
$1 \rightarrow 2$	0.8185	A	0.8185
$2 \rightarrow 3$	0.5220	B	-0.2965
$3 \rightarrow 4$	0.3065	C	-0.2155
$4 \rightarrow 5$	0.1500	D	-0.1565
$5 \rightarrow 6$	0.0425	E	-0.1075

The derived set of functions produces the voltage-charge characteristic of the desired nonlinear capacitance. This characteristic must now be incorporated into a circuit which will yield the voltage-current characteristic of the capacitance.

The complete equivalent circuit of the nonlinear capacitance can be presented by studying a simple *RC* network as shown in Fig. 5.28. Figure 5.28b contains the ECAP equivalent circuit of the network in which the nonlinear capacitance, Fig. 5.28a, is replaced by its model. The model of the voltage-charge characteristic is represented by branches 3, 4, 5, 6, and 7. The currents flowing in these branches are determined by the voltage e at node 1, which is the voltage across the capacitance. The magnitude of the charge on the capacitance is equal to the element current in branch 8.

The current flowing in branch 8 is converted to a voltage of equal magnitude at node 3 by transferring the current through a dependent current source T3 to a $1 \, \Omega$ resistance R_{10} between node 3 and ground. To prevent loading of the $1 \, \Omega$ resistance by the capacitance in branch 11, any current drawn from the node by the capacitance is returned to the node via dependent current source T4.

Then, since all current from the dependent current source T3 flows in the $1 \, \Omega$ resistance, the voltage at node 3 is equal in magnitude to the charge Ce on the nonlinear capacitance. Also, since the current that flows in the capacitance C_{11} is equal to the time rate of change of the voltage at node 3, it is also proportional to the time rate of change of the charge Ce. The current flowing in the capacitance C_{11} is

$$i = C_{11} \frac{de_3}{dt} = C_{11} \frac{d}{dt} (Ce) \qquad (5.23)$$

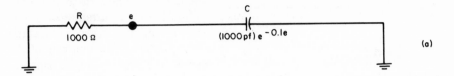

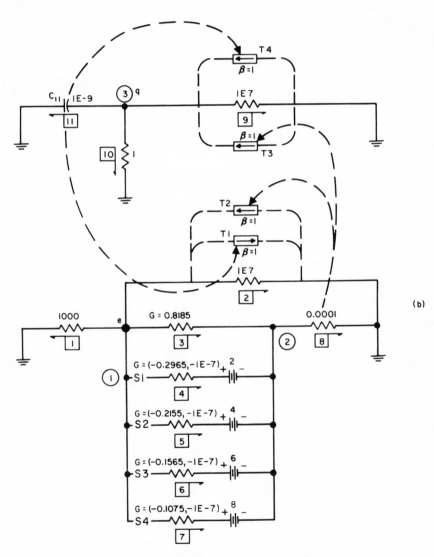

Fig. 5.28. *RC* circuit with nonlinear capacitance.

```
C       NONLINEAR CAPACITOR MODEL EXAMPLE
        TR
    B1  N(1,0),R=1000
    B2  N(1,0),R=1E7
    B3  N(1,2),G=0.8185
    B4  N(1,2),G=(-0.2965,-1E-7),E=-2
    B5  N(1,2),G=(-0.2155,-1E-7),E=-4
    B6  N(1,2),G=(-0.1565,-1E-7),E=-6
    B7  N(1,2),G=(-0.1075,-1E-7),E=-8
    B8  N(2,0),R=0.0001
    B9  N(3,0),R=1E7
   B10  N(3,0),R=1
   B11  N(3,0),C=1000E-12,F0=-3.679
    T1  B(11,2),BETA=1
    T2  B(8,2),BETA=-1
    T3  B(8,9),BETA=-1
    T4  B(11,9),BETA=-1
    S1  B=4,(4),OFF
    S2  B=5,(5),OFF
    S3  B=6,(6),OFF
    S4  B=7,(7),OFF
        TI=0.01E-6
        OU=2
        FI=2E-6
        PRINT,NV,CA
        EX
```

Fig. 5.29. Input data for nonlinear capacitor response calculations.

where Ce is expressed in nanocoulombs. Then the instantaneous current i in coulombs per second is equal to the product of the time rate of the change of charge, expressed in nanocoulombs per second (nanoamps), and the capacitance C_{11}. The current flowing in the network is scaled to coulombs per second, or amperes, by multiplying the derivative by 10^{-9}, or 1000 pf.

The dependent current source T1 removes a current equal to the current flowing in C_{11}, which simulates the current flowing in the nonlinear capacitance. To prevent the loading effect of the voltage-charge characteristic network at node 1, a

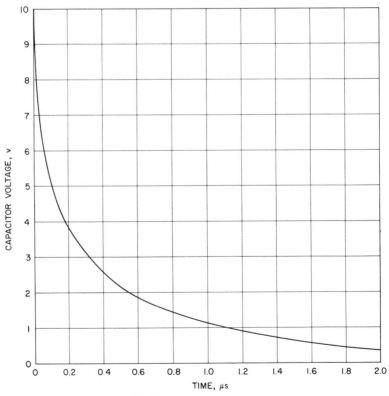

Fig. 5.30. Nonlinear capacitor response.

current equal to the current drawn from node 1 by the network is returned to the node by the dependent current source T2.

As an example of the input data and response of the nonlinear capacitance model, the transient response of the circuit in Fig. 5.28 is presented. The initial voltage on the capacitance is 10.0 v. Note that the initial condition is entered as 3.679 nanocoulombs in the input data in Fig. 5.29. The resulting transient response is plotted in Fig. 5.30.

5.5 PHYSICAL ANALOGY: MATHEMATICAL MODELING

5.5.1 General Concepts

The Electronic Circuit Analysis Program performs calculations in terms of voltages and currents only. Provisions for parameters, such as charge and magnetic flux, have not been included in the program and cannot be used directly. Section 5.4 describes one approach that may be used to indirectly operate in terms of charge, etc. Modeling problems frequently arise in which portions of the model are defined in parameters other than voltage and current, and where the elements used in the definition are not included in the standard list. If either of these conditions exists, it is necessary to devise methods to relate the new parameters to functions of voltage and current or to redefine the new elements into elements included in the standard list that will perform the required functions in terms of voltage and current.

In this section, the physical-analogy concept will be applied to the modeling of equations which are presented in terms of nonstandard parameters or elements or both. The concept itself is quite simple and easy to understand. The application of the concept to the development of a model requires imagination more than anything else.

If the set of equations that defines an element characteristic possesses an analogy in terms of voltage and current, generally the element can be modeled. This is similar to the solution of mechanical and thermal problems by electrical analogy techniques. Establishing the analogy is the first step in the derivation of an equivalent circuit using a physical analogy to model the equations that define an element characteristic. The equations must relate the parameters not standard to ECAP to those which are available. Once the analogy has been established, the second step is linking the derived analog network to the remainder of the model which includes the element terminals. The element terminal characteristics, to be compatible with the remainder of the circuit, must always be defined in terms of voltage and current. Linking the network to the model is accomplished with dependent current sources which isolate the analog network from the model yet provide the necessary coupling described by the equations between parts of the model.

The charge-control transistor model described in Appendix A.2 is an excellent example of the application of physical analogy techniques in modeling the charge equations which govern the transistor response. The complete transistor model also utilizes the piecewise linear approximation technique to model the base-emitter and base-collector diode characteristics.

5.5.2 Transistor Model

The transistor equivalent circuit developed in Appendix A.2 represents an analytic model of the transistor and as such can be used directly to calculate the

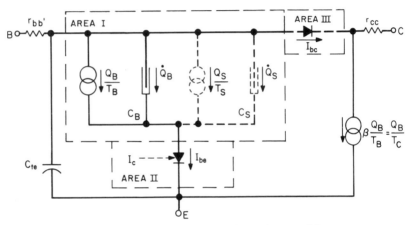

Fig. 5.31. Basic charge control transistor model.

response of any transistor. Unfortunately, its form does not lend itself to solutions using ECAP. However, the equivalent circuit can be broken into three separate areas coupled to each other via dependent current sources, and each area can be easily modeled for ECAP applications. These three areas are indicated in Fig. 5.31. The first area is the base region, which will be modeled using the physical analogy concepts discussed in this section. The second and third areas can be modeled as diodes using the piecewise linear modeling techniques described in Sec. 5.2.

To simplify construction of the base region of the transistor model, the development will be conducted in two steps. The first step describes construction of the basic charge-control transistor model with the storage time constant T_S assumed negligible. The second step describes the additional development necessary assuming that the storage time constant is not negligible. The basic configuration for the charge-control transistor model is shown in Fig. 5.31.

BASIC CHARGE CONTROL TRANSISTOR MODEL $(T_S \rightarrow 0)$

The charge-control transistor model in Fig. 5.31 can be simplified if the recombination rate T_s of the excess stored charge Q_s can be assumed negligible.

Area I is effectively the charge model of the transistor base region. Since no voltage drop exists across the charge-control network in this region, the base region can be removed from the basic transistor network and modeled separately as a charge network with the necessary coupling between charge network and transistor.

At this point the concept for components using physical analogies to model mathematical relationships must be applied. It is often more practical to model devices or specific portions of devices using a reference basis different from the conventional voltage-current reference. The charge reference basis will be referred to as the *charge domain*. In the charge domain the magnitude of charge is analogous to voltage level, and a charge drop across a component is analogous to a voltage drop in the voltage domain.

Consider first the current generator Q_B/T_B in the charge-control network illustrated in Fig. 5.32. The current flowing in that branch is a function of the charge level and the recombination rate of carriers in the base region. The constant of proportionality between charge and current in the charge domain is the recombination rate. An analogous constant of proportionality between voltage and current is resistance in the voltage domain. Therefore, in the charge domain, the base time constant T_B can be treated as a resistance since the number of recombinations occurring in unit time is directly proportional to the number

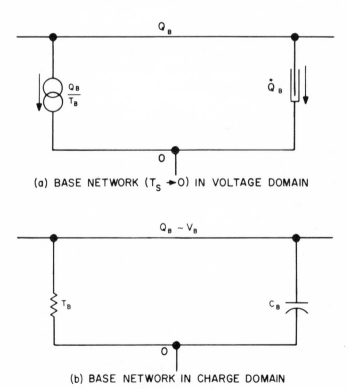

(a) BASE NETWORK ($T_S \to 0$) IN VOLTAGE DOMAIN

(b) BASE NETWORK IN CHARGE DOMAIN

Fig. 5.32. Voltage-charge analogies.

of excess minority carriers present and inversely proportional to the effective lifetime of carriers in the base region, or

$$I_R = \frac{Q_B}{T_B} \qquad (5.24)$$

where I_R = recombination current
Q_B = base control charge
T_B = recombination time constant (base)

If charge storage is occurring in the base region, then the charge density Q_B will be increasing with time at a rate which is a function of the current I or charge per unit time entering the base region. Since

$$\frac{dQ_B}{dt} = \frac{I}{C_B} \qquad (5.25)$$

it appears feasible to treat the base storage constant C_B as a charge term analogous to capacitance in the voltage domain. In the charge domain

$$\frac{dQ_B}{dt} = I \qquad (5.26)$$

so that C_B must be equal to one unit equivalent to 1 farad in the voltage domain.*
The time constant τ of the network in the charge domain can be defined as

$$\tau = T_B C_B = T_B \cdot 1 = T_B \qquad (5.27)$$

We can now show that the only important consideration in selecting values for elements in the charge network is that $\tau = T_B$. The absolute value of Q_B has no importance other than to maintain the ratio

$$I = \frac{Q_B}{T_B} \qquad (5.28)$$

This is demonstrated in the following development. If we let

$$I = \frac{Q_B}{k_1 R_B} \qquad (5.29)$$

where k_1 = arbitrary constant

$$R_B = T_B/k_1 \qquad (5.30)$$

and let

$$\frac{dQ_B}{dt} = \frac{I}{k_2 C_1} \qquad (5.31)$$

where $\qquad k_2 = \text{constant} = 1/k_1 \qquad (5.32)$

then $\qquad\qquad T_B = R_B C_1 \qquad (5.33)$

---

*Practical limitations may prevent use of the value $C_B = 1$ in the charge network because of the ECAP nodal impedance ratio limitations.

We now see that

$$\frac{Q_B}{T_B} = I = \frac{Q'_B}{R_B} \quad \text{where} \quad Q'_B = k_1 Q_B \tag{5.34}$$

and that

$$\frac{d(Q'_B)}{dt} = \frac{I}{C_1} \tag{5.35}$$

which justifies the assumption that the absolute value of Q_B is not important. Using these results

$$T_B = R_B C_1 \simeq \frac{1.22(\beta + 1)}{\omega_n} \simeq \frac{1.22}{\omega_\beta} \tag{5.36}$$

where $\omega_\beta = 2\pi f_\beta =$ beta cutoff frequency

An identical development is possible with the excess charge storage elements Q_S/T_S and dQ_S/dt.

The next consideration in the construction of the model of area I is communication between the basic transistor operating in the voltage domain and the base charge network in the charge domain.

The emitter and collector current equations (A.15 and A.17) developed in Appendix A.2 describe the coupling between the domains. These are

$$I_{be} = \frac{Q_B}{T_B} + \dot{Q}_B + \frac{Q_S}{T_S} + \dot{Q}_S \tag{5.37}$$

$$I_{bc} = \frac{Q_S}{T_S} + \dot{Q}_S \tag{5.38}$$

$$I_c = \beta \frac{Q_B}{T_B} \tag{5.39}$$

Using these equations, the base region (area I) of the transistor can be modeled as illustrated in Fig. 5.33.

The current I_{bc} does not contribute to the collector current of the transistor but contributes only to Q_S in the base region. Because of this, I_{bc} is returned to

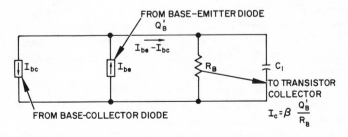

Fig. 5.33. Base region of transistor ($T_s \rightarrow 0$).

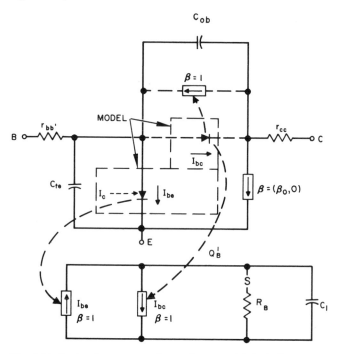

Fig. 5.34. Basic charge control transistor model including stray capacitances assuming $(T_S \to 0)$.

the base region in the model via the indicated dependent current source where it comprises part of I_{be} (illustrated in Fig. 5.34).

When the charge Q_B is less than zero, the dependent current source which generates I_c is set to zero to prevent reverse collector current $[-\beta(Q_B/T_B)]$ from flowing. The switch in series with R_B controls the beta value of the transistor.

Area II consists of the model of the base-emitter diode. When modeling the *E-I* characteristic of this diode, the effects of the collector current in the transistor must be accounted for. Figure 5.35 shows a typical *E-I* characteristic of a silicon planar epitaxial transistor; it illustrates the effect of collector current on the base-emitter diode characteristics.

In a saturated transistor, the error encountered when selecting the base-emitter diode characteristic corresponding to the collector saturation current is minimal and can usually be neglected. This greatly simplifies the selection of the appropriate *E-I* characteristic for the diode.

Area III consists of the model of the base-collector diode. This diode characteristic, shown in Fig. 5.35, is not a function of two independent variables as was the base-emitter diode characteristic.

If the excess stored charge recombination rate T_S is negligible, the transistor model including the parasitic elements f_β, C_{ob}, and C_{te} can be constructed as shown in Fig. 5.34.

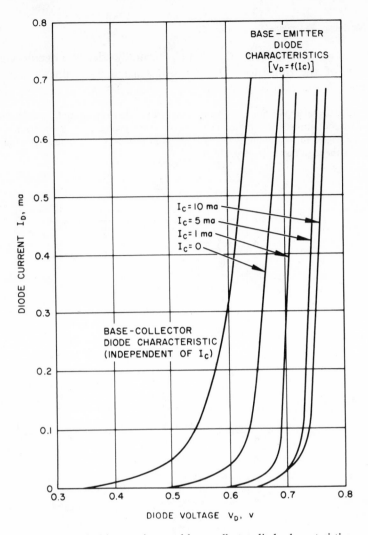

Fig. 5.35. Typical base-emitter and base-collector diode characteristics.

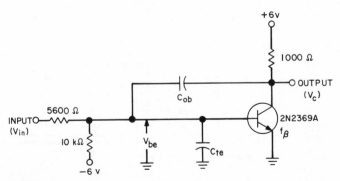

Fig. 5.36. Transistor circuit model used to illustrate C_{te}, C_{ob}, and f_β effects.

The values selected for f_β, C_{ob}, and C_{te} naturally affect the operational characteristics of the transistor. The parasitic capacitance values are generally available from the transistor specification sheets.* The value of either the gain-

*For large-signal applications, the base-emitter transition capacitance C_{te} can be considered constant without appreciable error, and the effect of the change in collector-base voltage on $C_{b'c}$ can be neglected. Assuming that $C_{b'c}$ is constant generally adds a negligible amount of error unless the load resistance R_L is high.

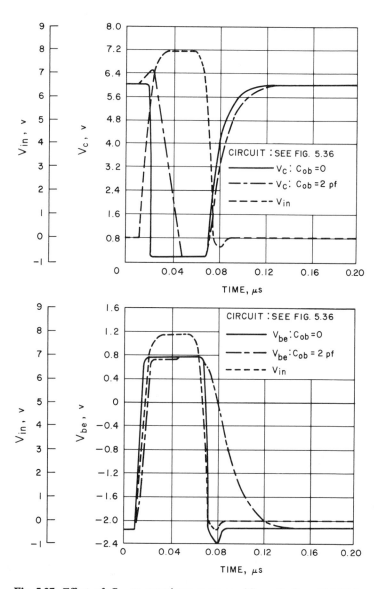

Fig. 5.37. Effect of C_{ob} on transistor response ($C_{te} = 0$, $f_\beta = 21$ MHz).

bandwidth product f_T or the alpha cutoff frequency f_n is also given in the transistor specifications. Since beta is not constant during the switching transition, it is reasonably accurate to compute f_β using the average value of forward beta β_{av} existing during the switching time. If the transistor is overdriven, then the "forced" beta must be considered. For example, if a transistor has a DC beta of 70 and is used as a switch with a forced beta of 20, the β_{av} value, which is a function of the drive level during the transition, might equal 30. The approxi-

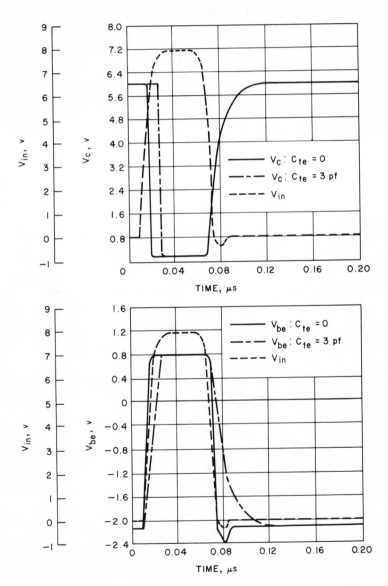

Fig. 5.38. Effect of C_{te} on transistor response ($C_{ob} = 0$, $f_\beta = 21$ MHz).

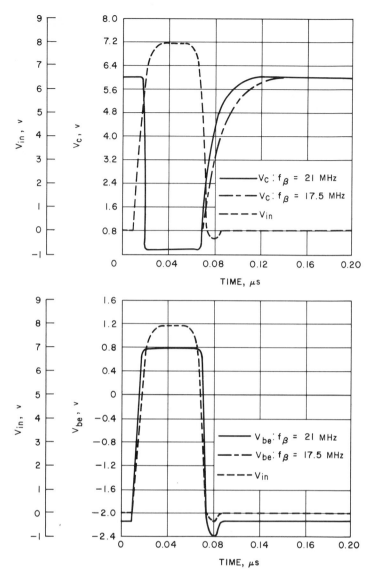

Fig. 5.39. Effect of f_β on transistor response ($C_{ob} = C_{te} = 0$).

mate conversion of f_T(or f_n) to f_β or $f_{\beta_{av}}$ is simply f_T/β_{av} (or f_n/β_{av}). These three elements are the most important response-limiting parameters in this transistor model. In some applications where transient response is not critical, however, these values may be unimportant and may be neglected.

To illustrate the effects of the parasitics on the transistor response, an ECAP model of the circuit in Fig. 5.36 (p. 141) was constructed using the charge-control transistor model. The circuit responses shown in Figs. 5.37, 5.38, 5.39, and 5.40 were calculated using ECAP to illustrate the effects of C_{ob}, C_{te}, f_β, and a composite of C_{ob} and C_{te}, respectively.

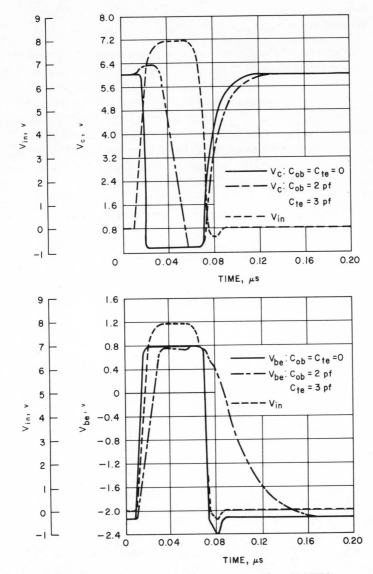

Fig. 5.40. Effect of C_{ob} on transistor response ($f_\beta = 21$ MHz).

The values selected for these illustrations are almost identical to the nominal values of C_{ob} and C_{te} of a high-frequency NPN switching transistor (2N2369A).

BASIC CHARGE CONTROL MODEL ($T_S \neq 0$)

In the development of the charge model, the storage time constant T_S was assumed negligible. In many of the applications using high-frequency switching transistors, or transistors operating in a nonsaturating mode, this assumption is valid. However, there are applications in which the storage characteristics are important.

In this section, the additions to the transistor model, required when the recombination time constant of the stored charge T_S is not negligible, are developed. The mechanics of the development in this section are basically quite simple. However, the representation of the physics of the device using the elements available in ECAP tends to make the model appear somewhat complicated. An extra effort will be made here to explain fully the function of each network as it is being constructed to aid in understanding the ECAP model details.

The model, based on the assumption that the stored charge cannot be neglected, consists of the basic model already developed and a few additions. The first addition is a charge storage network which models the behavior of the excess stored charge. The basic network is shown in Fig. 5.41.

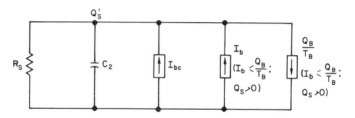

Fig. 5.41. Charge storage network.

The values of R_S and C_2 are selected so that $T_S = R_S C_2$, where

$$R_S = k_3 T_S \tag{5.40}$$

$$C_2 = \frac{C_S}{k_3} \tag{5.41}$$

$$k_3 = \text{arbitrary constant}$$

The development of the relationship between T_S and $R_S C_2$ is the same as the relationship between T_B and $R_B C_1$ developed earlier. The selection of the value of k_3 may be limited by the ECAP nodal impedance ratio restrictions.

Two of the dependent current sources in the Q_S network of Fig. 5.41 contribute to the operation of the circuit only if the magnitude of excess stored charge Q_S is greater than zero and the active recombination current Q_B/T_B is greater than the base current I_b. This defines the condition that exists when the transistor is recovering from saturation. To present the clearest picture of the operation of the Q_S network and its relationship to the transistor state, the description follows a transistor operation from cutoff to saturation and back again.

The mechanics of turning on the transistor are very simple. They are basically the same as those used when $T_S \to 0$ was assumed with the addition of a Q_S network to store the charge generated by I_{bc}. The following description is primarily concerned with the turn-off period, during which the transistor is recovering from saturation.

During the time the transistor is in saturation and the base current I_b is less than Q_B/T_B, Q_S must be removed from the charge storage network to sustain the collector current while Q_B is held constant. The remaining paragraphs in this section discuss the mechanization of this operation. The two dependent current sources, I_b and Q_B/T_B, are used in the charge storage network to remove the stored charge from the network. The dependent current source labeled I_b removes the charge from capacitance C_2 if the base current I_b is less than Q_B/T_B. This simulates the removal of stored charge by reverse drive. However, in the region of operation where $Q_B/T_B > I_b > 0$, the charge removed from the excess charge network is equal to the difference between Q_B/T_B and I_b. Therefore, the third dependent source Q_B/T_B must be added to the network to satisfy this condition.

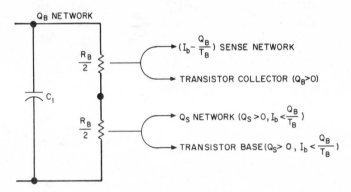

Fig. 5.42. Control charge network ($T_S \neq 0$).

The excess charge Q_S stored in the base region is removed by three methods: (1) recombination controlled by the excess charge time constant $R_S C_2$; (2) recombination to sustain the emitter current Q_B/T_B; and (3) reverse drive. During the time the charge Q_S is providing the drive to sustain the collector current, the base control charge will remain constant at the level $Q_B = I_c T_B/\beta$.

In ECAP, if a dependent current source is given two values, it cannot be changed from one to the other (upon actuation of a switch) unless the switch controls the *from*-branch associated with the dependent current source. R_B is divided into two resistances in the control charge network in Fig. 5.42 because the branch containing R_B is used as the *from*-branch for several dependent current sources which are switched under two different conditions. To eliminate the possible interaction of the switches, each of the two series resistances controls some of the dependent current sources.

In the basic transistor network it is also necessary to add two dependent current sources. The first source removes the reverse base current I_b from the transistor input only when $Q_S > 0$ and $I_b < Q_B/T_B$. During this time interval, the collector control current Q_B/T_B is applied by the second source to the base of the transistor to maintain the V_{be} drop and to sustain the level of the control charge in the Q_B network. The current Q_B/T_B input at the base terminal will

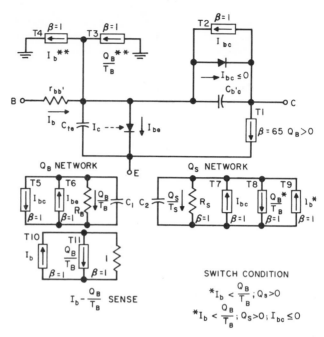

Fig. 5.43. Charge control transistor model ($T_S \neq 0$).

be transferred to the Q_B network since $I_{bc} = 0$ when $I_b < Q_B/T_B$. When the model is assembled, the reverse base current will, in effect, be drawn directly from the excess stored charge network until the excess charge is removed. The placement of the dependent current sources is illustrated in Fig. 5.43.

The switches required to mechanize the stored charge control are special types because they must select one of four possible states rather than the two of four states provided by standard ECAP switch combinations. The transistor model switch states are summarized in Table 5.4. The special set of series switches

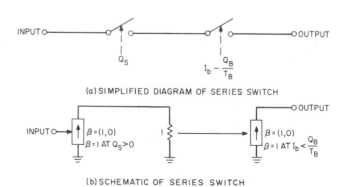

Fig. 5.44. Series switch.

Table 5.4.   Transistor switch states.

Q_S	$I_b - \dfrac{Q_B}{T_B}$	I_b return (T4)	$\dfrac{Q_B}{T_B} \to$ base (T3)	$\dfrac{Q_B}{T_B} \to Q_S$ net (T9)	$I_b \to Q_S$ net (T9)	Q_B	Transistor beta (T1)
$+$	$-$	1	1	1	1	$+$	65
$-$	$+$	0	0	0	0	$-$	0
$+$	$+$	0	0	0	0	$-$	0
$-$	$-$	0	0	0	0	$-$	0

is shown in Fig. 5.44. The current at the input is not transferred to the output unless both dependent current sources have a BETA value equal to unity.

The complete charge-control transistor model is illustrated in Fig. 5.43. The value of T_S cannot be accurately determined, and it is advantageous to base the selection of the storage time constant on lab data using reasonable tolerances in the selection of T_S for circuit analysis. Several methods of measurement to determine the value of T_S have been summarized very well by R. P. Nanavati[4], whose conclusions are supported by other experts.

PROBLEMS

5.1. Construct an equivalent network which simulates the *minimum* forward voltage-current characteristic for the 1N3600 diode (Fig. 5.3). The network should be formed by at least five linear segments.

5.2. Derive a general equivalent circuit for an ideal transformer with three windings. The equivalent circuit should be valid in the DC, AC, and transient analysis programs.

5.3. Repeat prob. 5.2. for a four-winding ideal transformer.

5.4. Derive a general equivalent circuit for a three-winding transformer with a coupling coefficient k_{ij} less than 0.999995 which does not use M cards to specify inductive coupling. Verify the model by comparing its frequency response with the response of the equivalent transformer using M cards to specify mutual coupling. Let the winding inductance $L_1 = L_2 = L_3 = 1$ mh, and the coupling coefficient $k_{12} = k_{13} = k_{23} = 0.75$ for the verification.

5.5. Repeat prob. 5.4. for a four-winding nonideal transformer using

[4]"Prediction of Storage Time Constants in Junction Transistors," *IRE Transactions on Electron Devices*, **ED**-7 (January 1960), 9–15.

Fig. p5.1

the inductance and coupling coefficient values specified in prob. 5.4. plus $L_4 = 1$ mh and $k_{14} = k_{24} = k_{34} = 0.75$ for the model verification.

5.6. Construct a piecewise linear network (Fig. p5.1) to approximate the characteristic $E_o = (E_i)^2$ for $0 \leq E_i \leq 10$ v. Use at least four linear segments in the piecewise linear network. The approximating characteristic should have $E_o = 0$ at $E_i = 0$.

5.7. Draw a simple equivalent network for the integrated circuit resistor illustrated in Fig. p5.2.

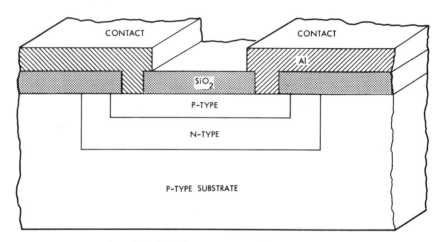

CROSS SECTIONAL VIEW OF DIFFUSED RESISTOR

Fig. p5.2

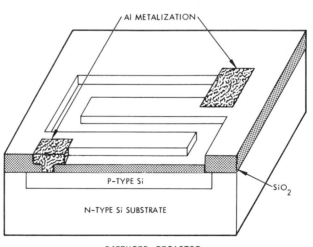

DIFFUSED RESISTOR

Fig. p5.3

5.8. Determine the transient response of an integrated circuit resistance (Fig. p5.3) that has a total resistance $R = 1000\ \Omega$ and a junction capacitance $C = 10$ pf to a -1 v step input. Assume a load resistance $R_L = 1000\ \Omega$ and that the effects of the semiconductor junction diode are negligible. Compare the transient response of the lumped element RC network (≥ 10 sections) with the responses of one-, two-, three-, and four-section networks with lumped values for R and C to determine the relative accuracies of the lumped element models to a distributed RC network.

5.9. Construct a general equivalent circuit for a uniform lossy transmission line when the characteristic impedance Z_o, the line inductance L_T, and the line resistance R_T are specified. (The practical limit to the number of sections in the ECAP equivalent circuit is about 40 nodes or about 20 sections to allow for input and output circuitry.)

5.10. Construct an equivalent circuit for a complex network whose transfer function is

$$\frac{E_o(s)}{E_i(s)} = G(s) = \frac{100(s + 5)}{(s + 1)(s + 10)}$$

The input resistance equals $1000\ \Omega$ and is independent of frequency. Determine the frequency response of the equivalent circuit to verify the model.

5.11. Construct an equivalent circuit for a complex network whose transfer function is

$$\frac{E_o(s)}{E_i(s)} = G(s) = \frac{10^5(s + 5 \times 10^3)}{(s + 10^3)(s + 10^6)}$$

and whose input impedance is

$$\frac{E_i(s)}{I_i(s)} = Z_i(s) = \frac{10^7}{s + 10^4}$$

Verify the equivalent circuit using the AC analysis program.

5.12. Determine the amount of compensation required in the network shown in Fig. p5.4 to make the voltage at node 1 equal to zero when the input signal equals zero. All switches are to be in their initially assumed states.

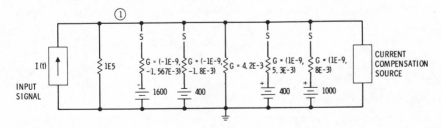

Fig. p5.4

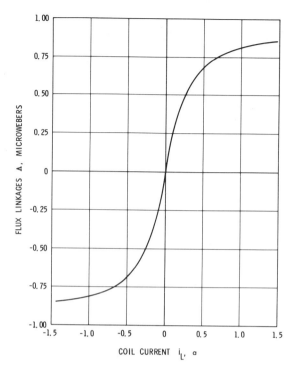

Fig. p5.5

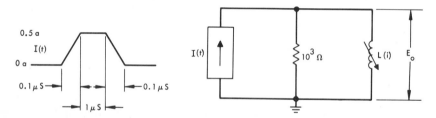

Fig. p5.6

5.13. Construct an equivalent circuit for a nonlinear saturating inductance with the flux-current characteristic shown in Fig. p5.5. The value of Φ should be equal to zero at $i = 0$. The piecewise linear model should contain at least five linear segments. (*Hint:* $i_L = (1/L) \int e\, dt$ and $Li_L = N\Phi$.)

5.14. Determine the voltage response of the nonlinear inductance model (prob. 5.13) in the circuit shown in Fig. p5.6. (*Note:* The value selected for TI must be small enough to define the response $e = Ldi/dt$.)

5.15. Construct an equivalent circuit for the Ebers-Moll transistor model shown in Fig. p5.7. The diode and junction capacitance characteristics shown

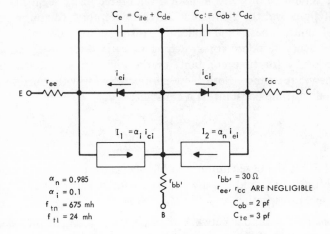

$C_e = C_{ite} + C_{de}$ $C_c = C_{ob} + C_{dc}$

i_{ei} i_{ci}

r_{ee} r_{cc}

E   C

$I_1 = \alpha_i i_{ci}$ $I_2 = \alpha_n i_{ei}$

$\alpha_n = 0.985$
$\alpha_i = 0.1$
$f_{tn} = 675$ mh
$f_{ti} = 24$ mh

$r_{bb'}$

$r_{bb'} = 30\,\Omega$
r_{ee}, r_{cc} ARE NEGLIGIBLE
$C_{ob} = 2$ pf
$C_{te} = 3$ pf

B

(a) EBERS–MOLL TRANSISTOR MODEL

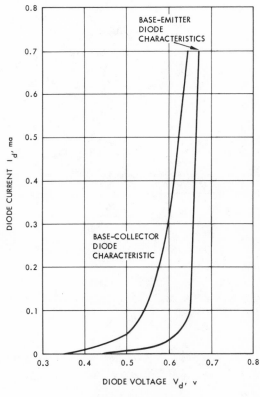

BASE–EMITTER
DIODE
CHARACTERISTICS

BASE–COLLECTOR
DIODE
CHARACTERISTIC

DIODE CURRENT I_d, ma

DIODE VOLTAGE V_d, v

(b) DIODE CHARACTERISTICS

Fig. p5.7

in the figure should be modeled in no less than three linear segments. Assume the values of alpha and the gain-bandwidth product, forward and inverse, and the values of emitter, base, and collector resistance are constant. *Note:* The transistor frequency response (or switching time) is determined primarily by the 3-db frequency (or time constant) of the diffusion capacitance and the junction dynamic resistance. Both these parameters vary with junction current, but their product (time constant) is a constant:

$$r_e C_{de} = 1/2\pi f_{Tn}$$

$$r_c C_{dc} = 1/2\pi f_{Tn}$$

where r_e and r_c are the base-emitter and base-collector junction dynamic resistances, respectively; C_{de} and C_{dc} are the base-emitter and base-collector diffusion capacitances, respectively.

5.16. Construct a simple network to generate a 1000 Hz sine wave. The amplitude of the 1000 Hz signal should be 2 v p-p. (*Hint:* $y = \cos(wt + \theta)$ is the solution to the equation $(d^2y/dt^2 + w^2y = 0.)$

6

PRACTICAL
EXAMPLES

6.1 INTRODUCTION

Chapters 1 through 4 of the text explained the program fundamentals and the basic techniques necessary for effective application of the DC analysis, AC analysis, and transient analysis programs. The most important approaches to the development of element models frequently required in the analysis programs were discussed in Chapter 5 using techniques developed in the first four chapters. With this background, we will now demonstrate how ECAP can be used effectively as a tool in the analysis of electronic circuits.

Up to this point, the principles of ECAP have been explained using simple examples in order to present the information clearly without irrelevant details. As a result, no illustrations have been presented demonstrating the use of ECAP in the solution of practical circuit analysis problems.

This chapter contains a collection of typical electronic circuit analyses. The selected examples yield a great amount of useful information without unnecessary complexity. The example circuits and analyses to be performed are listed in Table 6.1. Each analysis is discussed in a separate section to minimize the time required to isolate a particular analysis for reference.

Table 6.1. Practical analysis example problems.

Section no.	Circuit	Analysis type	Comments
6.2	Composite filter	AC	Multisection, passive-element, filter-design example
6.3	Composite filter	Transient	
6.4	Digital logic inverter	Transient	Application of charge-control transistor model, standard initial condition solution
6.5	10 MHz crystal oscillator	DC	Two-transistor, differential amplifier with a high-Q series resonant feedback (crystal)
6.6	10 MHz crystal oscillator	AC	Response of high-Q resonant circuit
6.7	10 MHz crystal oscillator	Transient	Solution technique where no steady-state solution exists

6.2 COMPOSITE FILTER: AC ANALYSIS

The design of passive filter networks is one of the most powerful, yet one of the simplest, applications of ECAP. In this example, values of the components in the filter are calculated, the network is assembled, and the resulting network is analyzed to determine the accuracy of the calculations and the frequency response of the filter when terminated in its characteristic impedance R_o. The response of the filter is also calculated when terminated in a resistance R_L equal to one-half the characteristic impedance and in a resistance equal to twice the characteristic impedance. No component modeling is required in this application since all the components are standard ECAP elements.

Suppose it is necessary to design a low-pass composite filter to work into a 100 Ω resistive load and to have a cutoff at $f_o = 12.0$ kHz. The filter must consist of two terminating half-sections, one prototype section, and one m-derived section with $f_\infty = 14.0$ kHz. The resulting filter must be analyzed to determine the frequency and phase response when terminated in $R_L = 100, 50,$ and 200 Ω.

For the prototype filter section

$$C = \frac{1}{(\pi f_o R_o)} \text{ f}$$

$$C = \frac{1}{\pi(12 \times 10^3)(100)} = 0.265 \ \mu\text{f}$$

$$L = \frac{R_o}{\pi f_o}$$

$$L = \frac{100}{\pi(12 \times 10^3)} = 2.65 \text{ mh}$$

For the m-derived sections

$$m = \sqrt{1 - f_o^2/f_\infty^2}$$

$$m = \sqrt{1 - \left(\frac{12 \times 10^3}{14 \times 10^3}\right)^2} = 0.515$$

Using this value of m, the elements of the m-derived section can be found.

$$L_1' = mL = 0.515 \ (2.65 \text{ mh}) = 1.365 \text{ mh}$$

$$L_2' = \frac{1 - m^2}{4m} L = \frac{1 - (0.515)^2}{4(0.515)} (2.65 \text{ mh}) = 0.946 \text{ mh}$$

$$C_2' = mC = 0.515(0.265 \ \mu\text{f}) = 0.1365 \ \mu\text{f}$$

The T-section, from which the terminating half-sections are obtained, is found by using $m = 0.6$. These elements are

$$L_{1T} = mL = 0.6(2.65 \text{ mh}) = 1.59 \text{ mh}$$

$$L_{2T} = \frac{1 - m^2}{4m} L = \frac{1 - (0.6)^2}{4(0.6)} (2.65 \text{ mh}) = 0.707 \text{ mh}$$

$$C_{2T} = mC = 0.6(0.265 \ \mu\text{f}) = 0.159 \ \mu\text{f}$$

The T-section is then modified to make two terminating half-sections. The complete composite filter is shown in Fig. 6.1.

In actual construction, the filter would appear as in Fig. 6.2.

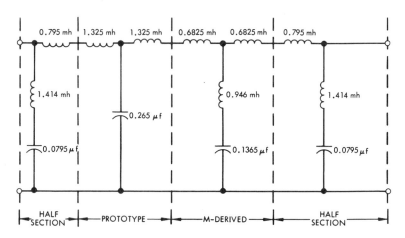

Fig. 6.1. Calculated composite filter for filter example.

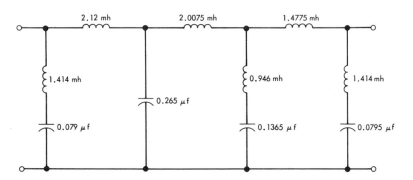

Fig. 6.2. Complete composite filter for filter example.

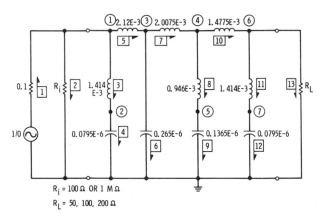

Fig. 6.3. Composite filter example ECAP equivalent circuit (AC) using ideal elements.

The ECAP equivalent circuit of the composite filter for the AC analysis appears in Fig. 6.3. Input data for this analysis (Fig. 6.4) are very simple since all components used are standard ECAP elements. Frequency- and phase-response analysis results are plotted in Figs. 6.5, 6.6, and 6.7, for R_L values of 100, 50, and 200 Ω, respectively. The input resistance value R_i had no effect on the response.

The AC analysis results verify calculated component values and provide the frequency- and phase-response required. The calculations and the circuit in Fig. 6.2 assume that all components used in the network are ideal (i.e., lossless). However, an inductor always has an associated series resistance that alters the

```
C       AC ANALYSIS OF COMPOSITE FILTER EXAMPLE
C       IDEAL COMPONENTS
        AC
C       RI=1 MEG.,RL=100
   B1 N(0,1),R=0.1,E=1/0
   B2 N(1,0),R=1E6
   B3 N(1,2),L=1.414E-3
   B4 N(2,0),C=0.0795E-6
   B5 N(1,3),L=2.12E-3
   B6 N(3,0),C=0.265E-6
   B7 N(3,4),L=2.0075E-3
   B8 N(4,5),L=0.946E-3
   B9 N(5,0),C=0.1365E-6
  B10 N(4,6),L=1.4775E-3
  B11 N(6,7),L=1.414E-3
  B12 N(7,0),C=0.0795E-6
  B13 N(6,0),R=100
       FREQ=5000
       PRINT,NV,CA
       EX
       MODIFY 1
       FREQ=1000(1.033)50E3
       EX
       MODIFY 2   RL=50   RI=1E6
  B13 R=50
       FREQ=1000(1.033)50E3
       EX
       MODIFY 3   RL=200 RI=1E6
  B13 R=200
       FREQ=1000(1.033)50E3
       EX
       MODIFY 4   RL=200 RI=100
   B2 R=100
       FREQ=1000(1.033)50E3
       EX
       MODIFY 5   RL=100 RI=100
  B13 R=100
       FREQ=1000(1.033)50E3
       EX
       MODIFY 6   RL=50   RI=100
  B13 R=50
       FREQ=1000(1.033)50E3
       EX
```

Fig. 6.4. Composite filter example input data (AC) using ideal elements.

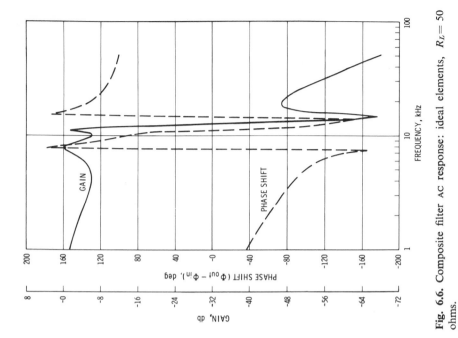

Fig. 6.6. Composite filter AC response: ideal elements, $R_L = 50$ ohms.

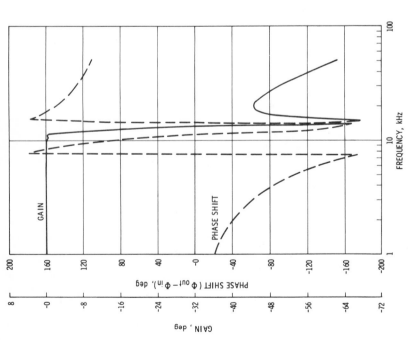

Fig. 6.5. Composite filter AC response: ideal elements, $R_L = 100$ ohms.

160

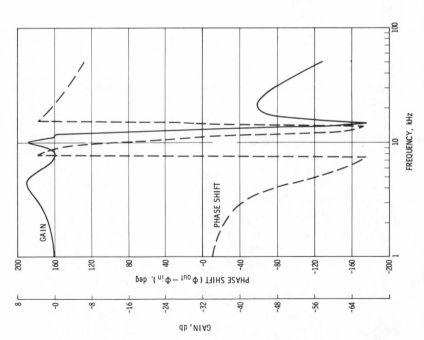

Fig. 6.8. Composite filter example ECAP equivalent circuit (AC) using nonideal elements.

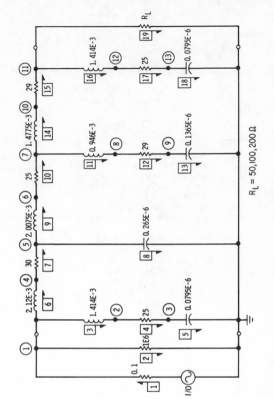

Fig. 6.7. Composite filter AC response: ideal elements, $R_L = 200$ ohms.

161

```
C       AC ANALYSIS OF COMPOSITE FILTER EXAMPLE
C       NONIDEAL COMPONENTS
        AC
C       RI=1 MEG, RL=50
   B1   N(0,1),R=0.1,E=1/0
   B2   N(1,0),R=1E6
   B3   N(1,2),L=1.414E-3
   B4   N(2,3),R=25
   B5   N(3,0),C=0.0795E-6
   B6   N(1,4),L=2.12E-3
   B7   N(4,5),R=30
   B8   N(5,0),C=0.265E-6
   B9   N(5,6),L=2.0075E-3
  B10   N(6,7),R=25
  B11   N(7,8),L=0.946E-3
  B12   N(8,9),R=29
  B13   N(9,0),C=0.1365E-6
  B14   N(7,10),L=1.4775E-3
  B15   N(10,11),R=29
  B16   N(11,12),L=1.414E-3
  B17   N(12,13),R=25
  B18   N(13,0),C=0.0795E-6
  B19   N(11,0),R=50
        FREQ=5000
        PRINT,NV,CA
        EX
        MO
        FREQ=1000(1.033)50E3
        EX
        MODIFY 2 RL=100
  B19   R=100
        FREQ=1000(1.033)50E3
        EX
        MODIFY 3 RL=200
  B19   R=200
        FREQ=1000(1.033)50E3
        EX
```

Fig. 6.9. Composite filter example input data (AC) using nonideal elements.

response of the network. Also, a capacitor equivalent circuit contains a shunt resistance because of the dielectric-leakage resistance of the component. In most cases, this leakage resistance is large enough that it can be neglected. Thus, to obtain good agreement between experimental results and the computer analysis of the filter, the loss elements in the network must be included in the equivalent circuit. The modified composite-filter ECAP circuit illustrated in Fig. 6.8 includes the resistances measured in a breadboard circuit. Input data used to compute the AC response of the network are shown in Fig. 6.9. The frequency- and phase-response calculations of the filter containing lossy elements are plotted in Figs. 6.10, 6.11, and 6.12. The results of the experimental gain measurements performed to verify the ECAP calculations are also plotted in these figures. The correlation between the calculated and the experimental frequency-response characteristics demonstrates the validity of the analysis.

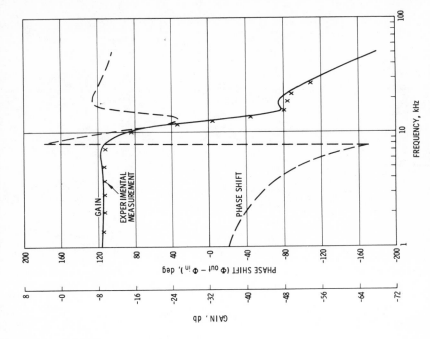

Fig. 6.11. Composite filter AC response: nonideal elements, $R_L = 50$ ohms.

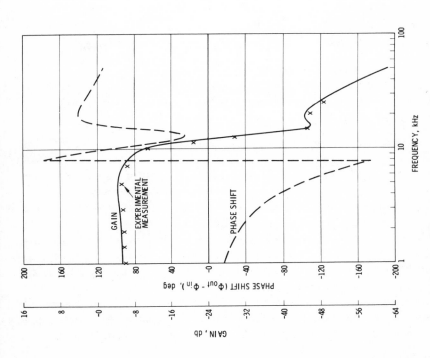

Fig. 6.10. Composite filter AC response: nonideal elements, $R_L = 100$ ohms.

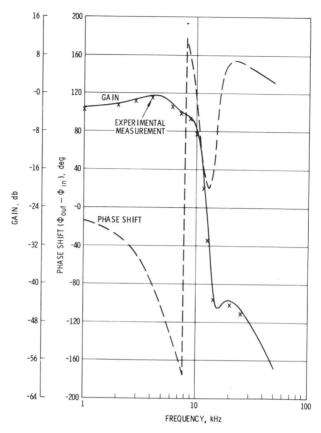

Fig. 6.12. Composite filter AC response: nonideal elements, $R_L = 200$ ohms.

6.3 COMPOSITE FILTER: TRANSIENT ANALYSIS

Application of ECAP techniques in the analysis of transient responses of filter networks is illustrated in this section. Values for the components used in the filter in this example were calculated in the first part of the AC analysis of the composite filter in Sec. 6.2. In this section, the filter network is analyzed to determine the transient pulse response of the filter when terminated in its characteristic impedance R_o, when terminated in one-half R_o, and when terminated in twice R_o. No component modeling is required in this application because all components used are standard, linear ECAP elements.

A transient analysis is also performed on the filter, assuming that nonideal (or lossy) components are used in the network. The results of this analysis are again compared with the experimental results obtained from the breadboard network assembled in the laboratory to demonstrate the validity of the analysis.

Since only standard elements are used, the input data for this analysis are very simple. The transient analysis equivalent circuit shown in Fig. 6.13 is the same as that used in the AC analysis example (see Fig. 6.3), except that the sinusoidal driving function in branch 1 has been replaced by a positive 1.0 v pulse source. The driving pulse has a duration of 1.0 msec and a rise and fall

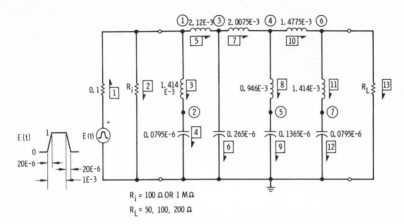

Fig. 6.13. Composite filter example ECAP equivalent circuit (transient analysis) using ideal elements.

time equal to 20.0 μsec. The input data required for the transient analysis using ideal components are presented in Fig. 6.14. Results of this transient analysis are plotted in Figs. 6.15, 6.16, and 6.17 (p. 168) for terminating resistance values of 100, 50, and 200 Ω. The assumed value of R_i was 1 MΩ.

To obtain an experimentally valid transient analysis of the filter network, it is necessary to insert into the network the series resistance present in each of the inductors. The resistance shunting the capacitors is generally negligible because of low dielectric-leakage current. The equivalent circuit of the nonideal filter containing the loss elements shown in Fig. 6.18 (p. 168) is a good approximation of the test network used in the laboratory. Input data representing this filter are in Fig. 6.19 (p. 169). The correlation between the experimental results and the calculated response demonstrates the validity of the equivalent circuit and the analysis. The experimental and calculated results are superimposed in Figs. 6.20, 6.21, and 6.22 (pp. 171–72) for comparison.

Transient analysis of the composite filter is not limited to problems in which linear terminations are used. For example, the filter network could have been terminated in a nonlinear reistance, such as a diode, without noticeably increasing the complexity of the ECAP input requirements. The advantage of the ECAP technique over calculating the transient response of the filter using conventional or classical methods is obvious when using a nonlinear terminating resistance.

```
C      TRANSIENT RESPONSE OF COMPOSITE FILTER NETWORK
C      IDEAL COMPONENTS
       TR--RL=100
  B1  N(0,1),R=0.1
  B2  N(1,0),R=1E6
  B3  N(1,2),L=1.414E-3
  B4  N(2,0),C=.0795E-6
  B5  N(1,3),L=2.12E-3
  B6  N(3,0),C=.265E-6
  B7  N(3,4),L=2.0075E-3
  B8  N(4,5),L=0.946E-3
  B9  N(5,0),C=.1365E-6
 B10  N(4,6),L=1.4775E-3
 B11  N(6,7),L=1.414E-3
 B12  N(7,0),C=.0795E-6
 B13  N(6,0),R=100
  E1  (1),0,1,1,1,1,1,1,1,1,1,1,1,1,1,1,1,1,1,1,1,1,1,1,1,1,1,1,1,
     *1,1,1,1,1,1,1,1,1,1,1,1,1,1,1,1,1,1,1,1,1,1,1,1,1,1,1,1,1,0
       TI=20E-6
       OU=1
       FI=2E-3
       PRINT,NV,CA
       EX
       TR--RL=50
  B1  N(0,1),R=.1
  B2  N(1,0),R=1E6
  B3  N(1,2),L=1.414E-3
  B4  N(2,0),C=.0795E-6
  B5  N(1,3),L=2.12E-3
  B6  N(3,0),C=.265E-6
  B7  N(3,4),L=2.0075E-3
  B8  N(4,5),L=0.946E-3
  B9  N(5,0),C=.1365E-6
 B10  N(4,6),L=1.4775E-3
 B11  N(6,7),L=1.414E-3
 B12  N(7,0),C=.0795E-6
 B13  N(6,0),R=50
  E1  (1),0,1,1,1,1,1,1,1,1,1,1,1,1,1,1,1,1,1,1,1,1,1,1,1,1,1,1,1,
     *1,1,1,1,1,1,1,1,1,1,1,1,1,1,1,1,1,1,1,1,1,1,1,1,1,1,1,1,1,0
       TI=20E-6
       OU=1
       FI=2E-3
       PRINT,NV,CA
       EX
       TR--RL=200
  B1  N(0,1),R=0.1
  B2  N(1,0),R=1E6
  B3  N(1,2),L=1.414E-3
  B4  N(2,0),C=.0795E-6
  B5  N(1,3),L=2.12E-3
  B6  N(3,0),C=.265E-6
  B7  N(3,4),L=2.0075E-3
  B8  N(4,5),L=.946E-3
  B9  N(5,0),C=.1365E-6
 B10  N(4,6),L=1.4775E-3
 B11  N(6,7),L=1.414E-3
 B12  N(7,0),C=.0795E-6
 B13  N(6,0),R=200
  E1  (1),0,1,1,1,1,1,1,1,1,1,1,1,1,1,1,1,1,1,1,1,1,1,1,1,1,1,1,1,
     *1,1,1,1,1,1,1,1,1,1,1,1,1,1,1,1,1,1,1,1,1,1,1,1,1,1,1,1,1,0
       TI=20E-6
       OU=1
       FI=2E-3
       PRINT,NV,CA
       EX
```

Fig. 6.14. Composite filter example input data transient analysis using ideal elements.

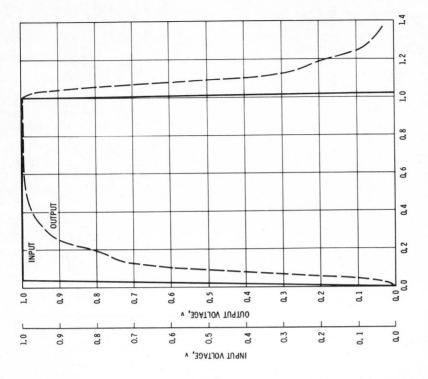

Fig. 6.15. Composite filter transient response: ideal elements, R_L = 100 ohms.

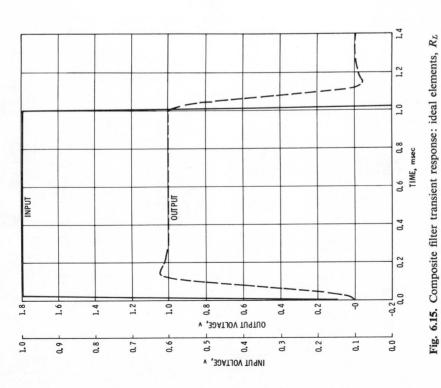

Fig. 6.16. Composite filter transient response: ideal elements, R_L = 50 ohms.

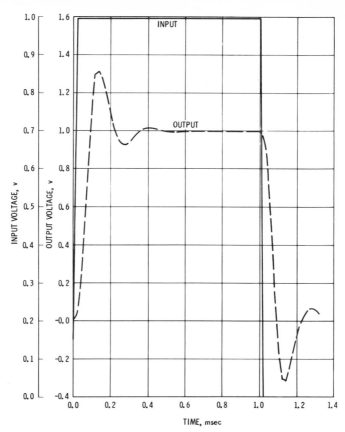

Fig. 6.17. Composite filter transient response: ideal elements, $R_L = 200$ ohms.

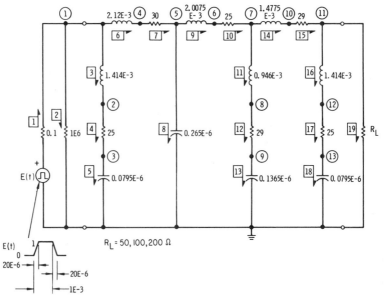

Fig. 6.18. Composite filter example ECAP equivalent circuit (transient analysis) using nonideal elements.

168

```
C       TRANSIENT ANALYSIS OF COMPOSITE FILTER
C       NONIDEAL COMPONENTS
        TR      RL=100
   B1  N(0,1),R=0.1
   B2  N(1,0),R=1F6
   B3  N(1,2),L=1.414E-3
   B4  N(2,3),R=25
   B5  N(3,0),C=0.0795E-6
   B6  N(1,4),L=2.12E-3
   B7  N(4,5),R=30
   B8  N(5,0),C=0.265E-6
   B9  N(5,6),L=2.0075E-3
  B10  N(6,7),R=25
  B11  N(7,8),L=0.946E-3
  B12  N(8,9),R=29
  B13  N(9,0),C=0.1365E-6
  B14  N(7,10),L=1.4775E-3
  B15  N(10,11),R=29
  B16  N(11,12),L=1.414E-3
  B17  N(12,13),R=25
  B18  N(13,0),C=0.0795E-6
  B19  N(11,0),R=100
   E1  (1),0,1,1,1,1,1,1,1,1,1,1,1,1,1,1,1,1,1,1,1,1,1,1,1,1,1,1,1,1,
      *1,1,1,1,1,1,1,1,1,1,1,1,1,1,1,1,1,1,1,1,1,1,1,1,1,1,1,1,0
        TI=20E-6
        FI=2E-3
        OU=1
        PRINT,NV,CA
        EXECUTE
        TR      RL=50
   B1  N(0,1),R=0.1
   B2  N(1,0),R=1E6
   B3  N(1,2),L=1.414E-3
   B4  N(2,3),R=25
   B5  N(3,0),C=0.0795E-6
   B6  N(1,4),L=2.12E-3
   B7  N(4,5),R=30
   B8  N(5,0),C=0.265E-6
   B9  N(5,6),L=2.0075E-3
  B10  N(6,7),R=25
  B11  N(7,8),L=0.946E-3
  B12  N(8,9),R=29
  B13  N(9,0),C=0.1365E-6
  B14  N(7,10),L=1.4775E-3
  B15  N(10,11),R=29
  B16  N(11,12),L=1.414E-3
  B17  N(12,13),R=25
  B18  N(13,0),C=0.0795E-6
  B19  N(11,0),R=50
   E1  (1),0,1,1,1,1,1,1,1,1,1,1,1,1,1,1,1,1,1,1,1,1,1,1,1,1,1,1,1,1,
      *1,1,1,1,1,1,1,1,1,1,1,1,1,1,1,1,1,1,1,1,1,1,1,1,1,1,1,1,0
        TI=20E-6
        OU=1
        FI=2E-3
        PRINT,NV,CA
        EX
```

Fig. 6.19. Composite filter example input data (transient analysis) using nonideal elements. (*Continued on next page.*)

```
       TR      RL=200
   B1  N(0,1),R=0.1
   B2  N(1,0),R=1F6
   B3  N(1,2),L=1.414E-3
   B4  N(2,3),R=25
   B5  N(3,0),C=0.0795E-6
   B6  N(1,4),L=2.12E-3
   B7  N(4,5),R=30
   B8  N(5,0),C=0.265E-6
   B9  N(5,6),L=2.0075E-3
  B10  N(6,7),R=25
  B11  N(7,8),L=0.946E-3
  B12  N(8,9),R=29
  B13  N(9,0),C=0.1365E-6
  B14  N(7,10),L=1.4775E-3
  B15  N(10,11),R=29
  B16  N(11,12),L=1.414E-3
  B17  N(12,13),R=25
  B18  N(13,0),C=0.0795E-6
  B19  N(11,0),R=200
   F1  (1),0,1,1,1,1,1,1,1,1,1,1,1,1,1,1,1,1,1,1,1,1,1,1,1,1,1,
      *1,1,1,1,1,1,1,1,1,1,1,1,1,1,1,1,1,1,1,1,1,1,1,1,1,0
       TI=20E-6
       OU=1
       FI=2E-3
       PRINT,NV,CA
       EX
```

Fig. 6.19 (*continued*).

6.4   DIGITAL LOGIC-INVERTER:
TRANSIENT ANALYSIS

This example has two primary functions. First, it demonstrates that the transient analysis program can serve reliably as an electronic circuit simulator to predict the performance of large-signal, or nonlinear, circuits. This capability is illustrated by analyzing the transient response of a digital inverter circuit to a specified input signal. The second function of the example is to demonstrate the application of the nonlinear transistor model, which was developed in Chapter 5 from the large-signal transistor equivalent circuit described in Appendix A.2, to the solution of nonlinear circuit analysis problems. This particular example illustrates all the important details involved in the application of the charge-control transistor model. Results of this transient analysis will be compared with experimental measurements to demonstrate the transistor model accuracy.

The transient response of the digital logic-inverter circuit in Fig. 6.23 to the input signal defined in the same figure is calculated in this example. The first step in the solution of an analysis problem involving very high-speed transient signals is to define the parasitic elements existing in the circuit as precisely as is necessary to obtain the required degree of accuracy (e.g., component lead

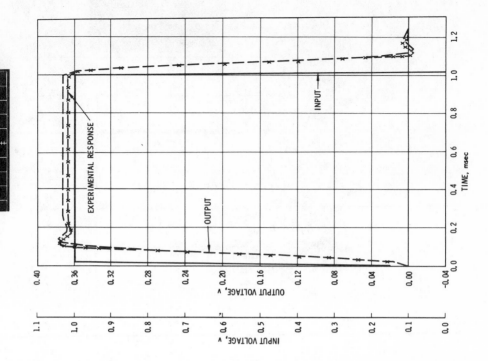

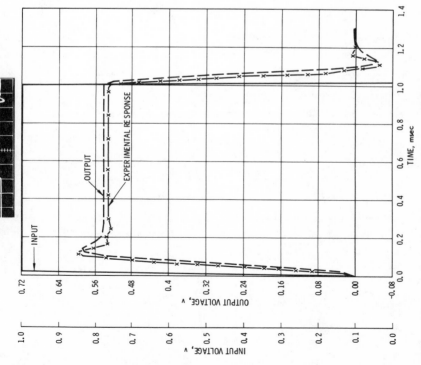

Fig. 6.20. Composite filter transient response: nonideal elements, $R_L = 100$ ohms. **Fig. 6.21.** Composite filter transient response: nonideal elements, $R_L = 50$ ohms.

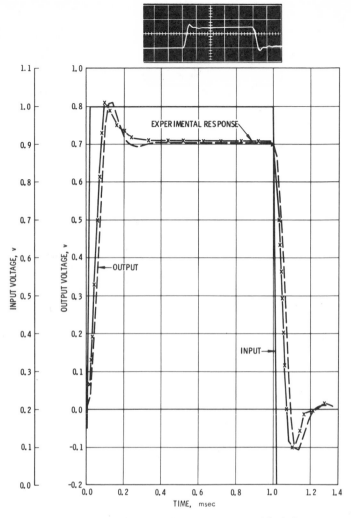

Fig. 6.22. Composite filter transient response: nonideal elements, $R_L = 200$ ohms.

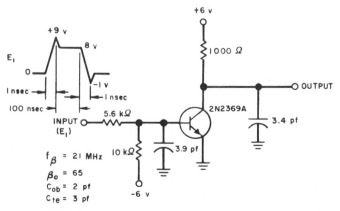

Fig. 6.23. Test circuit for experimental verification.

172

inductance can be neglected in a problem involving only audio frequencies, but it becomes very important in the circuit response at ultra-high frequencies). The second step, of equal importance, is the determination of the parameters of the nonlinear components in the circuit. In this analysis, the only nonlinear component present is the transistor, for which the large-signal, charge-control transistor model (Appendix A.2) is selected as an equivalent circuit.

The only parasitic elements having an important effect on the response of this circuit are the stray capacitances associated with the base and the collector of the transistor. The stray capacitance values measured in the breadboard circuit are

$$C_{\text{stray-}b} = 3.9 \text{ pf}$$

$$C_{\text{stray-}c} = 3.4 \text{ pf}$$

The component layout used makes it possible to assume that the parasitic inductances are negligible. The values for C_{ob} (output capacitance) and C_{te} (emitter-transition capacitance) can generally be directly obtained from the manufacturer's transistor data sheet. The data sheet* for the 2N2369A transistor used in this example is shown in Fig. 6.24. The values used for C_{ob} and C_{te} in this example are measured so that the correlation between the calculated and the experimental transient response can be checked. (Note that the values measured for use in the example are reasonably close to the values stated in the specification sheet. If a worst case transistor had been required, the min-max values from the specification could have been used.)

The value measured for DC beta for the test transistor is the same as the typical value given in the specification sheet at $I_c = 10$ ma. The typical value of f_T from the data sheet is 675 MHz. Using this value and assuming the forward average beta during the switching interval

$$\beta_{\text{AV}} \doteq 31.5$$

$$f_\beta = \frac{f_T}{\beta_{\text{AV}}} = \frac{675 \text{ MHz}}{31.5} = 21.4 \text{ MHz}$$

$$T_B \simeq \frac{1.22}{2\pi f_\beta} = 9.1 \text{ nsec}$$

To minimize the error in current balance at the node connecting R_B and C_1, arbitrarily set $R_B = 3850 \ \Omega$ and select

$$C_1 = \frac{T_B}{R_B} = \frac{9.1 \text{ nsec}}{3.85 \text{ k}\Omega} = 2.36 \text{ pf}$$

As a check, the equivalent resistance value of C_1 is

$$\frac{\Delta t}{C_1} = \frac{1 \text{ nsec}}{2.36 \text{ pf}} = 433 \ \Omega$$

which is sufficiently close to R_B to minimize the nodal current unbalance.

*Courtesy of Fairchild Semiconductor, a division of Fairchild Camera and Instrument Corp., Mountain View, Calif.

DATA SHEET AND PROCUREMENT SPECIFICATIONS · DECEMBER 1963

■ FAIRCHILD NPN DIFFUSED SI SWITCH

2N2369A

FAIRCHILD NPN DIFFUSED SILICON PLANAR EPITAXIAL TRANSISTOR

HIGH-SPEED SATURATED SWITCH

I_C Collector Current (10 μsec Pulse)   500 mA
I_c DC Collector Current   200 mA

ELECTRICAL CHARACTERISTICS (25°C free air temperature unless otherwise noted)

SYMBOL	CHARACTERISTIC	MIN.	TYP.	MAX.	UNITS	TEST CONDITIONS
h_{FE}	DC Pulse Current Gain [Note 5]	40	66	120		$I_C = 10$ mA, $V_{CE} = 1.0$ V
h_{FE}	DC Pulse Current Gain [Note 5]	40	63	120		$I_C = 10$ mA, $V_{CE} = 0.35$ V
h_{FE}	DC Pulse Current Gain [Note 5]	30	71			$I_C = 30$ mA, $V_{CE} = 0.4$ V
h_{FE}	DC Pulse Current Gain [Note 5]	20	50			$I_C = 10$ mA, $V_{CE} = 0.35$ V
h_{FE}	DC Pulse Current Gain [Note 5]	20				$I_C = 100$ mA, $V_{CE} = 1.0$ V
h_{FE} (−55°C)	DC Pulse Current Gain [Note 5]	0.7	0.8			$I_C = 10$ mA, $V_{CE} = 1.0$ V
V_{BE} (sat)	Base Saturation Voltage	0.59	0.9	0.85	Volts	$I_C = 10$ mA, $I_B = 1.0$ mA
V_{BE} (sat)	Base Saturation Voltage (−55°C to +125°C)		0.9	1.02	Volts	$I_C = 10$ mA, $I_B = 1.0$ mA
V_{BE} (sat)	Base Saturation Voltage		1.1	1.15	Volts	$I_C = 30$ mA, $I_B = 3.0$ mA
V_{BE} (sat)	Base Saturation Voltage		1.1	1.6	Volts	$I_C = 10$ mA, $I_B = 1.0$ mA
V_{CE} (sat)	Collector Saturation Voltage		0.14	0.2	Volts	$I_C = 100$ mA, $I_B = 10$ mA
V_{CE} (sat)	Collector Saturation Voltage		0.17	0.25	Volts	$I_C = 10$ mA, $I_B = 1.0$ mA
V_{CE} (sat)	Collector Saturation Voltage (125°C)		0.19	0.3	Volts	$I_C = 30$ mA, $I_B = 3.0$ mA
V_{CE} (sat)	Collector Saturation Voltage		0.28	0.5	Volts	$I_C = 10$ mA, $I_B = 1.0$ mA
h_{fe}	High Frequency Current Gain (f = 100 mc)	5.0	6.75			$V_{CE} = 10$ V
C_{ob}	Output Capacitance		2.3	4.0	pf	$V_{CB} = 5.0$ V, $I_E = 0$
I_{CBO} (150°C)	Collector Reverse Current		0.05	0.4	μA	$V_{CB} = 20$ V, $I_E = 0$
	Collector Cutoff Current		10	30		$V_{BE} = 0$
BV_{CES}	Collector to Emitter Breakdown Voltage	40			Volts	$I_C = 10$ μA, $I_B = 0$
BV_{CBO}	Collector to Base Breakdown Voltage	40			Volts	$I_C = 10$ μA, $I_E = 0$
V_{CEO} (sust)	Collector to Emitter Sustaining Voltage [Notes 4 and 5]	15			Volts	$I_C = 10$ mA, $I_B = 0$ (pulsed)
BV_{EBO}	Emitter to Base Breakdown Voltage	4.5			Volts	$I_E = 10$ μA, $I_C = 0$
τ_s	Charge Storage Time Constant [Note 6]		6.0	13	nsec	$I_C \cong 10$ mA, $I_{B1} \cong 10$ mA
t_{on}	Turn On Time [Note 6]		9.0	12	nsec	$I_C \cong 10$ mA, $I_{B1} \cong I_{B2} = 3.0$ mA
t_{off}	Turn Off Time [Note 6]		13	18	nsec	$I_C \cong 10$ mA, $I_{B1} \cong 3.0$ mA, $I_{B2} \cong -1.5$ mA

NOTES:

(1) These ratings are limiting values above which the serviceability of any individual semiconductor device may be impaired.
(2) These are steady state limits. The factory should be consulted on applications involving pulsed or low duty cycle operations.
(3) These ratings give a maximum junction temperature of 200°C and junction-to-case thermal resistance of 146°C/watt (derating factor of 6.85 mW/°C). Junction-to-ambient thermal resistance of 486°C/watt (derating factor of 2.06 mW/°C).
(4) Ratings refers to a high-current point where collector-to-emitter voltage is lowest. For more information send for Fairchild Publication APP-4.
(5) Pulse Conditions: length = 300 μsec; duty cycle ≤ 2%.
(6) See switching circuits for exact value of I_C, I_{B1} and I_{B2}.

EMITTER TRANSITION AND OUTPUT CAPACITANCES VERSUS REVERSE BIAS VOLTAGE

CONTOURS OF CONSTANT GAIN BANDWIDTH PRODUCT (f_t)

Fig. 6.24. 2N2369A transistor specification. Left: Typical data sheet. Right: C_{te}, C_{ob}, and f_t characteristics.

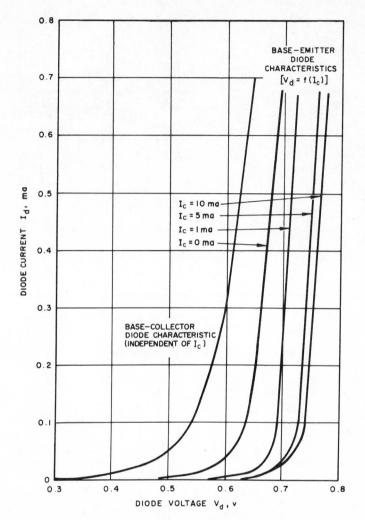

Fig. 6.25. Base-emitter and base-collector diode characteristics typical.

Again, the minimum or maximum values for any or all of these parameters can be used to study the circuit transient response using a worst case transistor.

The base-emitter and base-collector diode characteristics in Fig. 6.25 are also obtained by direct measurement. The piecewise linear approximations of the characteristics are shown in Fig. 6.26. The base-emitter diode used in this analysis is an average of the four V_{be} characteristics at low emitter current values; at higher emitter currents ($I_d > 25$ μa), the V_{be} characteristic at $I_c = 5$ ma is used.

If the assumption is made that the storage-time constant is negligible, the ECAP equivalent circuit for the digital inverter will appear as in Fig. 6.27. This

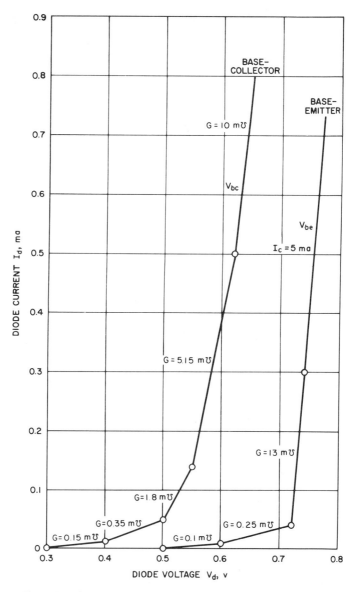

Fig. 6.26. Piecewise linear approximations of base-emitter and base-collector diode characteristics (2N2369A transistor).

equivalent circuit is much simpler than the one required if the storage time cannot be neglected; in the latter case, the equivalent circuit appears as in Fig. 6.28. The storage-time constant T_S assumed in Fig. 6.28 is equal to the base-time constant T_B. The value selected for T_S is usually determined experimentally because of the difficulty in establishing it analytically.

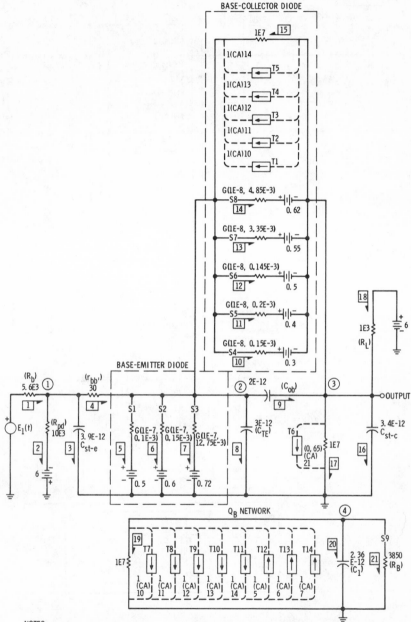

Fig. 6.27. Equivalent ECAP circuit ($T_S \longrightarrow 0$).

NOTES:
1) BRANCHES 15, 17, AND 19 CONTAIN DUMMY RESISTORS USED AS TO−BRANCHES FOR THE DEPENDENT CURRENT SOURCES.
2) THE NOTATION OF THE FORM A(CA)N ON THE DEPENDENT CURRENT SOURCES LITERALLY MEANS "THE CURRENT IN THE DEPENDENT CURRENT SOURCE IS A TIMES THE ELEMENT CURRENT IN BRANCH N".

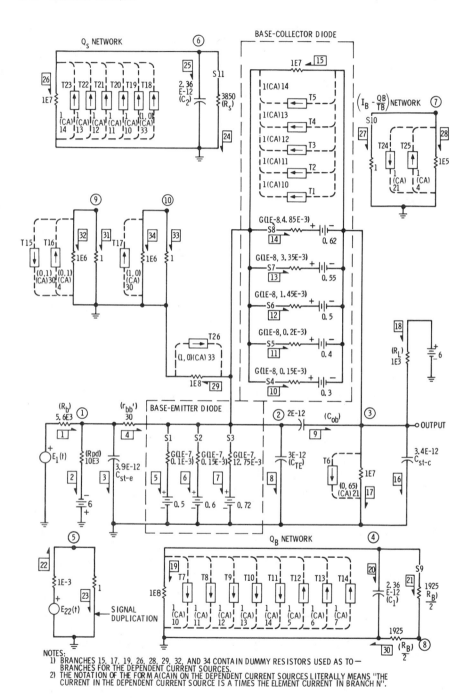

Fig. 6.28. Equivalent ECAP circuit ($T_S = T_B$).

```
C      TRANSIENT ANALYSIS OF DIGITAL LOGIC INVERTER
C      USING CHARGE CONTROL TRANSISTOR MODEL
C      ASSUMING TS=0
C      EQUILIBRIUM SOLUTION
       TR
  B1 N(0,1),R=5.6E3
  B2 N(1,0),R=10E3,E=6
  B3 N(1,0),C=3.9E-12
  B4 N(1,2),R=30
  B5 N(2,0),G=(1E-7,.1E-3),E=-.5
  B6 N(2,0),G=(1E-7,.15E-3),E=-.6
  B7 N(2,0),G=(1E-7,12.75E-3),E=-.72
  B8 N(2,0),C=3E-12
  B9 N(2,3),C=2E-12
 B10 N(2,3),G=(1E-8,.15E-3),E=-.3
 B11 N(2,3),G=(1E-8,.2E-3),E=-.4
 B12 N(2,3),G=(1E-8,1.45E-3),E=-.5
 B13 N(2,3),G=(1E-8,3.35E-3),E=-.55
 B14 N(2,3),G=(1E-8,4.85E-3),E=-.62
 B15 N(3,2),R=1E7
 B16 N(3,0),C=3.4E-12
 B17 N(3,0),R=1E7
 B18 N(0,3),R=1E3,E=6
 B19 N(4,0),R=1E7
 B20 N(4,0),C=2.36E-12
 B21 N(4,0),R=3850
  T1 B(10,15),BETA=1
  T2 B(11,15),BETA=1
  T3 B(12,15),BETA=1
  T4 B(13,15),BETA=1
  T5 B(14,15),BETA=1
  T6 B(21,17),BETA=(0,65)
  T7 B(10,19),BETA=1
  T8 B(11,19),BETA=1
  T9 B(12,19),BETA=1
 T10 B(13,19),BETA=1
 T11 B(14,19),BETA=1
 T12 B( 5,19),BETA=-1
 T13 B( 6,19),BETA=-1
 T14 B( 7,19),BETA=-1
  S1 B=5,(5),OFF
  S2 B=6,(6),OFF
  S3 B=7,(7),OFF
  S4 B=10,(10),OFF
  S5 B=11,(11),OFF
  S6 B=12,(12),OFF
  S7 B=13,(13),OFF
  S8 B=14,(14),OFF
  S9 B=21,(21),OFF
     TI=1E-9
     OU=1
     FI=200E-9
     EQUILIBRIUM
     PRINT,NV,CA
     EXECUTE
```

Fig. 6.29. Digital inverter circuit example initial condition input data ($T_S = 0$).

```
STEADY STATE SOLUTION


NODES                 NODE VOLTAGES

 1-  4        -0.21435003E 01    -0.21434202E 01     0.59971724E 01   -0.15168735E-02

BRANCHES              ELEMENT CURRENTS

 1-  4         0.38276803D-03     0.38564986D-03    -0.21435002D-06   -0.26674863D-05
 5-  8        -0.26434200D-06    -0.27434200D-06    -0.28634199D-06   -0.21434202D-06
 9- 12        -0.81405706D-06    -0.84405911D-07    -0.85405911D-07   -0.86405910D-07
13- 16        -0.86905910D-07    -0.87605910D-07     0.38332951D-06    0.59971704D-06
17- 20         0.59971704D-06     0.28275523D-05     0.39414475D-06   -0.15168732D-09
21- 21        -0.39399284D-06
```

Fig. 6.30. Initial conditions for digital inverter circuit ($T_S = 0$).

```
C      TRANSIENT ANALYSIS OF DIGITAL LOGIC INVERTER
C      USING CHARGE CONTROL TRANSISTOR MODEL
C      ASSUMING TS=0
       TR
 B1   N(0,1),R=5.6E3
 B2   N(1,0),R=10E3,E=6
 B3   N(1,0),C=3.9E-12,E0=2.1435
 B4   N(1,2),R=30
 B5   N(2,0),G=(1E-7,.1E-3),E=-.5
 B6   N(2,0),G=(1E-7,.15E-3),E=-.6
 B7   N(2,0),G=(1E-7,12.75E-3),E=-.72
 B8   N(2,0),C=3E-12,E0=2.1434
 B9   N(2,3),C=2E-12,E0=8.1406
B10   N(2,3),G=(1E-8,.15E-3),E=-.3
B11   N(2,3),G=(1E-8,.2E-3),E=-.4
B12   N(2,3),G=(1E-8,1.45E-3),F=-.5
B13   N(2,3),G=(1E-8,3.35E-3),E=-.55
B14   N(2,3),G=(1E-8,4.85E-3),E=-.62
B15   N(3,2),R=1E7
B16   N(3,0),C=3.4E-12,E0=-5.99717
B17   N(3,0),R=1E7
B18   N(0,3),R=1E3,E=6
B19   N(4,0),R=1E7
B20   N(4,0),C=2.36E-12,E0=0.15169E-2
B21   N(4,0),R=3850
 E1   (1),0,0,0,0,0,9,8,8,8,8,8,8,8,8,8,8,8,8,8,8,8,8,8,8,8,8,8,8,8,8,8,
      *8,8,8,8,8,8,8,8,8,8,8,8,8,8,8,8,8,8,8,8,8,8,8,8,8,8,8,8,8,
      *8,8,8,8,8,8,8,8,8,8,8,8,8,8,8,8,8,8,8,8,8,8,8,8,8,8,8,8,
      *8,8,8,8,8,8,8,8,8,8,8,8,8,8,8,8,8,8,8,8,8,8,8,8,8,-1,0
 T1   B(10,15),BETA=1
 T2   B(11,15),BETA=1
 T3   B(12,15),BETA=1
 T4   B(13,15),BETA=1
 T5   B(14,15),BETA=1
 T6   B(21,17),BETA=(0,65)
 T7   B(10,19),BETA=1
 T8   B(11,19),BETA=1
 T9   B(12,19),BETA=1
T10   B(13,19),BETA=1
T11   B(14,19),BETA=1
T12   B( 5,19),BETA=-1
T13   B( 6,19),BETA=-1
T14   B( 7,19),BETA=-1
 S1   B=5,(5),OFF
 S2   B=6,(6),OFF
 S3   B=7,(7),OFF
 S4   B=10,(10),OFF
 S5   B=11,(11),OFF
 S6   B=12,(12),OFF
 S7   B=13,(13),OFF
 S8   B=14,(14),OFF
 S9   B=21,(21),OFF
      TI=1E-9
      OU=1
      FI=200E-9
      PRINT,NV,CA
      EXECUTE
```

Fig. 6.31. Digital inverter transient analysis example input data
($T_S = 0$).

180

Initial conditions for the equivalent circuit in Fig. 6.27 are calculated using the transient analysis program EQUILIBRIUM solution option since a steady-state solution exists. Input data and results of the initial condition calculations appear in Figs. 6.29 and 6.30. The input data incorporating the initial conditions to determine the inverter response to an input-pulse width of 100 nsec are shown in Fig. 6.31. Input data in Fig. 6.31 for the solution plotted in Fig. 6.34 differ from the data for the solution in Fig. 6.33 only in the driving-function pulse width and the designated final time.

The initial condition solution for the equivalent circuit in Fig. 6.28 is almost identical to the solution for the circuit in which $T_S = 0$ was assumed. Figure 6.32 contains the input data for the transient analysis assuming $T_S = T_B$ of the digital logic-inverter circuit response to the 100 nsec driving pulse. The results differ slightly from the analysis conducted under the assumption $T_S = 0$, indicating that, for transistors exhibiting more prominent storage effects, $T_S > T_B$ should be used in their analysis.

```
C       TRANSIENT ANALYSIS OF DIGITAL LOGIC INVERTER
C       USING CHARGE CONTROL TRANSISTOR MODEL
C       ASSUMING TS=TB
        TR
   B1   N(0,1),R=5.6E3
   B2   N(1,0),R=10E3,F=6
   B3   N(1,0),C=3.9E-12,EO=2.1434
   B4   N(1,2),R=30
   B5   N(2,0),G=(1E-7,.1E-3),E=-.5
   B6   N(2,0),G=(1E-7,.15E-3),E=-.6
   B7   N(2,0),G=(1E-7,12.75E-3),F=-.72
   B8   N(2,0),C=3E-12,EO=2.1433
   B9   N(2,3),C=2E-12,EO=8.1405
   B10  N(2,3),G=(1E-8,.15E-3),E=-.3
   B11  N(2,3),G=(1E-8,.2E-3), E=-.4
   B12  N(2,3),G=(1E-8,1.45E-3),E=-.5
   B13  N(2,3),G=(1E-8,3.35E-3),F=-.55
   B14  N(2,3),G=(1E-8,4.85E-3),E=-.62
   B15  N(3,2),R=1E7
   B16  N(3,0),C=3.4E-12,EO=-5.9972
   B17  N(3,0),R=1F7
   B18  N(0,3),R=1F3,E=6
   B19  N(4,0),R=1E8
   B20  N(4,0),C=2.36E-12,FO=0.15173F-2
   B21  N(4,8),R=1925
   B22  N(0,5),R=1E-3
   B23  N(5,0),R=1
   B24  N(6,0),R=3850
   B25  N(6,0),C=2.36E-12,FO=0.1657E-2
   B26  N(6,0),R=1E7
   B27  N(7,0),R=1
   B28  N(7,0),R=1E5
   B29  N(2,0),R=1F8
   B30  N(8,0),R=1925
   B31  N(9,0),R=1
   B32  N(9,0),R=1F6
   B33  N(10,0),R=1
   B34  N(10,0),R=1F6
   E1   (1),0,0,0,0,0,0,0,9,8,8,8,8,8,8,8,8,8,8,8,8,8,8,8,8,8,8,8,8,8,8,8,8,
       *8,8,8,8,8,8,8,8,8,8,8,8,8,8,8,8,8,8,8,8,8,8,8,8,8,
       *8,8,8,8,8,8,8,8,8,8,8,8,8,8,8,8,8,8,8,8,8,8,8,8,8,
       *8,8,8,8,8,8,8,8,8,8,8,8,8,8,8,8,8,8,8,8,8,8,8-1,0
   E22  (1),0,0,0,0,0,0,0,9,8,8,8,8,8,8,8,8,8,8,8,8,8,8,8,8,8,8,8,8,8,8,8,8,
       *8,8,8,8,8,8,8,8,8,8,8,8,8,8,8,8,8,8,8,8,8,8,8,8,8,
       *8,8,8,8,8,8,8,8,8,8,8,8,8,8,8,8,8,8,8,8,8,8,8,8,8,
       *8,8,8,8,8,8,8,8,8,8,8,8,8,8,8,8,8,8,8,8,8,8,8-1,0
```

Fig. 6.32. Digital inverter transient analysis example input data ($T_S = T_B$). (*Continued on next page.*)

```
T1  B(10,15),BETA=1
T2  B(11,15),BETA=1
T3  B(12,15),BETA=1
T4  B(13,15),BETA=1
T5  B(14,15),BETA=1
T6  B(21,17),BETA=(0,65)
T7  B(10,19),BETA=1
T8  B(11,19),BETA=1
T9  B(12,19),BETA=1
T10 B(13,19),BETA=1
T11 B(14,19),BETA=1
T12 B( 5,19),BETA=-1
T13 B( 6,19),BETA=-1
T14 B( 7,19),BETA=-1
T15 B(30,32),BETA=(0,1)
T16 B( 4,32),BETA=(0,-1)
T17 B(31,34),BETA=(-1,0)
T18 B(33,26),BETA=-1
T19 B(10,26),BETA=-1
T20 B(11,26),BETA=-1
T21 B(12,26),BETA=-1
T22 B(13,26),BETA=-1
T23 B(14,26),BETA=-1
T24 B(21,28),BETA=1
T25 B(4,28), BETA=-1
T26 B(33,29),BETA=(1,0)
S1  B=5,(5),OFF
S2  B=6,(6),OFF
S3  B=7,(7),OFF
S4  B=10,(10,33),OFF
S5  B=11,(11),OFF
S6  B=12,(12),OFF
S7  B=13,(13),OFF
S8  B=14,(14),OFF
S9  B=21,(21),OFF
S10 B=27,(31),OFF
S11 B=24,(4,30),OFF
    TI=1E-9
    OU=1
    FI=200E-9
    PRINT,NV,CA
    EXECUTE
```

Fig. 6.32 (*continued*).

Table 6.2. **Comparison between predicted and experimental results assuming $T_S = 0$ (digital logic-inverter).**

Parameter	Results	Calculated	Measured	Difference	Per cent error
V_{ce}	Delay time (t_d), nsec	30	28	2	6.7
	Rise time (t_r), nsec	18	18	0	0
	Fall time (t_f), nsec	27	27	0	0
	Peak overshoot, v	6.3	6.25	0.05	0.8
	Saturation voltage (V_{ces}), v	0.15	0.17	0.02	13
V_{be}	Rise time (t_r), nsec	20	20	0	0
	Fall time (t_f), nsec	65	70	5	7
	Saturation voltage (V_{bes}), v	0.79	0.77	0.02	2.5

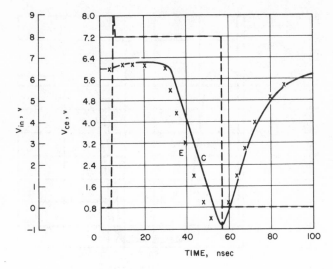

INPUT PULSE WIDTH = 50 nsec RC_Q = 57 nsec
INPUT PULSE T_R = T_F = I nsec β_0 = 65
C_{ob} = 2 pf f_β = 21 MHz
$C_{te.}$ = 3 pf
C_{st-e} = 3.9 pf
C_{st-c} = 3.4 pf

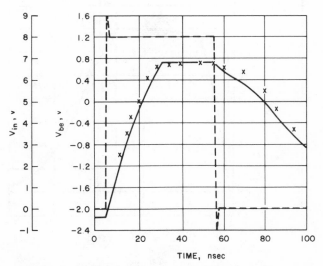

– – – – INPUT PULSE (V_{in})
———— CALCULATED CIRCUIT RESPONSE (T_S = T_B ; $T_S \leftarrow$ 0)
×××××× EXPERIMENTAL CIRCUIT RESPONSE
CIRCUIT: SEE FIG. 6.23

Fig. 6.33. Digital inverter transient response (input pulse width = 50 nsec).

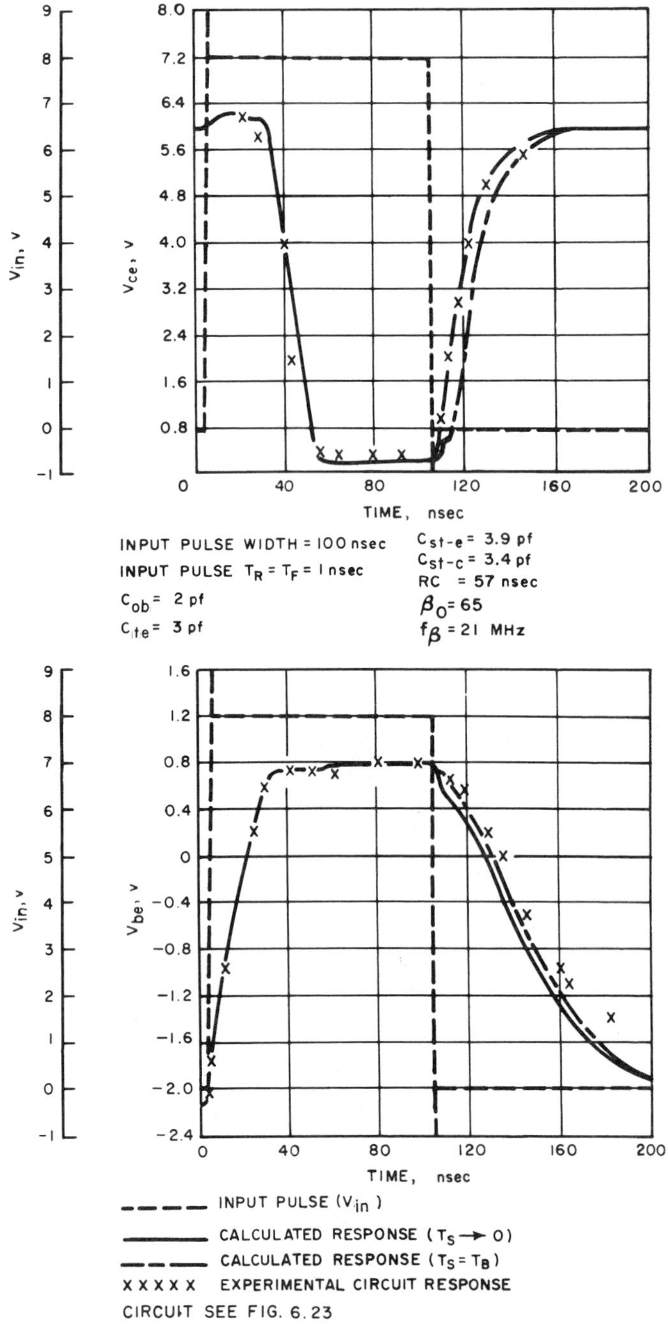

Fig. 6.34. Digital inverter transient response (input pulse width = 100 nsec).

Results of both analyses ($T_S = 0$ and $T_S = T_B$) are plotted in Fig. 6.33 for the condition in which the input pulse is too narrow to permit the transistor to saturate, and in Fig. 6.34 for the condition in which the transistor is saturated. The experimental results are superimposed on the calculated time-response to demonstrate the accuracy of the charge-control transistor model and the capabilities of the transient analysis program when applied to the analysis of non-linear circuits. Table 6.2 (p. 182) compares the calculated and the experimental results to further illustrate the accuracy of the calculations.

6.5 10 MHz CRYSTAL OSCILLATOR :
DC ANALYSIS

A DC, an AC, and a transient analysis of the two-transistor, 10 MHz crystal oscillator in Fig. 6.35 are performed, respectively. This particular circuit is used as an example because it provides, with a minimum of circuit complexity, a great amount of useful information about ECAP techniques not described previously in the text. The basic oscillator circuit consists of a differential amplifier with a high-Q, series-resonant, positive feedback network (crystal) to provide, over a narrow frequency band, adequate feedback to sustain oscillation.

In this section, the DC characteristics of the differential amplifier used in the oscillator are investigated. The crystal feedback network, consisting of the crystal and a 510 Ω resistor, has no effect on the circuit under DC conditions and can be neglected in the analysis. (The frequency response and the transient response of the complete circuit are studied in the following two sections.)

The DC analysis we will perform on this circuit consists primarily of

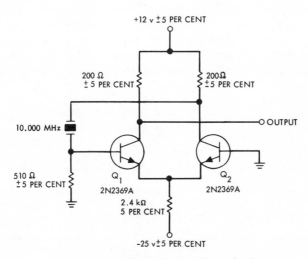

Fig. 6.35. 10 MHz crystal oscillator circuit.

investigating the DC operating point of the amplifier. The DC amplifier output-voltage level and the signal level are important design variables that must be considered when using direct coupling to the following stage. The required power ratings of the components, the nominal input impedance, and the DC gain of the circuit will also be calculated in the analysis.

The first step in the analysis is the reduction of the circuit in Fig. 6.35 to its ECAP DC equivalent. When the oscillator is operating properly, the transistors are overdriven and function as an emitter-coupled switch (i.e., the emitter current I_e flows in either Q_1 or Q_2 except during the switching transition). The DC output level, or quiescent level, then occurs when the emitter currents I_{e_1} and I_{e_2} are equal. The 510 Ω resistor, used as part of the feedback network, is removed and replaced with a small resistance (0.01 Ω) to minimize the amplifier DC output voltage offset due to the presence of the resistor.

The transistor model used in this analysis is the DC version of the hybrid-π transistor equivalent circuit described in Appendix A.1. Most of the parameters used in the model construction are listed in the 2N2369A data sheet in Fig. 6.24. To illustrate the techniques used in handling matched or related parameters, two extra requirements will be added to these parameters:

$$(1) \quad |h_{fe_1} - h_{fe_2}| \leq 0.1 h_{fe_1}$$

$$(2) \quad |V_{be_1} - V_{be_2}| \leq 10 \text{ mv}$$

The transistor model input impedance h_{ie} as a function of h_{fe} can be computed from

$$h_{ie} = r_{bb'} + (h_{fe} + 1)r_e \tag{6.1}$$

where $r_{bb'}$ is assumed equal to 30 Ω and I_e is assumed equal to 5.0 ma.

The calculated values of h_{ie} are tabulated in Table 6.3 with the corresponding values of h_{fe}. Since the 2N2369A transistor is specified for operation as a switch, the output conductance h_{oe} is not given. It is then necessary to either assume or measure a value for h_{oe}. In this case, the assumed value is 10^{-4} $\mho$.

The base-emitter voltage drop $V_{be(ON)}$ when the transistor is not operated as a saturated switch is also not provided. Here the value of $V_{be(ON)}$ is measured

Table 6.3. 2N2369A transistor parameters for 10 MHz oscillator DC analysis.

Parameter/Limit	Min	Nom	Max
h_{fe}	40	65	120
h_{ie}	244 Ω	373 Ω	660 Ω
h_{oe}		100/$\mu\mho$	
$V_{be(ON)}$	0.65 v	0.70 v	0.75 v

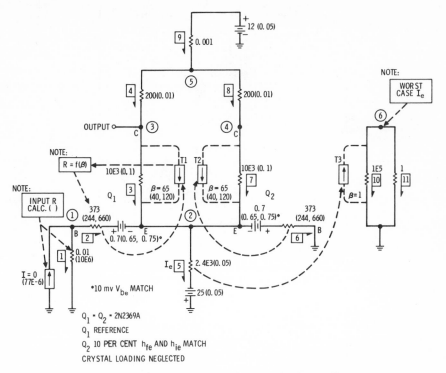

Fig. 6.36. DC ECAP equivalent circuit—10 MHz oscillator.

for a large sample of transistors since the value has a great effect on the amplifier operating point. The results of these measurements are listed in Table 6.3. The equivalent circuit for the DC analysis of the oscillator appears in Fig. 6.36.

A limitation has already been placed on the allowable h_{fe} difference between the transistors. Even though the value of h_{fe} in both transistors can be between 40 and 120, the difference cannot exceed $0.1(h_{fe_1})$. For a realistic analysis, the worst case calculations must be repeated for individual limit values of h_{fe} while the 10 per cent match is maintained since the analysis is performed using the input data parameter limit values. In this problem

$$0.9\,h_{fe_1} \le h_{fe_2} \le 1.1\,h_{fe_1}$$

The value of h_{fe_1} is fixed and the tolerance limit is placed on h_{fe_2} as shown in Fig. 6.37. Analyses are then performed using the upper and the lower h_{fe} limits and the nominal values. Since input resistance h_{ie} is directly related to h_{fe}, the h_{ie} value must correspond to the h_{fe_1} value used in each analysis step.

The tolerance on V_{be} is handled in a different manner. The base-emitter voltage drops are related by

$$|V_{be_1} - V_{be_2}| \le 10 \text{ mv}$$

```
C       DC ANALYSIS OF 1OMHZ OSCILLATOR
C       CRYSTAL REMOVED
        DC
 B1   N(1,0),R=0.01,I=0
 B2   N(1,2),R=373,E=-0.7(-0.705,-0.695)
 B3   N(3,2),R=10E3(.1)
 B4   N(5,3),R=200(.01)
 B5   N(2,0),R=2.4E3(.05),E=25(.05)
 B6   N(0,2),R=373(.1),E=-0.7(-0.705,-0.695)
 B7   N(4,2),R=10E3(.1)
 B8   N(5,4),R=200(.05)
 B9   N(0,5),R=0.001,E=12
B10   N(6,0),R=1E5
B11   N(6,0),R=1
 T1   B(2,3),BETA=65
 T2   B(6,7),BETA=65(.1)
 T3   B(5,10),BETA=-1
        SE
        ST
        WO
        PRINT,NV,CA,BP,ST,SE,WO
        EX
        MODIFY 1  EMITTER RESISTANCE VARIATION
 B5   R=560(10)3600
        EX
        MODIFY 2  LOAD RESISTANCE VARIATION
 B4   P=51(10)820
 B5   R=2.4E3
        EX
        MODIFY 3    WORST CASE (MIN. BETA)
 B2   R=244
 B4   R=200
 B6   R=244(.1)
 T1   BETA=40
 T2   BETA=40(.1)
        WO
        ST
        EX
        MODIFY 4    WORST CASE (MAX. BETA)
 B2   R=660
 B6   R=660(.1)
 T1   BETA=120
 T2   BETA=120(.1)
        WO
        ST
        EX
        MODIFY 5  WORST CASE (MIN.VBE,MAX.BETA)
 B2   R=660,E=-0.655(-0.66,-0.65)
 B6   R=660(.1),E=-0.655(-0.66,-0.65)
 T1   BETA=120
 T2   BETA=120(.1)
        WO
        ST
        EX
        MODIFY 6  WORSTCASE (MAX.VBE,MAX.BETA)
 B2   E=-0.745(-0.75,-0.74)
 B6   E=-0.745,-0.74)
        WO
        ST
        EX
        MODIFY 7  INPUT RESISTANCE
 B1   R=10E6,I=58.87E-6
 B2   R=373
 B6   R=373
 T1   BETA=65
 T2   BETA=65
        PRINT,NV,CA,MI
        EX
```

Fig. 6.37. 10 MHz crystal oscillator circuit DC analysis input data.

Since no ratio relationship exists, the tolerance can be divided so that the maximum difference between V_{be_1} and V_{be_2} is 10 mv as demonstrated in the DC analysis input data (Fig. 6.37).

The input data in Fig. 6.37 contain the necessary information to perform a DC analysis of the 10 MHz oscillator including power ratings of the components, DC gain, and a complete worst case analysis of the output operating point.

This analysis also demonstrates a method to calculate the input resistance of the circuit. In the initial calculations and the first six modify routines of the analysis, the amplifier input was connected to ground through a small series resistance to prevent an offset in output voltage. To measure the input impedance of the amplifier, it is necessary to "open" the input terminal to the amplifier. This, of course, causes biasing, or operating-point, problems. If it is necessary to avoid an offset in the amplifier operating point, the input terminal of the amplifier can be connected to ground through an arbitrarily large resistance while a bias current is simultaneously applied to the input terminal to maintain the required operating-point balance as shown in the input data (MODIFY 7) in Fig. 6.37. This is not necessary in this application because the impedance to ground from any node is directly available from the nodal impedance matrix obtained as miscellaneous output.

With this background, results of the DC analysis in Fig. 6.38 are studied.

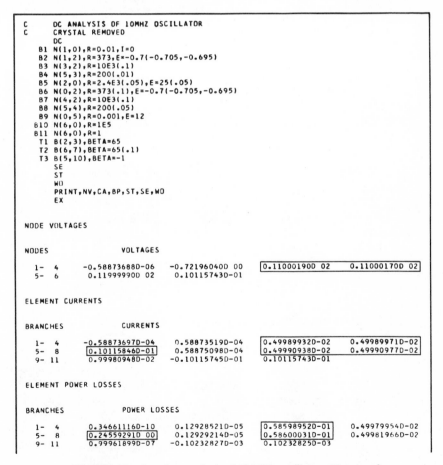

Fig. 6.38. Results of DC analysis of 10 MHz oscillator. (*Continued on next seven pages.*)

```
PARTIAL DERIVATIVES AND SENSITIVITIES OF NODE VOLTAGES

WITH RESPECT TO RESISTANCES

BRANCH    NODE      PARTIALS              SENSITIVITIES

  1        1       -0.58872869D-04       -0.58872835E-08
  1        2       -0.29384554D-04       -0.29384541E-08
  1        3        0.10070149D-02        0.10070141E-06
  1        4       -0.10046218D-02       -0.10046210E-06
  1        5        0.11965381D-10        0.11965376E-14
  2        3        0.10070123D-02        0.37561546E-02
  2        4       -0.10046192D-02       -0.37472283E-02
  2        5        0.11965364D-10        0.44630785E-10
  2        6       -0.12243406D-07       -0.45667885E-07
  3        3        0.11683472D-04        0.11683470E-02
  4        3       -0.49491693D-02       -0.98983385E-02
  5        3        0.41148630D-03        0.99476688E-02
  5        6       -0.42098395D-05       -0.10103614E-03
  6        3       -0.10046662D-02       -0.37473289E-02
  6        4        0.10070395D-02        0.37562556E-02
  6        5        0.11965982D-10        0.44633089E-10
  6        6       -0.12244063D-07       -0.45670337E-07
  7        3       -0.11301114D-04       -0.11301110E-02
  8        3       -0.48195485D-04       -0.96390955E-04
  9        3       -0.99948335D-02       -0.99948295E-07
 10        3        0.0                   0.0
 11        3        0.0                   0.0

WITH RESPECT TO BETAS

BETA      NODE      PARTIALS              SENSITIVITIES

  1        1        0.44508452D-08        0.28930478E-08
  1        2        0.16602101D-03        0.10791360E-03
  1        3       -0.58679269D-02       -0.38141506E-02
  2        3        0.56760531D-02        0.36894325E-02
  3        3        0.0                   0.0

WITH RESPECT TO VOLTAGE SOURCES

BRANCH    NODE      PARTIALS              SENSITIVITIES

  2        1       -0.13428268D-04       -0.93997812E-07
  2        2        0.49911207D 00        0.34937821E-02
  2        3       -0.17104677D 02       -0.11973262E 00
  2        4        0.17064029D 02        0.11944813E 00
  2        5       -0.20323827D-06       -0.14226669E-08
  2        6        0.20796120D-03        0.14557272E-05
  5        3       -0.40973973D-01       -0.10243490E-01
  6        3        0.17064029D 02        0.11944813E 00
  6        4       -0.17104678D 02       -0.11973262E 00
  6        5       -0.20324372D-06       -0.14227051E-08
  6        6        0.20796678D-03        0.14557663E-05
  9        3        0.99967383D 00        0.11996078E 00
```

Fig. 6.38 (*continued*).

Unimportant portions of the output are omitted, and other portions are reduced to graphic form to eliminate unnecessary bulk.

The first portion of the output in Fig. 6.38 contains the node voltages, element currents, and element power losses. The node voltage calculations (NV3, NV4) indicate the amplifier is balanced, which can also be verified by observing the magnitude of the collector currents (CA4, CA8). The collector current in each transistor is equal to 5.0 ma, which justifies the assumptions made when constructing the transistor model. The voltage level at node 3 is the nominal DC (or average) output level of the oscillator since the average level occurs when the collector currents are equal. The output signal level is calculated from

STANDARD DEVIATIONS OF NODE VOLTAGES

NODE	STD. DEV.	NOM.-STD. DEV.	NOMINAL	NOM.&STD. DEV.
1	0.39406920E-07	-0.62814377E-06	-0.58873688D-06	-0.54932991E-06
2	0.14682387E-02	-0.72342861E 00	-0.72196040D 00	-0.72049212E 00
3	0.50352816E-01	0.10949837E 02	0.11000190D 02	0.11050543E 02
4	0.52989010E-01	0.10947181E 02	0.11000170D 02	0.11053159E 02
5	0.23799112E-06	0.11999990E 02	0.11999990D 02	0.11999990E 02
6	0.24171032E-03	0.98740309E-02	0.10115743D-01	0.10357451E-01

WORST CASE SOLUTIONS FOR NODE VOLTAGES

NODE	WCMIN	NOMINAL	WCMAX
1	-0.87732894E-06	-0.58873684E-06	-0.29673623D-06
2	-0.73321182E 00	-0.72196037E 00	-0.71157440D 00
3	0.10618425E 02	0.11000190E 02	0.11378959D 02
4	0.10526979E 02	0.11000170E 02	0.11371874D 02
5	0.11999989E 02	0.11999990E 02	0.11999991D 02
6	0.91361254E-02	0.10115739E-01	0.11198584D-01

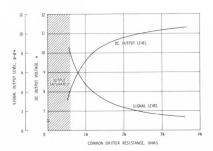

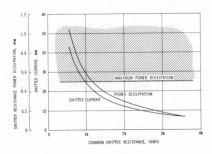

Fig. 6.38 (*continued*).

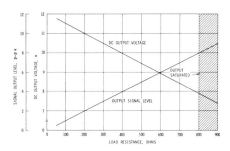

```
       MODIFY 3       WORST CASE (MIN. BETA)
     B2 R=244
     B4 R=200
     B6 R=244(.1)
     T1 BETA=40
     T2 BETA=40(.1)
        WO
        ST
        EX


NODE VOLTAGES


NODES               VOLTAGES

   1-  4      -0.94744744D-06    -0.72311862D 00     0.11007418D 02      0.11007388D 02
   5-  6       0.11999990D 02     0.10115260D-01


ELEMENT CURRENTS


BRANCHES            CURRENTS

   1-  4      -0.94744758D-04     0.94744574D-04     0.49628359D-02      0.49628577D-02
   5-  8       0.10115363D-01     0.94748457D-04     0.49629881D-02      0.49630100D-02
   9- 11       0.99258678D-02    -0.10115262D-01     0.10115260D-01


ELEMENT POWER LOSSES


BRANCHES            POWER LOSSES

   1-  4       0.89765678D-10     0.21902750D-05     0.58216729D-01      0.49259924D-02
   5-  8       0.24556947D 00     0.21904545D-05     0.58218364D-01      0.49262947D-02
   9- 11       0.98522851D-07    -0.10231851D-03     0.10231848D-03
```

Fig. 6.38 (*continued*).

```
STANDARD DEVIATIONS OF NODE VOLTAGES


NODE     STD. DEV.            NOM.-STD. DEV.      NOMINAL             NOM.&STD. DEV.

 1       0.57679156E-07      -0.10051263E-05     -0.94744744D-06     -0.88976827E-06
 2       0.14055767E-02      -0.72452414E 00     -0.72311862D 00     -0.72171301E 00
 3       0.45263793E-01       0.10962154E 02      0.11007418D 02      0.11052682E 02
 4       0.48309192E-01       0.10959078E 02      0.11007388D 02      0.11055696E 02
 5       0.16911940E-06       0.11999990E 02      0.11999990D 02      0.11999990E 02
 6       0.17339003E-03       0.99418685E-02      0.10115260D-01      0.10288648E-01


WORST CASE SOLUTIONS FOR NODE VOLTAGES


NODE     WCMIN               NOMINAL             WCMAX

 1       -0.13253311E-05     -0.94744740E-06     -0.54111038D-06
 2       -0.73282379E 00     -0.72311860E 00     -0.71370847D 00
 3        0.10710276E 02      0.11007418E 02      0.11326754D 02
 4        0.10610617E 02      0.11007387E 02      0.11330587D 02

 5   PARTIAL W.R.T.   R   1   HAS CHANGED SIGN AT MIN


 5   PARTIAL W.R.T.   R   2   HAS CHANGED SIGN AT MIN


 5   PARTIAL W.R.T.   E   2   HAS CHANGED SIGN AT MIN


 5   PARTIAL W.R.T.   R   6   HAS CHANGED SIGN AT MAX


 5   PARTIAL W.R.T.   E   6   HAS CHANGED SIGN AT MAX

 5        0.11999990E 02      0.11999990E 02      0.11999991D 02
 6        0.95918253E-02      0.10115258E-01      0.10638811D-01

         MODIFY 4      WORST CASE (MAX. BETA)
    B2  R=660
    B6  R=660(.1)
    T1  BETA=120
    T2  BETA=120(.1)
        WO
        ST
        EX


NODE VOLTAGES


NODES              VOLTAGES

 1-  4     -0.32119287D-06     -0.72119902D 00      0.10994803D 02      0.10994792D 02
 5-  6      0.11999990D 02      0.10116060D-01


ELEMENT CURRENTS


BRANCHES           CURRENTS

 1-  4     -0.32119292D-04      0.32119263D-04      0.50259107D-02      0.50259323D-02
 5-  8      0.10116163D-01      0.32119750D-04      0.50259679D-02      0.50259896D-02
 9- 11      0.10051922D-01     -0.10116062D-01      0.10116060D-01


ELEMENT POWER LOSSES


BRANCHES           POWER LOSSES

 1-  4      0.10316487D-10      0.68088708D-06      0.58883581D-01      0.50520002D-02
 5-  8      0.24560831D 00      0.68090771D-06      0.58884194D-01      0.50521153D-02
 9- 11      0.10104113D-06     -0.10233469D-03      0.10233467D-03
```

Fig. 6.38 (*continued*).

```
STANDARD DEVIATIONS OF NODE VOLTAGES


NODE    STD. DEV.         NOM.-STD. DEV.      NOMINAL          NOM.&STD. DEV.

 1      0.20833106E-07    -0.34202594E-06     -0.32119287D-06   -0.30035972E-06
 2      0.13732789E-02    -0.72257227E 00     -0.72119902D 00   -0.71982574E 00
 3      0.49032010E-01     0.10945771E 02      0.10994803D 02    0.11043835E 02
 4      0.51827274E-01     0.10942964E 02      0.10994792D 02    0.11046618E 02
 5      0.17195407E-06     0.11999990E 02      0.11999990D 02    0.11999990E 02
 6      0.17340819E-03     0.99426508E-02      0.10116060D-01    0.10289468E-01



WORST CASE SOLUTIONS FOR NODE VOLTAGES


NODE    WCMIN              NOMINAL            WCMAX

 1      -0.45520869E-06    -0.32119283E-06     -0.17626088D-06
 2      -0.73053396E 00    -0.72119898E 00     -0.71214122D 00
 3       0.10679153E 02     0.10994802E 02      0.11336117D 02
 4       0.10575781E 02     0.10994791E 02      0.11335175D 02
 5       0.11999989E 02     0.11999990E 02      0.11999990D 02
 6       0.95926598E-02     0.10116059E-01      0.10639564D-01

        MODIFY 5  WORST CASE (MIN.VBE,MAX.BETA)
    B2  R=660,E=-0.655(-0.66,-0.65)
    B6  R=660(.1),E=-0.655(-0.66,-0.65)
    T1  BETA=120
    T2  BETA=120(.1)
        WO
        ST
        EX


NODE VOLTAGES


NODES          VOLTAGES

 1-  4    -0.32235408D-06    -0.67627561D 00     0.10992955D 02    0.10992943D 02
 5-  6     0.11999990D 02     0.10134778D-01


ELEMENT CURRENTS


BRANCHES          CURRENTS

 1-  4    -0.32235413D-04    0.32235333D-04     0.50351619D-02    0.50351751D-02
 5-  8     0.10134881D-01    0.32235822D-04     0.50352193D-02    0.50352326D-02
 9- 11     0.10070408D-01   -0.10134780D-01     0.10134778D-01


ELEMENT POWER LOSSES


BRANCHES          POWER LOSSES

 1-  4     0.10391217D-10    0.68581705D-06     0.58756463D-01    0.50705988D-02
 5-  8     0.24651806D 00    0.68583783D-06     0.58757076D-01    0.50707145D-02
 9- 11     0.10141311D-06   -0.10271374D-03     0.10271372D-03
```

Fig. 6.38 (*continued*).

$$E_{\text{SIG}} = I_{et} \times R_L = (10.116 \text{ ma}) \times (200 \ \Omega) = 2.0232 \text{ v} \qquad (6.2)$$

where $I_{et} = $ common-emitter current (CA5)
$\quad\quad\ \ R_L = $ load resistance (B4)

The element power losses determine the required power ratings of the individual components. The nominal power dissipated in each component is less than 5 mw with the exception of that in branches 3, 5, and 7. The highest nominal power

```
STANDARD DEVIATIONS OF NODE VOLTAGES

NODE    STD. DEV.          NOM.-STD. DEV.      NOMINAL            NOM.&STD. DEV.

  1     0.20841952E-07     -0.34319601E-06     -0.32235408D-06     -0.30151210E-06
  2     0.13738638E-02     -0.67764944E 00     -0.67627561D 00     -0.67490172E 00
  3     0.49052805E-01      0.10943901E 02      0.10992955D 02      0.11042007E 02
  4     0.51857185E-01      0.10941086E 02      0.10992943D 02      0.11044800E 02
  5     0.17195407E-06      0.11999990E 02      0.11999990D 02      0.11999990E 02
  6     0.17340819E-03      0.99613667E-02      0.10134778D-01      0.10308184E-01


WORST CASE SOLUTIONS FOR NODE VOLTAGES

NODE    WCMIN              NOMINAL             WCMAX

  1     -0.45644737E-06     -0.32235408E-06     -0.17734457D-06
  2     -0.68561566E 00     -0.67627561E 00     -0.66721270D 00
  3      0.10677123E 02      0.10992954E 02      0.11334450D 02
  4      0.10573649E 02      0.10992943E 02      0.11333594D 02
  5      0.11999989E 02      0.11999990E 02      0.11999990D 02
  6      0.96113756E-02      0.10134775E-01      0.10658284D-01

        MODIFY 6  WORST CASE (MAX.VBE,MAX.BETA)
     B2 E=-0.745(-0.75,-0.74)
     B6 E=-0.745(-0.75,-0.74)
        WO
        ST
        EX

NODE VOLTAGES

NODES           VOLTAGES

  1- 4     -0.32003396D-06     -0.76612243D 00     0.10996652D 02     0.10996640D 02
  5- 6      0.11999990D 02      0.10097342D-01

ELEMENT CURRENTS

BRANCHES         CURRENTS

  1- 4     -0.32003401D-04      0.32003281D-04     0.50166700D-02     0.50166893D-02
  5- 8      0.100974450-01      0.32003766D-04     0.50167271D-02     0.50167463D-02
  9-11      0.100334360-01     -0.10097344D-01     0.10097342D-01

ELEMENT POWER LOSSES

BRANCHES         POWER LOSSES

  1- 4      0.10242175D-10      0.67597863D-06     0.59009958D-01     0.50334353D-02
  5- 8      0.24470025D 00      0.67599911D-06     0.59010571D-01     0.50335498D-02
  9-11      0.10066983D-06     -0.10195634D-03     0.10195631D-03
```

Fig. 6.38 (*continued*).

dissipation calculated is

$$BP5 = 0.2456 \text{ w}$$

Allowing a 50 per cent power derating for a safety factor requires that the 2.4 kΩ resistor in branch 5 be rated at 0.5 w or greater. The power dissipations in branches 3 and 7 are effectively equal to the power dissipated in the transistors. The currents CA3 and CA7 equal the collector currents, and the voltages

```
STANDARD DEVIATIONS OF NODE VOLTAGES

NODE     STD. DEV.          NOM.-STD. DEV.      NOMINAL            NOM.&STD. DEV.

 1       0.20824302E-07    -0.34085821E-06     -0.32003396D-06    -0.29920960E-06
 2       0.13726971E-02    -0.76749510E 00     -0.76612243D 00    -0.76474971E 00
 3       0.49011324E-01     0.10947640E 02      0.10996652D 02     0.11045663E 02
 4       0.51797464E-01     0.10944842E 02      0.10996640D 02     0.11048437E 02
 5       0.17195407E-06     0.11999990E 02      0.11999990D 02     0.11999990E 02
 6       0.17340819E-03     0.99239312E-02      0.10097342D-01     0.10270748E-01

WORST CASE SOLUTIONS FOR NODE VOLTAGES

NODE      WCMIN              NOMINAL             WCMAX

 1       -0.45397229E-06    -0.32003391E-06     -0.17517718D-06
 2       -0.77545226E 00    -0.76612240E 00     -0.75706975D 00
 3        0.10681184E 02     0.10996652E 02      0.11333778D 02
 4        0.10577913E 02     0.10996640E 02      0.11336757D 02
 5        0.11999989E 02     0.11999990E 02      0.11999990D 02
 6        0.95739439E-02     0.10097340E-01      0.10620844D-01

      MODIFY 7  INPUT RESISTANCE
    B1 R=10E6,I=58.87E-6
    B2 R=373
    B6 R=373
    T1 BETA=65
    T2 BETA=65
       PRINT,NV,CA,MI
       EX

NODAL CONDUCTANCE MATRIX

ROW      COLS

 1       1 - 4      0.26810650D-02-0.26809650D-02 0.0              0.0
 1       5 - 6      0.0             0.0
 2       1 - 4     -0.17694367D 00 0.35450400D 00-0.99999990D-04-0.99999990D-04
 2       5 - 6      0.0             0.0
 3       1 - 4      0.17426270D 00-0.17436270D 00 0.50999989D-02 0.0
 3       5 - 6     -0.49999990D-02 0.0
 4       1 - 4      0.0            -0.17436270D 00 0.0             0.50999989D-02
 4       5 - 6     -0.49999990D-02 0.0
 5       1 -.4      0.0             0.0            -0.49999990D-02-0.49999990D-02
 5       5 - 6      0.10000100D 04 0.0
 6       1 - 4      0.0            -0.41666650D-03 0.0             0.0
 6       5 - 6      0.0             0.10000100D 01

EQUIVALENT CURRENT VECTOR

NODE NO.    CURRENT

    1       0.20561886D-02
    2      -0.27406260D 00
    3       0.12982565D 00
    4       0.12982565D 00
    5       0.12000000D 05
    6       0.10416660D-01

NODAL IMPEDANCE MATRIX

ROW      COLS

 1       1 - 4      0.74463211D 03 0.57420961D 01 0.11259124D 00 0.11259124D 00
 1       5 - 6      0.11259009D-05 0.0
 2       1 - 4      0.37165987D 03 0.54423103D 01 0.11259544D 00 0.11259544D 00
 2       5 - 6      0.11259429D-05 0.0
 3       1 - 4     -0.12736863D 05 0.12087564D 00 0.19608180D 03 0.33313027D-02
 3       5 - 6      0.98041566D-03 0.0
 4       1 - 4      0.12706594D 05 0.19632350D 03 0.38504798D 01 0.19992895D 03
 4       5 - 6      0.10188868D-02 0.0
 5       1 - 4     -0.15133978D-03 0.98221186D-03 0.99965121D-03 0.99965121D-03
 5       5 - 6      0.10000000D-02 0.0
 6       1 - 4      0.15485667D 00 0.23926044D-02 0.46914277D-04 0.46914277D-04
 6       5 - 6      0.46913798D-09 0.99999000D 00
```

Fig. 6.38 (*continued*).

developed across the branches are the collector-to-emitter voltages of the transistors. Hence

$$P_{Q_1} = V_{ce_1} \times I_{c_1} = (\text{NV3} - \text{NV2}) \times \text{CA3} = 58.6 \text{ mw}$$

and $\qquad P_{Q_2} = V_{ce_2} \times I_{c_2} = (\text{NV4} - \text{NV2}) \times \text{CA7} = 58.6 \text{ mw}$

The next section of the output consists of partial derivatives and sensitivity coefficients for each element in the circuit. This section is valuable in determining the component tolerances required to maintain a specified DC output-level tolerance. When considering the effect of resistor tolerance on the output level, the sensitivity column is generally the most useful because resistor tolerance and degradation are usually given in per cent of nominal value. From these calculations, the tolerance of h_{ie} (B6) is important in determining the stability of the output operating point. Unfortunately, manufacturing tolerances of the transistor do not permit a tighter tolerance specification of this parameter. However, when two transistors with similar characteristics (matched h_{fe}, etc.) are used in the circuit configuration specified for this problem, the variation due to the h_{ie_2}(B6) tolerance is compensated for by h_{ie_1}(B2). Notice that the sign of the partial derivative with respect to B2 is opposite that with respect to B6 and that the magnitudes are almost equal. Then, since the parameters will track (i.e., drift in the same direction) with temperature, emitter current, etc., the effective sensitivity coefficient will be approximately

$$\text{SE(B2)} + \text{SE(B6)} = (0.1007012 - 0.1004646)10^{-2}$$

$$= (0.236)10^{-5}$$

which is much smaller than either of its components. This, of course, does not include the 10 per cent h_{fe}-match tolerance and its effect on h_{ie} for which the sensitivity coefficient is determined by combining the effects of h_{fe} and the effects of h_{ie}.

The circuit configuration also compensates for the effect of h_{fe} on the DC output level and aids the stability of the circuit. The net effect of h_{fe_1} and h_{fe_2} is approximately

$$\text{SE(T1)} + \text{SE(T2)} = (0.586793 - 0.567605)10^{-2}$$

$$= (0.19188)10^{-3}$$

because the transistor h_{fe} values track with temperature, I_e, etc. Note: When the resistance or the dependent current source values, or both, are changed, the matrix defining the circuit is also changed. For small changes, the partial derivative remains relatively constant, but for a large component value change, such as 50 per cent, the matrix may be modified enough that the partial derivative will change appreciably.

The effects of the base-emitter voltage drops V_{be_1} and V_{be_2} are easier to compute using the partial derivatives since the partial is given in volts per volt.

With a 10 mv V_{be} match ($|V_{be_1} - V_{be_2}| \leq 10$ mv) the operating-point toler-
ance is increased approximately ± 85 mv. The absolute tolerance of V_{be} is
± 0.05 v. However, since V_{be_1} and V_{be_2} track within the 10 mv V_{be} match
specification, the net partial derivative is

$$\text{PD(E2)} + \text{PD(E6)} = (0.1710468 - 0.1706403)10^2$$
$$= 0.04065 \text{ v/v}$$

The DC gain of the amplifier is immediately available since a change in the base-
emitter voltage drop is equivalent to a change in the input signal. The partial
derivative with respect to E2 or E6 yields the gain directly; in the example
DC gain = 17.1.

The remaining two calculations in the nominal solution are the standard
deviation and worst case calculations using the nominal values of h_{ie}, h_{fe}, and
V_{be} with only the matching tolerances considered.

The first modify routine calculates the DC output level as a function of the
common-emitter resistor R_e in branch 5. The nominal DC output level is obviously not a linear function of R_e, as illustrated by the change in magnitude of the
partial derivative over the range of emitter resistance values. The output signal
level is again calculated from Eq. (6.2) and is plotted with the DC output level.
For very small values of the emitter resistor, the amplifier output level is very
sensitive to changes in R_e and the amplifier may saturate. For large values of
R_e, the output level as a function of R_e becomes more stable but the signal level
decreases so that the DC output level tolerance becomes critical.

The first modify routine also calculates the power dissipated in R_e.
Assuming that 0.5 w is the maximum allowable power dissipation in the emitter
resistor because of the heat generated in the circuit, the minimum value of R_e
would have to be approximately 1.2 kΩ.

The second modify routine calculates the DC output level as a function of
the load resistor R_L. The output signal level determined from Eq. (6.2) is
superimposed to illustrate the maximum value of R_L that can be used without
saturating the amplifier.

Up to this point, only nominal characteristics have been used in computing
acceptable values for the amplifier components. The next step in the analysis
determines the values acceptable under any worst case conditions present in
the circuit. Modify routines 3 through 6 calculate the information necessary to
perform a worst case analysis of the circuit. In considering the circuit for use
as an oscillator, the DC output level is important to guarantee not only that the
operating point is within allowable limits for biasing the following stage but
also that the combined effects of the component tolerances cannot cause one
of the transistors to be biased in a cutoff condition and thus prevent sustained
oscillation.

MODIFY 3 and MODIFY 4 set h_{fe} to its minimum and maximum values, respectively, to determine which value has the greatest effect on the worst case
emitter current and the DC operating point. In the worst case emitter-current

calculations, the difference is negligible, but in operating-point calculations, the maximum h_{fe} produces the greatest change. Also, in calculations for the worst case solution for node voltage 5 under minimum h_{fe} conditions (MODIFY 3), the partial derivatives with respect to R_1, R_2, R_6, E_2, and E_6 change signs. However, since the values of these partial derivatives in the nominal solution are very small, their sign changes are of little significance (see sec. 2.6).*

MODIFY 5 performs a standard deviation and worst case analysis on the amplifier under the conditions of minimum V_{be} and maximum h_{fe}. MODIFY 6 repeats the analysis under maximum V_{be} and minimum h_{fe} conditions. Results of those two routines are illustrated in Fig. 6.39. The worst case common-emitter current is shown in Fig. 6.39a. The product of this current and the load resistor R_L defines the maximum and minimum output signals shown with the

*If the value of the partial derivative is very small, the sign change could be due to inherent machine-accuracy limitations. In any case, the change can generally be neglected.

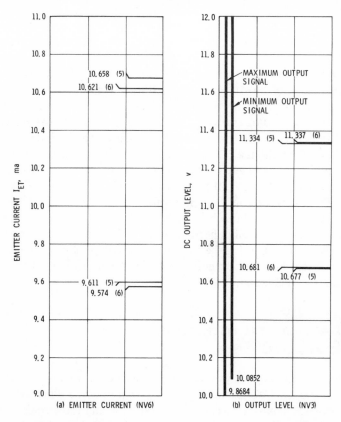

Fig. 6.39. Worst case emitter current and output levels (10 MHz oscillator DC analysis).

DC output level in Fig. 6.39b. (The output signal varies with time between $+12.0$ v and $+9.8684$ v or $+10.0852$ v under worst case conditions.) This figure illustrates the limits of the DC operating point when it is being used as an amplifier; it also demonstrates that sustained oscillation will exist with the addition of the crystal feedback network because the circuit will not be biased with one transistor in a cutoff condition. Increasing the value of the resistor in branch 1 will increase the DC offset of the amplifier and eventually will bias Q_1 in a cutoff state. Assuming that the V_{be} threshold of the transistor is 0.5 and that

$$\frac{\partial(\text{NV3})}{\partial R_1} = -0.5887 \times 10^{-4}$$

is linear, the limiting value of R_1 will be equal to the difference between the lower-limit DC output level and the minimum signal level minus the base-emitter threshold voltage, all divided by the partial derivative with respect to R; that is,

$$\frac{(\text{DC}_{\text{op. pt.}} - E_{\text{sig}}) - V_{be(\text{thresh})}}{\partial(\text{NV3})/\partial R_1} = \frac{(10.677 - 10.085) - 0.5}{0.5887 \times 10^{-4}} \simeq 1.6 \text{ k}\Omega$$

The 150 Ω resistor used in the feedback network should then be acceptable in the oscillator circuit.

The input resistance of the amplifier can be read directly from the nodal impedance matrix of MODIFY 7. The resistance from any node (m) to ground is equal to the resistance on the diagonal of the nodal impedance matrix at (m, m) (see sec. 2.16). For example, the input resistance of the amplifier is equal to the resistance from node 1 to ground. The resistance at the intersection of row 1 and column 1 of the matrix is 744.632 Ω. (Do not load this node with any resistance not to be considered in the input resistance. If the value of R_1 had not been changed from 0.01 Ω to 10 MΩ, the calculated input resistance would have been 0.01 Ω, which, of course, is not valid.)

The current source in branch 1 equal to 58.87 μa is used only to demonstrate a method of keeping the amplifier balanced. It does not affect conductance or impedance matrices. Omitting the current source will produce nonsensical values for node voltages and element currents, but these are ignored in the nodal impedance matrix calculation.

6.6 10 MHz CRYSTAL OSCILLATOR: AC ANALYSIS

In this section, an AC analysis is performed on the two-transistor 10 MHz oscillator in Fig. 6.35. Here, the frequency response of the differential amplifier (crystal feedback removed) used in the oscillator and the frequency response of the complete oscillator are calculated. The gain and phase shift characteristics of the amplifier are studied first to determine the amplifier gain response and the amount of phase shift present in the amplifier which will shift the oscillator frequency from the natural series resonant frequency of the crystal.

The only component requiring a model in the analysis of the circuit, assuming temporarily that the crystal has been removed, is the transistor. The small-signal, hybrid-π transistor equivalent circuit (see Appendix A.1) is appropriate for this application. The value of $r_{b'e}$ calculated in Sec. 6.5 is

$$r_{b'e} = (h_{fe} + 1)r_e = 343 \ \Omega$$

for an emitter current equal to 5.0 ma. The value of f_T from the 2N2369A transistor data sheet (see Fig. 6.24) is typically 675 MHz.

The emitter-junction capacitance $C_{b'e}$, calculated using the transistor data and Eq. (A.3), is

$$C_{b'e} = \frac{1}{2\pi f_T r_e} = \frac{1}{2\pi (675 \times 10^6)5.2} = 45.5 \ \text{pf}$$

The resistance $r_{b'c}$ can be neglected at the frequencies of interest in this analysis. The assumed common-emitter output resistance r_{ce} is

$$r_{ce} = \frac{1}{h_{oe}} = 10 \ \text{k}\Omega$$

as in the DC analysis in Sec. 6.5. The forward current gain h_{fe} and the output capacitance $C_{b'c}$ are assumed equal to the typical values from the data sheet. Then

$$C_{b'c} = C_{ob} = 2.0 \ \text{pf}$$

and

$$h_{fe} = 65$$

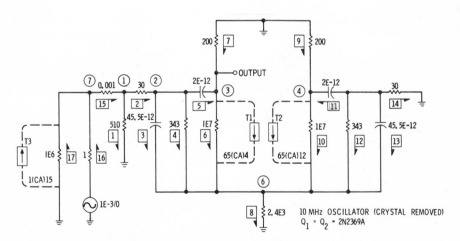

Fig. 6.40. ECAP equivalent circuit (AC analysis) 10 MHz oscillator (crystal removed) $Q_1 = Q_2 = $ 2N2369A.

The AC equivalent circuit of the oscillator (amplifier) without the crystal is in Fig. 6.40.

The network consisting of branches 15, 16, and 17 and the dependent current source T3 at the circuit input are added to demonstrate a means of applying a constant-voltage deriving function to the input terminal of the amplifier without adding any loading effects. Briefly, by injecting a current equal to CA15 into node 7 via T3, the net current flowing in branch 16 equals zero. Hence, the voltage at node 7 is equal to the constant voltage driving function in branch 16. Alternatively, in this example the resistance in series with the signal source could have been decreased to approximately 10^{-3} Ω and connected directly to node 1. However, nodal current unbalance problems frequently arise in circuits with reactive input elements that require a constant input voltage independent of loading and frequency.

Input data for the AC analysis of the amplifier are shown in Fig. 6.41. The first modify routine is used to obtain the amplitude and phase response of the amplifier over a wide frequency range from 1 MHz to 50 MHz. This analysis yields the wide-band amplifier response plotted in Fig. 6.42. The second modify routine obtains the frequency response on a linear frequency scale over

```
            AC ANALYSIS OF 10MHZ OSCILLATOR
     C      CRYSTAL REMOVED
     B1  N(1,0),R=510
     B2  N(1,2),R=30
     B3  N(2,6),C=45.5E-12
     B4  N(2,6),R=343
     B5  N(2,3),C=2E-12
     B6  N(3,6),R=1E7
     B7  N(0,3),R=200
     B8  N(6,0),R=2.4E3
     B9  N(0,4),R=200
    B10  N(4,6),R=1E7
    B11  N(5,4),C=2E-12
    B12  N(5,6),R=343
    B13  N(5,6),C=45.5E-12
    B14  N(5,0),R=30
    B15  N(7,1),R=0.001
    B16  N(0,7),R=1,E=1E-3/0
    B17  N(0,7),R=1E6
     T1  B(4,6),BETA=65
     T2  B(12,10),BETA=65
     T3  B(15,17),BETA=1
         FREQ=10E6
         PRINT,NV,CA
         EX
         MO
         FREQ=1E6(1.04)50E6
         EX
         MO
         FREQ=9.940E6(+100)10.060E6
         EX
```

Fig. 6.41. AC analysis of 10 MHz crystal oscillator input data (crystal removed).

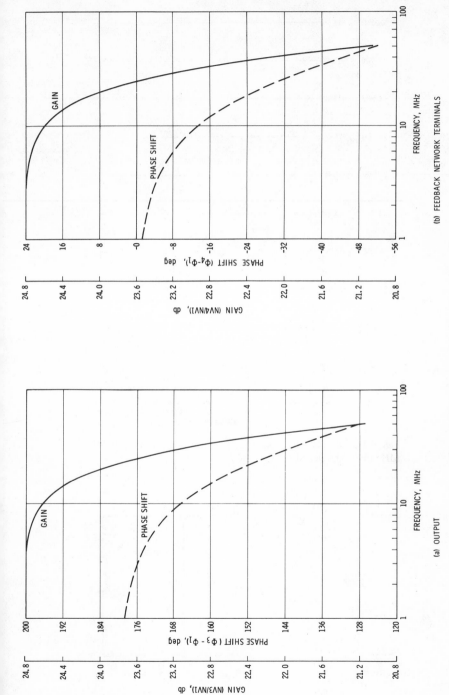

Fig. 6.42. Amplifier frequency- and phase-response.

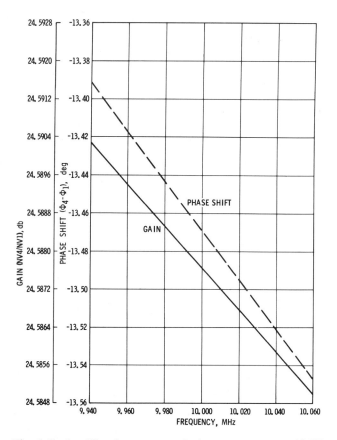

Fig. 6.43. Amplifier frequency- and phase-response near 10.000 MHz (feedback network terminals).

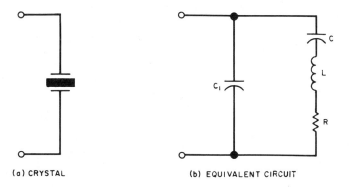

(a) CRYSTAL   (b) EQUIVALENT CIRCUIT

Fig. 6.44. Crystal equivalent circuit.

the narrow frequency band from 9.940 MHz to 10.060 MHz (plotted in Fig. 6.43). This set of calculations magnifies the frequency scale near 10.000 MHz. In the next step of the AC analysis, in which the crystal is included in the oscillator, the response in this frequency band is very important.

Before proceeding with the AC analysis of the 10 MHz oscillator including the crystal feedback network, an equivalent circuit for the crystal used in the circuit must be derived. We select the commonly used equivalent circuit shown in Fig. 6.44 for a vibrating quartz crystal. In this circuit, the resonant frequency is defined by

$$f_r = \frac{1}{2\pi\sqrt{LC}} \tag{6.3}$$

and the network Q is

$$Q = \frac{2\pi f_r L}{R} \tag{6.4}$$

The holder, or shunt, capacitance C_1 is a function of the crystal and the mounting technique.

For the crystal used in this oscillator, the following specifications are given:

$$\text{Series } f_r = 10.000 \text{ MHz}$$

$$\text{Series } R \doteq 25 \, \Omega \text{ max}$$

$$C_1 = 5 \text{ pf}$$

$$Q = 250{,}000$$

From these values,

$$L = \frac{QR}{2\pi f_r} = \frac{(2.5 \times 10^5)25}{2\pi(10^7)} = 100 \text{ mh}$$

$$C = \frac{1}{(2\pi f_r)^2 L} = \frac{1}{4\pi^2(10^{14})0.1} = 2.53299 \times 10^{-15} \text{ f}$$

The equivalent circuit for the 10.000 MHz quartz crystal used in the oscillator analysis is shown in Fig. 6.45a.

The input data used to compute the crystal response are contained in Fig. 6.45b. The first modify routine in the AC analysis computes the amplitude- and phase-response of the crystal over a 120 kHz frequency range centered at 10.000 MHz. The second modify routine narrows the range to 30 kHz with the same number of calculations. The results of these calculations are plotted in Figs. 6.46 and 6.47. The results of the frequency-response calculations for the crystal equivalent circuit are a good approximation of actual crystal characteristics.

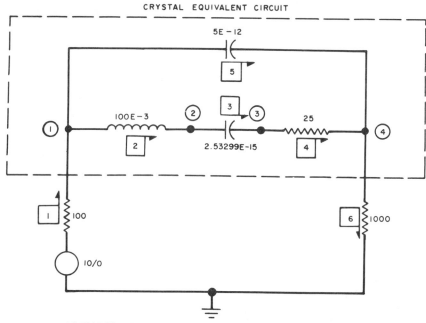

(a) ECAP EQUIVALENT CIRCUIT FOR CRYSTAL RESPONSE CALCULATIONS

```
       AC ANALYSIS OF 10.0 MHZ CRYSTAL RESPONSE
B1  N(0,1),R=100,E=10/0
B2  N(1,2),L=100.000E-3
B3  N(2,3),C=2.53299E-15
B4  N(3,4),R=25
B5  N(1,4),C=5E-12
B6  N(4,0),R=1000
    FREQUENCY=10E6
    PR,NV
    EX
    MO
    FREQ=9.940E6(+100)10.060E6
    EX
    MO
    FREQ=9.990E6(+100)10.020E6
    EX
```

(b) INPUT DATA FOR CRYSTAL RESPONSE CALCULATIONS

Fig. 6.45. Equivalent circuit and input data for 10 MHz crystal frequency response calculations.

The frequency- and phase-response differences between the crystal and its equivalent circuit are minor and will be neglected.

AC equivalent circuits are now available for all components used in the oscillator. The ECAP equivalent circuit for the complete 10 MHz oscillator is shown in Fig. 6.48. This circuit is the same as the circuit in Fig. 6.40 with the

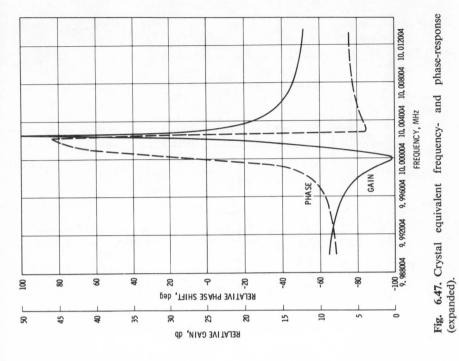

Fig. 6.47. Crystal equivalent frequency- and phase-response (expanded).

Fig. 6.46. Crystal equivalent circuit frequency- and phase-response.

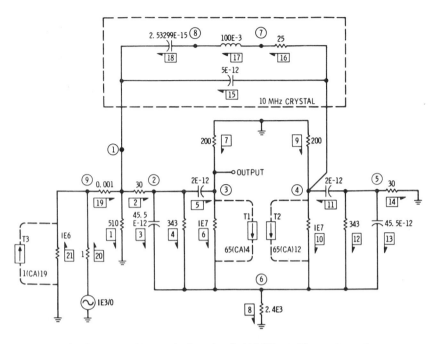

Fig. 6.48. AC ECAP equivalent circuit 10 MHz oscillator. $Q_1 = Q_2$ = 2N2369A.

addition of the crystal feedback element. Input data for the AC response calculations of the complete oscillator circuit are given in Fig. 6.49. The AC analysis of this circuit is conducted in the same manner as the analysis with the crystal removed, except that an additional analysis over a 40 kHz frequency band around the 10.000 MHz resonant frequency of the crystal is perfomed to further expand the view of the response in this area.

The best way to illustrate the reasons for calculating the frequency response over the expanded regions is to study the results of the first modify routine, which are plotted in Fig. 6.50. The frequency-response calculations obtained are identical to the frequency-response calculations when the crystal was removed from the circuit. (Authors' note: We added the thin lines appearing in the figure from additional data.) There are no differences because there are no calculations performed within the frequency spectrum in which the high-Q crystal has any effect, even though the increments between the frequency steps used in the solution are very small. Portions of the calculations for the two frequency steps nearest 10.000 mHz are shown in Fig. 6.51 to illustrate the frequency-step problem which must be considered in the use of high-Q circuits. An insufficient number of points in the frequency range of interest would have been calculated even if a finer frequency step had been used. Figure 6.52 shows the results of the AC analysis over a 120 kHz frequency range centered at 10.000 MHz. This, in effect, expands the region of interest near the resonant frequency of the oscillator and provides a more accurate picture of response characteristics in the region. It also clarifies the reason for calculating the response of the differential amplifier over this narrow frequency band. The response plotted in Fig. 6.43 guarantees that there are no undesirable resonances or other frequency- or phase-distortions in the region near 10.000 MHz in the amplifier. The effect of the crystal resonance still occurs over a relatively small portion of the AC response calculations. Since there are only 100 frequency steps used in this series of calculations, there is still a reasonable probability of inaccuracy in the graphs of the oscillator response.

The frequency spectrum, over which the calculations are performed, are further limited to effectively expand the frequency scale; in this way more accurate representation of the frequency- and phase-responses of the oscillator are obtained. Figure 6.53 (p. 212) shows the results of the AC analysis of MODIFY 3

```
        AC ANALYSIS OF 10MHZ CRYSTAL OSCILLATOR
  B1  N(1,0),R=510
  B2  N(1,2),R=30
  B3  N(2,6),C=45.5E-12
  B4  N(2,6),R=343
  B5  N(2,3),C=2E-12
  B6  N(3,6),R=1E7
  B7  N(0,3),R=200
  B8  N(6,0),R=2.4E3
  B9  N(0,4),R=200
  B10 N(4,6),R=1E7
  B11 N(5,4),C=2E-12
  B12 N(5,6),R=343
  B13 N(5,6),C=45.5E-12
  B14 N(5,0),R=30
  B15 N(4,1),C=5E-12
  B16 N(4,7),R=25
  B17 N(7,8),L=100E-3
  B18 N(8,1),C=2.53299E-15
  B19 N(9,1),R=0.001
  B20 N(0,9),R=1,E=1E-3/0
  B21 N(0,9),R=1E6
   T1  B(4,6),BETA=65
   T2  B(12,10),BETA=65
   T3  B(19,21),BETA=1
        FREQ=10E6
        PRINT,NV,CA
        EX
        MO
        FREQ=1E6(1.04)50E6
        EX
        MO
        FREQ=9.940E6(+100)10.060E6
        EX
        MO
        FREQ=9.980E6(+100)10.02E6
        EX
```

Fig. 6.49. 10 MHz crystal oscillator—AC analysis input data.

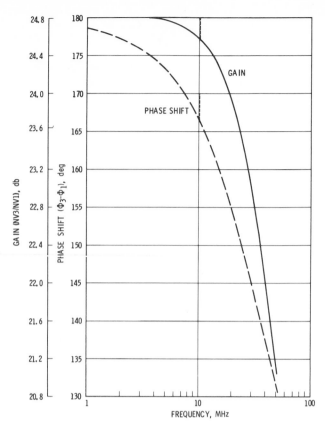

Fig. 6.50. 10 MHz crystal oscillator frequency- and phase-response output.

```
FREQ =   0.97258480E 07

        NODES            NODE VOLTAGES

MAG     1-  4   0.99999621E-03   0.94423164E-03   0.16968898E-01   0.16853970E-01
PHA             0.90965812E-04  -0.56364288E 01   0.16689987E 03  -0.16361908E 02


MAG     5-  8   0.10826177E-03   0.50105737E-03   0.16853951E-01   0.29483157E 00
PHA             0.54074356E 02  -0.28870928E 00  -0.16365784E 02  -0.17323334E 02


MAG     9-  9   0.99999877E-03
PHA             0.0


FREQ =   0.10114881E 08

        NODES            NODE VOLTAGES

MAG     1-  4   0.99999621E-03   0.94302208E-03   0.16933683E-01   0.16816273E-01
PHA             0.85238644E-04  -0.58473959E 01   0.16639554E 03  -0.16889053E 02


MAG     5-  8   0.11175114E-03   0.50112419E-03   0.16816322E-01   0.68595159E 00
PHA             0.54304184E 02  -0.30280638E 00  -0.16879639E 02   0.16204575E 03


MAG     9-  9   0.99999877E-03
PHA             0.0
```

Fig. 6.51. Frequency-response calculations nearest 10.000 MHz from AC response routine (MODIFY1) using logarithmic frequency steps.

over the narrow frequency range of 40 kHz centered at the natural resonant frequency of the crystal. The results differ from those in Fig. 6.52 because of the increased accuracy obtained from a finer frequency step. Also, the resonance characteristic is expanded enough to present some of the important details concerning the resonant frequency of the oscillator. For example, the circuit will oscillate at a frequency at which the closed-loop gain of the circuit is equal to unity and at which the phase shift between output (node 4) and input (node 1) is equal to zero. In this circuit, resonant frequency is between 10.00000 MHz and 10.00040 MHz. The phase shift at 10.00000 MHz is approximately $-55.8°$ and at 10.00040 MHz is approximately $+8.31°$. Frequency increments used in the calculations were not sufficiently fine to determine frequency where $\Phi_4 - \Phi_1 = 0$ and NV4/NV1 $\simeq 1$. Calculation of the exact frequency of oscillation is left as an exercise.

Using an analysis of this type, the effect of each element in the circuit on the frequency of the oscillator can be determined; this is difficult, if not impossible,

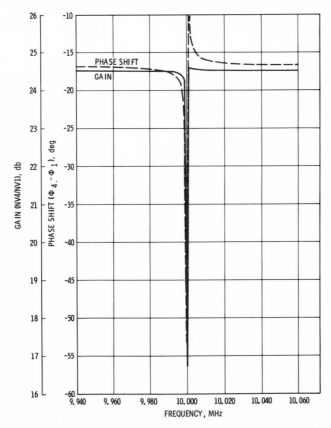

Fig. 6.52. 10 MHz crystal oscillator frequency- and phase-response over a 120 kHz frequency range centered at 10.000 MHz (feedback network terminals).

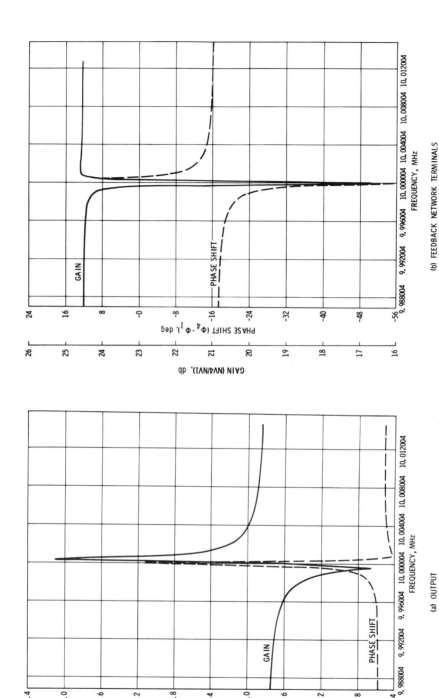

Fig. 6.53. 10 MHz crystal oscillator frequency- and phase-response over a 24 kHz frequency range centered at 10.000 MHz.

in the laboratory. In concluding this section, it should be emphasized that solution control parameters can often significantly affect the results. In this example, appropriate frequency step selection was required to provide the proper frequency response for the high-Q network.

6.7 10 MHz CRYSTAL OSCILLATOR: TRANSIENT ANALYSIS

Time response, or transient behavior, of the 10 MHz crystal oscillator is studied in this section. During the analysis, several problems that may occur in the application of the transient analysis program are discussed. Three of these problems are time step selection, initial condition solutions, and the use of high-Q resonant circuits. Much of the information used in this section has been derived in previous sections of this chapter. In these specific instances, references will be made to the source of information.

The first step in transient analysis is the selection of the proper transistor model to be used. Considering the circuit configuration and the gain of the amplifier at resonance, it is reasonable to assume that circuit operation will be nonlinear. Using this assumption as a guide, we select the charge-control transistor model developed in Appendix A.2.

The emitter current in the transistors can vary between 0.0 ma and 10.0 ma. From Fig. 6.25, the $I_c = 10$ ma curve should be used to model the base-emitter characteristic. In Sec. 6.4, the equivalent circuit for the 2N2369A transistor used the curve for $I_c = 5$ ma to model the base-emitter diode characteristic. Since the difference between the $I_c = 5$ ma and $I_c = 10$ ma curves is very slight, the same base-emitter diode model is used for this application. Error in the analysis due to this decision should be very small, especially considering the manufacturing spread of the V_{be} parameter. Other required transistor parameters are the same as those used in Sec. 6.4 (C_{ob}, C_{te}, h_{fe}) except for the base-time constant T_B.

In the previous example, the assumed average forward beta during the switching interval was equal to 31.5. In this example

$$\beta_{av} = 9.5$$

then

$$f_\beta = \frac{f_T}{\beta_{av}} = \frac{675\ \text{MHz}}{9.5} = 71.2\ \text{MHz}$$

and

$$T_B \simeq \frac{1.22}{2\pi f_\beta} = 2.71\ \text{nsec}$$

Using the value for C_1 calculated in Sec. 6.4

$$R_B = \frac{T_B}{C_1} = \frac{2.71\ \text{nsec}}{2.36\ \text{pf}} = 1150\ \Omega$$

The equivalent resistance of C_1 is

$$R_{\text{EQ}(C_1)} = \frac{\Delta t}{C_1} = \frac{0.1 \text{ nsec}}{2.36 \text{ pf}} = 43.3 \ \Omega$$

which is relatively close to R_B and satisfies the nodal impedance ratio requirements discussed in Chapter 4.

The results obtained when investigating the digital logic-inverter indicated it was possible to neglect the storage-time constant T_S in that particular application. The transistor is operated as an emitter-coupled switch in this analysis, which should further reduce storage effects. Therefore, T_S is assumed equal to zero.

The only remaining element in the transistor model which has not been considered is the base-collector diode. Since neither transistor in the oscillator circuit ever achieves a saturated or partially saturated condition, it can be assumed that the base-collector diode remains reverse-biased and can be neglected.

The crystal equivalent circuit used in the AC analysis of the oscillator is invalid in this analysis. The numerical-integration approximation techniques used in the transient analysis program make it difficult to use resonant networks, such as the crystal, with Q values exceeding approximately 250. This is partially a function of the time step selected for the analysis, but practical time-limit and other time-constant considerations usually prohibit use of time steps small enough to use high-Q networks. The Q of the crystal from the specification was $Q = 250,000$. The value of the elements used in the model for the AC analysis were $L = 100$ mh; $C = 0.00253299$ pf; $R = 25 \ \Omega$. To meet the Q requirements and still maintain the series resonant frequency, the crystal equivalent circuit values are scaled so that the new values are $L' = 100 \ \mu\text{h}$; $C' = 2.53299$ pf; $R = 25 \ \Omega$. Lowering the Q of the crystal will create some unavoidable second- or third-order effects in the circuit transient response. The computer analysis results are still in excellent agreement with the experimental results, as will be seen. The crystal holder capacitance ($C_1 = 1.21066$ pf) was decreased to shift the parallel resonant frequency above the region of interest. This result again is due to the low-Q crystal model used in the analysis.

The 10 MHz crystal oscillator equivalent circuit for a transient analysis can now be assembled using the component models that have been constructed. The complete circuit, shown in Fig. 6.54, does not contain any stray capacitances as did the inverter circuit in Sec. 6.4. Effects of the parasitic elements on the response of this circuit are considered negligible because of impedance levels and voltage changes involved in the response.

The equilibrium solution for this circuit is obtained by a completely different method than the one discussed throughout the text, which is one of the primary reasons for selecting this circuit as an example. In an oscillator circuit, a steady-state condition does not exist at any time. Therefore it is not valid to compute an equilibrium solution by substituting a short circuit (0.01 Ω) for the inductances

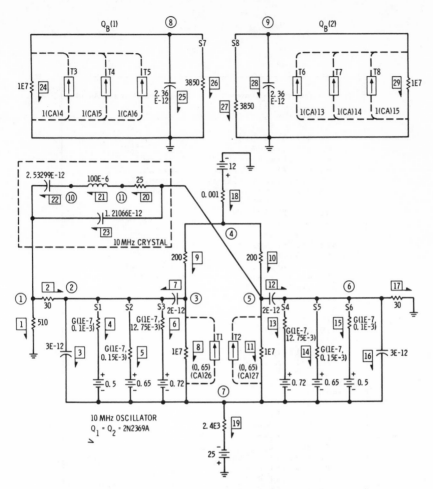

Fig. 6.54. Transient analysis ECAP equivalent circuit 10 MHz oscillator: $Q_1 = Q_2 = $ 2N2369A.

and an open circuit ($10^7\ \Omega$) for all capacitances. The results of this type of solution are obviously useless. Instead the pseudo-steady-state solution, or the initial solution, is obtained by starting the analysis without any specified initial conditions and allowing the analysis to run an arbitrary period of time until the desired oscillatory condition has become established and stable.* The input data used to obtain the pseudo-steady-state solution are contained in Fig. 6.55. Notice that this is a standard transient analysis with the FINAL TIME set arbitrarily at $t = 200$ nsec. The time selected to provide the initial conditions is arbitrary as long as the circuit response is stabilized (i.e., the voltages and currents associ-

---

*Improper time step selection or a value of Q that is too large, or both, will either prevent oscillation or produce a damped oscillation which is meaningless.

```
C       TRANSIENT ANALYSIS OF 10MHZ OSCILLATOR
C       EQUILIBRIUM SOLUTION
C       STEADY STATE DOES NOT EXIST
        TR
  B1    N(1,0),R=510
  B2    N(1,2),R=30
  B3    N(2,7),C=3E-12
  B4    N(2,7),G=(.1E-3,1E-7),E=-.5
  B5    N(2,7),G=(.15E-3,1E-7),E=-.65
  B6    N(2,7),G=(12.75E-3,1E-7),E=-.72
  B7    N(2,3),C=2E-12
  B8    N(3,7),R=1E7
  B9    N(4,3),R=200
  B10   N(4,5),R=200
  B11   N(5,7),R=1E7
  B12   N(5,6),C=2E-12
  B13   N(6,7),G=(1E-7,12.75E-3),E=-.72
  B14   N(6,7),G=(1E-7,.15E-3),E=-.65
  B15   N(6,7),G=(1E-7,.1E-3),E=-.5
  B16   N(6,7),C=3E-12
  B17   N(6,0),R=30
  B18   N(0,4),R=.001,E=12
  B19   N(7,0),R=2.4E3,E=25
  B20   N(5,11),R=25
  B21   N(11,10),L=100E-6
  B22   N(10,1), C=2.53299E-12
  B23   N(5,1),C=1.21066E-12
  B24   N(8,0),R=1E7
  B25   N(8,0),C=2.36E-12
  B26   N(8,0),R=1150
  B27   N(9,0),R=1150
  B28   N(9,0),C=2.36E-12
  B29   N(9,0),R=1E7
  T1    B(26,8),BETA=(20,0)
  T2    B(27,11),BETA=(0,20)
  T3    B(4,24),BETA=-1
  T4    B(5,24),BETA=-1
  T5    B(6,24),BETA=-1
  T6    B(13,29),BETA=-1
  T7    B(14,29),BETA=-1
  T8    B(15,29),BETA=-1
  S1    B=4,(4),ON
  S2    B=5,(5),ON
  S3    B=6,(6),ON
  S4    B=13,(13),OFF
  S5    B=14,(14),OFF
  S6    B=15,(15),OFF
  S7    B=26,(26),ON
  S8    B=27,(27),OFF
        TI=0.1E-9
        OU=20
        FINAL TIME=200E-9
        PRINT,NV,CA
        EXECUTE
```

Fig. 6.55. 10 MHz crystal oscillator equilibrium solution. Transient analysis input data.

```
 T =    0.1999970E-07


NODES                  NODE VOLTAGES

   1-  4      0.57509464E 00      0.55851907E 00      0.10033703E 02      0.11999988E 02
   5-  8      0.11660891E 02      0.19420420E-02     -0.19710290E 00      0.56852466E 00
   9- 11      0.14244097E-02      0.79189854E 01      0.11618888E 02

BRANCHES               ELEMENT CURRENTS

   1-  4      0.11276365D-02      0.55251865D-03     -0.51003615D-07      0.25562209D-04
   5-  8      0.15843311D-04      0.45418158D-03      0.56982556D-04      0.98884078D-02
   9- 12      0.98314231D-02      0.16954845D-02      0.25958139D-04     -0.10628821D-04
  13- 16     -0.52095479D-07     -0.45095482D-07     -0.30095483D-07     -0.75236266D-04
  17- 20      0.64734731D-04      0.115269080D-01     0.103345360D-01      0.16801441D-02
  21- 24      0.16801441D-02      0.16801441D-02      0.11085795D-07     -0.49553024D-03
  25- 28      0.11581176D-05      0.49436929D-03      0.123861710D-05     -0.13660461D-05
  29- 29      0.12742888D-06
```

Fig. 6.56. Initial conditions for 10 MHz crystal oscillator.

```
 C       TRANSIENT ANALYSIS OF 10MHZ OSCILLATOR
         TR
    B1  N(1,0),R=510
    B2  N(1,2),R=30
    B3  N(2,7),C=3E-12,EO=-0.7556220
    B4  N(2,7),G=(.1E-3,1E-7),E=-.5
    B5  N(2,7),G=(.15E-3,1E-7),E=-.65
    B6  N(2,7),G=(12.75E-3,1E-7),E=-.72
    B7  N(2,3),C=2E-12,EO=9.475184
    B8  N(3,7),R=1E7
    B9  N(4,3),R=200
   B10  N(4,5),R=200
   B11  N(5,7),R=1E7
   B12  N(5,6),C=2E-12,EO=-11.65895
   B13  N(6,7),G=(1E-7,12.75E-3),E=-.72
   B14  N(6,7),G=(1E-7,.15E-3),E=-.65
   B15  N(6,7),G=(1E-7,.1E-3),E=-.5
   B16  N(6,7),C=3E-12,EO=-0.1990450
   B17  N(6,0),R=30
   B18  N(0,4),R=0.001,E=12
   B19  N(7,0),R=2.4E3,E=25
   B20  N(5,11),R=25
   B21  N(11,10),L=100E-6,IO=0.1680144E-2
   B22  N(10,1),C=2.53299E-12,EO=-7.343891
   B23  N(5,1),C=5E-12,EO=-11.085796
   B24  N(8,0),R=1E7
   B25  N(8,0),C=2.36E-12,EO=-0.5685246
   B26  N(8,0),R=1150
   B27  N(9,0),R=1150
   B28  N(9,0),C=2.36E-12,EO=-0.1424410E-2
   B29  N(9,0),R=1E7
    T1  B(26,8),BETA=(20,0)
    T2  B(27,11),BETA=(0,20)
    T3  B(4,24),BETA=-1
    T4  B(5,24),BETA=-1
    T5  B(6,24),BETA=-1
    T6  B(13,29),BETA=-1
    T7  B(14,29),BETA=-1
    T8  B(15,29),BETA=-1
    S1  B=4,(4),ON
    S2  B=5,(5),ON
    S3  B=6,(6),ON
    S4  B=13,(13),OFF
    S5  B=14,(14),OFF
    S6  B=15,(15),OFF
    S7  B=26,(26),ON
    S8  B=27,(27),OFF
         TI=0.1E-9
         OU=20
         FI=200E-9
         PRINT,NV,CA
         EXECUTE
```

Fig. 6.57. 10 MHz crystal oscillator transient analysis input data.

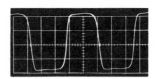

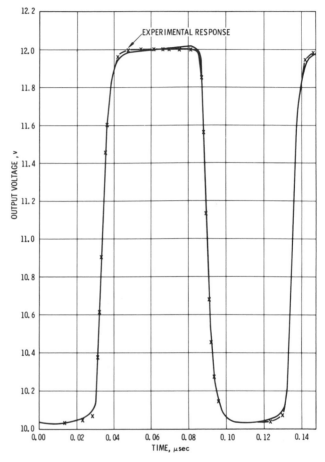

Fig. 6.58. 10 MHz crystal oscillator output voltage waveform.

ated with the reactive elements in the circuit which were not specified as initial conditions have become established and the circuit is functioning correctly). Since the circuit response is stable at $t = 20$ nsec and the oscillations in the circuit are not damped in the 200 nsec interval, indicating valid Q and TI selections, the initial condition information for the transient analysis solution can be obtained at $t = 20$ nsec. The output data for this solution are shown in Fig. 6.56.

The data from the $t = 20$ nsec solution (Fig. 6.56) are now incorporated into the input data in Fig. 6.55. The switch states from the pseudo-equilibrium

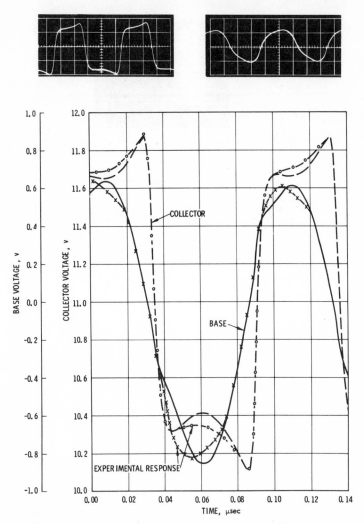

Fig. 6.59. 10 MHz crystal oscillator voltage waveforms at crystal terminals.

solution just described must agree with the initial switch states specified in the input data for the transient analysis solution. If the switch states in the initial condition solution ($t = 20$ nsec) do not agree with the initial switch states specified in the transient analysis, the initial conditions can be selected at a later time when the switch states do agree, or the initial switch states in the input data can be changed to satisfy the initial condition solution.

The input data for the transient analysis solution containing the initial conditions are illustrated in Fig. 6.57. The output voltage, crystal terminal voltage, and common-emitter voltage waveforms are shown in Figs. 6.58, 6.59, and 6.60.

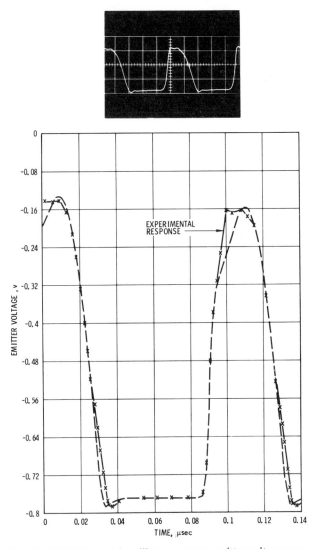

Fig. 6.60. 10 MHz crystal oscillator common emitter voltage wave-form.

In Fig. 6.58 the calculated oscillator output voltage waveform is superimposed on the experimentally measured output waveform to demonstrate the accuracy of the solution. Similarly, the voltage waveforms measured at the terminals of the crystal are superimposed on the calculated response in Fig. 6.59, and the calculated and measured common-emitter voltage waveforms are compared in Fig. 6.60. In each case, the correlation achieved was excellent. Photographs of the measured responses are included to further emphasize the accuracy of the calculations.

6.8 SUMMARY

In this chapter, a linear composite-filter network was designed in the first example using textbook equations and ideal components; then it was modified to include the loss elements in the components. An AC and a transient analysis were performed on the ideal and nonideal networks to determine the errors resulting from using lossy components and to illustrate the use of ECAP in this type of application. The results of each analysis were compared with laboratory data to verify their accuracy.

Second, a digital logic-inverter was analyzed to demonstrate the conversion of a schematic and a manufacturer's data sheet to an ECAP equivalent circuit. The use of the charge-control transistor model in nonlinear, or switching, applications was also explained in detail. The accuracy of the transistor model was verified by experimental measurements superimposed on the calculated response.

As a final example, a two-transistor crystal oscillator was analyzed in some detail. The DC analysis treated the circuit as a differential amplifier and determined the required power ratings of the elements and the effects of the tolerance of each component on the DC output level; it also performed a worst case analysis of the DC output level. An AC analysis was performed on the oscillator to determine the frequency-response characteristics and the frequency of oscillation of the circuit. In this analysis, the equivalent circuit of a quartz crystal was introduced.

In the transient analysis of the oscillator, the time response of the circuit was calculated and compared with experimental data measured to demonstrate the validity of the analysis.

PROBLEMS

6.1. Construct an equivalent circuit for the operational amplifier in Fig. p6.1. Neglect the frequency-response limitations of the amplifier.

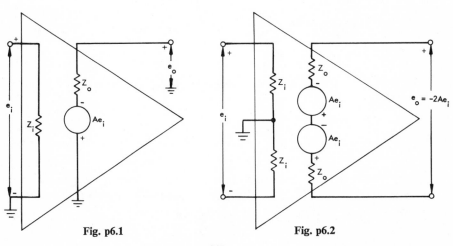

Fig. p6.1   Fig. p6.2

6.2. Construct an equivalent circuit for the operational amplifier in Fig. p6.2. Neglect the frequency-response limitations of the amplifier in the network.

6.3. Determine the transient response of the circuit in Fig. p6.3a to the input voltages specified. Assume the equivalent circuit of the amplifier is as shown in Fig. p6.3b.

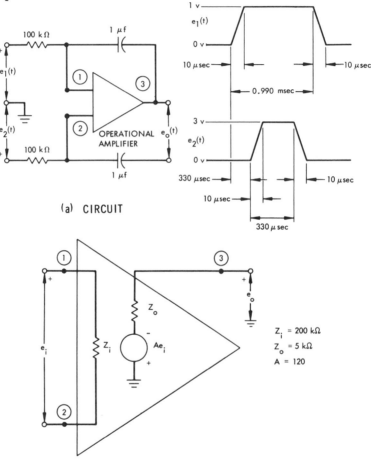

(a) CIRCUIT

(b) OPERATIONAL AMPLIFIER EQUIVALENT CIRCUIT

Fig. p6.3

6.4. Determine the frequency-response of the bandpass filter in Fig. p6.4 when the filter is terminated in $R_o = 600\ \Omega$.

6.5. Determine the frequency-response of the filter in Fig. p6.4 if the network is terminated in $R_o = 1200\ \Omega$. Compare the computed response with that obtained in prob. 6.4 and explain the differences.

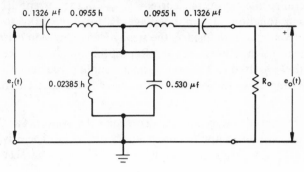

Fig. p6.4

6.6. Determine the response of the filter in Fig. p6.4 to the trapezoidal input voltage waveform in Fig. p6.5. Let the period of the input waveform $T = 1.35$ msec.

6.7. Repeat prob. 6.6 with $T = 2$ msec. Compare the computed transient response with the response obtained in prob. 6.6 and explain the differences observed. (*Note*: The input voltage waveform cannot be generated directly with a time-dependent source.)

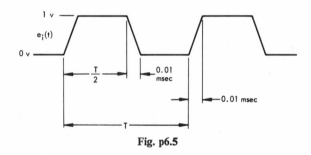

Fig. p6.5

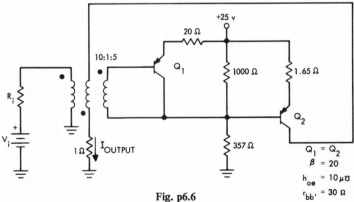

Fig. p6.6

6.8. Determine the DC output current I_{OUTPUT} of the current driver circuit in Fig. p6.6 as a function of input voltage V_i for $V_i = 5.0$ v to 10.0 v with $R_i = 300\ \Omega$. (*Note:* DC is defined as the steady-state condition existing immediately after the transient effects subside.) Repeat for input resistance $R_i = 100\ \Omega$ to $400\ \Omega$ with $V_i = 10.0$ v. The transformer coupling coefficient k_{ij} is greater than 0.999995 for all windings, and the primary inductance can be considered infinite.

6.9. Determine the voltage and the drive current waveforms at the input and termination of a digital computer core memory drive line. The drive line can be approximated by a transmission line with the following characteristics:

$$Z_o = 120\ \Omega$$

$$L_T = 5.5\ \mu\text{h}$$

$$R_T = 20\ \Omega$$

$$R_{\text{termination}} = 0.01\ \Omega$$

The equivalent circuit of the simplified memory current driver and drive line appears in Fig. p6.7. The specification for the diode is shown in Fig. 5.3.

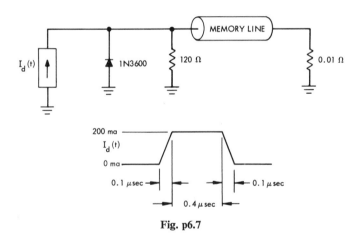

Fig. p6.7

6.10. Determine the range of T_s in the charge storage network in Fig. 5.41 to give maximum storage time of 20 to 40 nsec (reverse base current $I_{br} = 0$) with an overdrive current I_{bc} of sufficient duration to reach the maximum value of Q_s for that overdrive level.

6.11. Determine the base, emitter, and collector waveforms for the emitter-coupled monostable multivibrator shown in Fig. p6.8a when driven by the signal specified. The base-emitter and base-collector diode characteristics and the other necessary transistor parameters are shown in Fig. p6.8b.

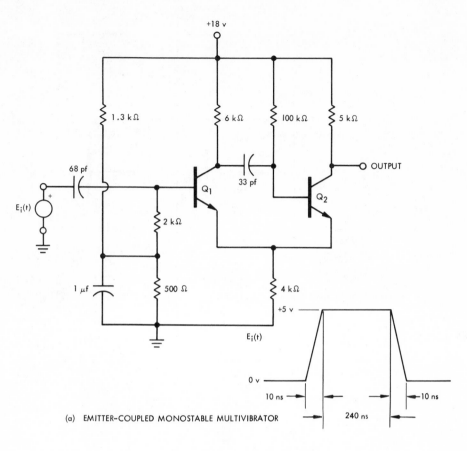

(a) EMITTER-COUPLED MONOSTABLE MULTIVIBRATOR

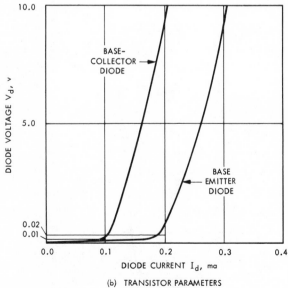

$Q_1 = Q_2$
$h_{fe} = 50$
$r_{bb'} = 200\ \Omega$
$C_{ob} = 2\ pf$
$C_{te} = 3\ pf$
$f_\beta = 19.4\ MHz$
$T_s = 0$

(b) TRANSISTOR PARAMETERS

Fig. p6.8

225

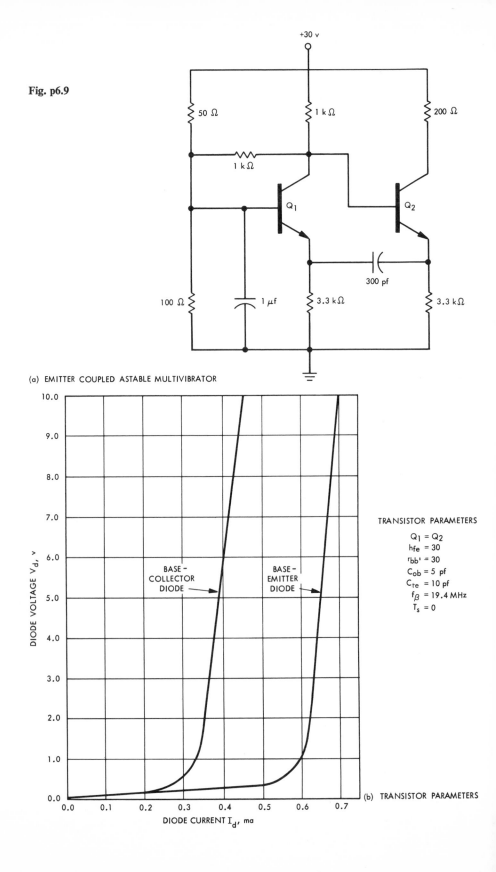

Fig. p6.9

+30 v

50 Ω

1 kΩ

1 kΩ

200 Ω

Q_1

Q_2

300 pf

100 Ω

1 μf

3.3 kΩ

3.3 kΩ

(a) EMITTER COUPLED ASTABLE MULTIVIBRATOR

DIODE VOLTAGE V_d, v

10.0

9.0

8.0

7.0

6.0

5.0

4.0

3.0

2.0

1.0

0.0

BASE−
COLLECTOR
DIODE

BASE−
EMITTER
DIODE

0.0   0.1   0.2   0.3   0.4   0.5   0.6   0.7

DIODE CURRENT I_d, ma

TRANSISTOR PARAMETERS

$Q_1 = Q_2$
$h_{fe} = 30$
$r_{bb'} = 30$
$C_{ob} = 5$ pf
$C_{te} = 10$ pf
$f_\beta = 19.4$ MHz
$T_s = 0$

(b) TRANSISTOR PARAMETERS

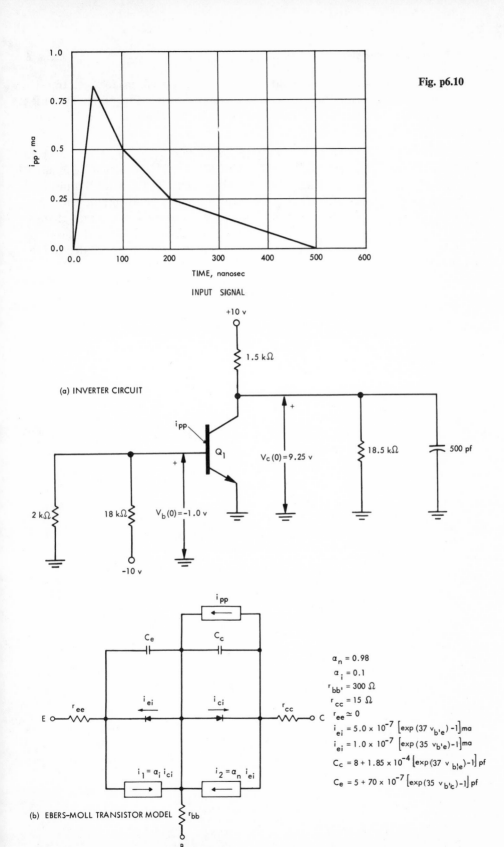

Fig. p6.10

INPUT SIGNAL

(a) INVERTER CIRCUIT

(b) EBERS–MOLL TRANSISTOR MODEL

$\alpha_n = 0.98$

$\alpha_i = 0.1$

$r_{bb'} = 300\ \Omega$

$r_{cc} = 15\ \Omega$

$r_{ee} \simeq 0$

$i_{ei} = 5.0 \times 10^{-7}\ \left[\exp\left(37\ v_{b'e}\right) - 1\right]$ ma

$i_{ei} = 1.0 \times 10^{-7}\ \left[\exp\left(35\ v_{b'e}\right) - 1\right]$ ma

$C_c = 8 + 1.85 \times 10^{-4}\ \left[\exp\left(37\ v_{b'e}\right) - 1\right]$ pf

$C_e = 5 + 70 \times 10^{-7}\ \left[\exp\left(35\ v_{b'c}\right) - 1\right]$ pf

6.12. Determine the base, emitter, and collector waveforms for both transistors of the emitter-coupled astable multivibrator shown in Fig. p6.9a. The base-emitter and base-collector diode characteristics and the other necessary transistor parameters are shown in Fig. p6.9b.

6.13. Determine the base- and collector-voltage response of the inverter circuit shown in Fig. p6.10a to the specified primary photo current i_{pp} induced in the transistor by a transient radiation environment. The Ebers-Moll equivalent circuit for the transistor, including the induced primary photocurrent generator, is shown in Fig. p6.10b. The nonlinear capacitance and conductance characteristics for the transistor junctions are specified mathematically with the other necessary parameters. The initial conditions for the circuit are included in Fig. p6.10a.

7

ANALYSIS
OF
PHYSICAL
SYSTEMS

7.1 INTRODUCTION

Up to this point, only the mastery of ECAP techniques and their utilization in the analysis of electronic circuits have been discussed. Several types of electrical networks were analyzed in the preceding chapter to illustrate the application of ECAP to practical circuit analysis problems and to demonstrate some of the program's capabilities. In this chapter, application of ECAP in the analysis of physical systems will be discussed. Physical systems may or may not have direct electrical analogies, but they all may be defined by transfer functions or by sets of simultaneous integral-differential equations.

The first part of this chapter describes the analysis of feedback control systems in which electrical networks equivalent to the function-block transfer characteristics can be formed. These networks, when interconnected according to the system block diagram, form an electrical network analogous to the control system. The ECAP simulation techniques developed here provide simple solutions to analysis problems in relatively short times.

The application of ECAP to heat transfer problems is the second topic discussed in this chapter. A direct electrical network analog exists for the thermal system, which makes the conversion of thermal quantities and parameters straightforward.

Mechanical and hydraulic systems, discussed in the last section of the chapter, are analyzed using two different techniques. The first utilizes a direct electrical network analog to obtain the solution to the problem. The second approach uses an ECAP simulation of an analog computer to solve the integral-differential equations defining the system.

A list of references is included at the end of the chapter for those interested in obtaining additional information on analog applications.

7.2 CONTROL SYSTEMS ANALYSIS

7.2.1 Introduction

It is often possible to analyze complex feedback control systems by conventional techniques described in texts devoted to the design and analysis of such systems. Sometimes, however, analysis problems become very difficult or even impossible to solve with any degree of accuracy using these techniques. If nonlinearities are present in the system, the problems become orders of magnitude more difficult to solve.

The standard approach to the system analysis problem, if the solution to the problem is difficult or very time-consuming, is to use the analog computer or, more recently, the digital computer. Analog computers in the past have been simpler to program and operate, and have yielded results in a convenient, easily interpretable, graphic form. The digital computer, on the other hand, is more versatile and accurate, but the programming problems involved in encoding the problem limit its effectiveness. Application-oriented digital computer programs[1] have now been written which have essentially eliminated the programming handicap.

Control-system performance can also be studied by analyzing an electrical network analogous to the system being investigated. That is, an electrical network defined by the same equations as the original system can be analyzed to obtain the system response.

Some functions used in systems analysis, $y(x) = \cos x$ and $z(x, y) = xy$ for example, require the use of nonlinear elements for their generation. These functions are, of course, only as accurate as the piecewise linear approximations to their nonlinear characteristics. Care should be taken in determining the applicability of ECAP to control-system-analysis problems to assure a realistic solution.

7.2.2   Simulation of Function Blocks

The first step in the analysis of a feedback control system, or any other system in which a function block is to be replaced by an electrical network simulating its transfer function, is the definition of the function block and its transfer characteristic.

Once a transfer function has been specified, it is generally possible to model the function with a combination of dependent current sources and simple RC networks. The most commonly used RC networks and their characteristics are listed in Table 7.1. The following procedure might be used in the synthesis of a transfer-function model:

1. Define the transfer characteristic of the function block. For example, consider

$$\frac{E_o(s)}{E_i(s)} = G(s) = \frac{K_1 s^2(\tau_1 s + 1)}{(\tau_2 s + 1)(\tau_3 s + 1)(\tau_4 s + 1)}$$

   to be an arbitrary transfer function which will be simulated by an electrical network.

2. Factor the transfer function into convenient terms that can be individually represented by elementary RC networks. Thus we have

$$\frac{E_o(s)}{E_i(s)} = G(s) = \frac{K_1}{\tau_2 \tau_3} \cdot \left(\frac{\tau_2 s}{\tau_2 s + 1}\right) \cdot \left(\frac{\tau_3 s}{\tau_3 s + 1}\right) \cdot \left(\frac{\tau_1 s + 1}{\tau_4 s + 1}\right)$$

[1] *System/360 Continuous System Modeling Program Users Manual* (360A-CX-16X), (White Plains, N.Y.: International Business Machines Corp., 1967). Also, H. E. Peterson and F. J. Sansom, "Mimic—A Digital Simulator Program," *SESCA Internal Memo 65–112* (Wright-Patterson Air Force Base, 1965). And, W. M. Syn and D. G. Wyman, "DSL-90 Digital Simulation Language—User's Guide," *SHARE*, Distribution #3358 (San Jose, Calif.: International Business Machines Corp., 1965).

Table 7.1. Commonly used RC networks and their characteristics.

NETWORK	TRANSFER FUNCTION	GAIN CHARACTERISTIC	POLE-ZERO CONFIGURATION
$\tau = RC$	$\dfrac{1}{\tau s + 1}$	$\dfrac{1}{\tau}$, -20 db/DEC	pole at $\dfrac{1}{\tau}$
$\tau = RC$	$\dfrac{\tau s}{\tau s + 1}$	$\dfrac{1}{\tau}$, $+20$ db/DEC	pole $-\dfrac{1}{\tau}$
$\tau = R_2C$, $k = R_2/(R_1 + R_2)$	$\dfrac{\tau s + 1}{\frac{\tau}{k}s + 1}$, $0<k<1$	$\dfrac{k}{\tau}$ $\dfrac{1}{\tau}$, -20 db/DEC	$-\dfrac{1}{\tau}$ $-\dfrac{k}{\tau}$
$\tau = R_1C$, $k = R_2/(R_1 + R_2)$	$\dfrac{k(\tau s + 1)}{k\tau s + 1}$, $0<k<1$	$\dfrac{1}{\tau}$ $\dfrac{1}{k\tau}$, $+20$ db/DEC	$-\dfrac{1}{k\tau}$ $-\dfrac{1}{\tau}$

3. Replace each of the factors with an RC network having the same transfer characteristic, using care to model the factors exactly. The factor

$$\frac{\tau_1 s + 1}{\tau_4 s + 1} = \frac{k(\tau_1 s + 1)}{k\tau_1 s + 1} \cdot \frac{1}{k}$$

assuming $\tau_4 < \tau_1$. If $\tau_4 > \tau_1$, a different RC network will be required. The multiplying factor k will affect the value of K_1, which is determined in the last step, as are τ_2 and τ_3.

4. The technique being used to construct the simulated transfer function is only valid if individual network functions obtained

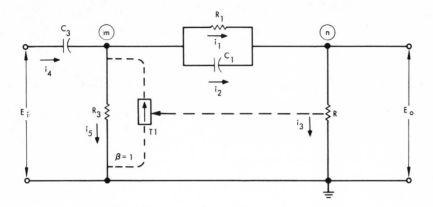

Fig. 7.1. Transfer function simulation network for

$$G_1(s) = \left(\frac{\tau_3 s}{\tau_3 s + 1}\right)\left(\frac{k(\tau_1 s + 1)}{k\tau_1 s + 1}\right)$$

by factoring do not load the previous network when the composite network is assembled. Consider the interconnection of the terms

$$G_1(s) = \frac{\tau_3 s}{\tau_3 s + 1}\left(\frac{k(\tau_1 s + 1)}{k\tau_1 s + 1}\right)$$

which form the network shown in Fig. 7.1. If the networks are directly connected, the first network $\tau_3 s/(\tau_3 s + 1)$ will be loaded by the second network, which will change the transfer function of the combined networks to something other than $G_1(s)$. The total load current $i_3 = i_2 + i_1$ can be returned to node m via the dependent current source T1 to compensate for loading. By using this current feedback technique, $i_4 = i_5$, and the loading on the first network due to the second network becomes negligible. This technique must be used to isolate each stage of the composite network from the succeeding stage. (Remember, the *from*-branch for a dependent current source using BETA must contain a resistance in the AC analysis program.)

5. The gain of the transfer function is the last element of the simulation network to be added. It must provide the gain specified in the transfer function plus compensate for any losses in individual *RC* networks. In this example

$$K_T = \frac{K_1}{k(\tau_2)\tau_3}$$

This term is mechanized as shown in Fig. 7.2. Before we discuss the gain term, observe that the network formed by $R_2 C_2$ is not

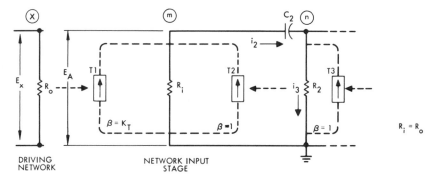

Fig. 7.2. Simulation network gain term

$$K_T = \frac{K_1}{k(\tau_2)\tau_3}$$

loaded by the following network because of the current feedback via T3. Also, since $i_2 = i_3$, the input node m is not loaded by R_2C_2 because of the current feedback via T2. The voltage at node x of the driving network is the input signal $E_i(s)$ for $G(s)$. The amplified input signal is developed across R_i at node m without loading the driving network by using dependent current source T1. The voltage developed at node m with respect to the voltage at node x is then

$$\frac{E_A}{E_x} = \beta \frac{R_i}{R_o} = K_T \frac{R_i}{R_o} \tag{7.1}$$

$$\frac{E_A}{E_x} = K_T \quad \text{if} \quad R_i = R_o \tag{7.2}$$

Hence, network gain is controlled by the value of BETA assigned to the current source linking the network being considered to the preceding network in a chain.

6. If a summing element such as that in Fig. 7.3 appears in the

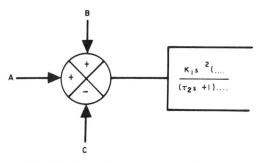

Fig. 7.3. Control system summing element.

problem, it is implemented (as shown in Fig. 7.4) in the same manner as it is in the single input term in step 5. The gain of each input is specified as before.

$$\frac{E_{iA}}{E_y} = \frac{\beta R_i}{R_{oy}} = K_T \quad \text{if} \quad R_i = R_{oy} \tag{7.3}$$

$$\frac{E_{iB}}{E_x} = \frac{\beta R_i}{R_{ox}} = K_T \quad \text{if} \quad R_i = R_{ox} \tag{7.4}$$

$$\frac{E_{iC}}{E_z} = -\frac{\beta R_i}{R_{oz}} = -K_T \quad \text{if} \quad R_i = R_{oz} \tag{7.5}$$

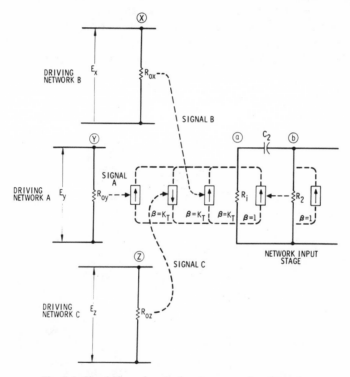

Fig. 7.4. Simulation of control system summing element.

Then, assuming $R_i = R_{ox} = R_{oy} = R_{oz}$, the input signal to the network equals

$$E_i = K_T(E_x + E_y - E_z) \tag{7.6}$$

which is desired. Note that the negative sign in the summing element is implemented by reversing the sign of the dependent current source associated with that signal. With this information, function blocks which do not have poles at the origin can be modeled. The techniques of manipulating this type of transfer function appear later in this section.

7.2.3   Simple Control-System Simulation: Example

The turbojet-engine speed-control system[2] illustrated in Fig. 7.5 is typical of the systems encountered in control problems. This simple system is used to demonstrate the application of ECAP techniques to the analysis of feedback control systems. The first part of the example demonstrates the analysis of the frequency response of the total closed-loop system and the inner-fuel-flow control loop. The second part of the example studies the transient response of the system to various reference-speed signal changes.

The values for the function-block gain terms and the system time constants are

$$K_1 = 4.550 \text{ lb/in.}^2 \text{ per radian/sec}$$

$$K_2 = 0.695 \text{ gal/min per lb/in.}^2$$

$$K_3 = 4.320 \text{ radians/sec per gal/min}$$

$$K_4 = 1.000 \text{ radian/sec per radian/sec}$$

$$K_5 = 0.531 \text{ lb/in.}^2 \text{ per gal/min}$$

$$\tau_1 = 0.0192 \text{ sec}$$

$$\tau_2 = 0.010 \text{ sec}$$

$$\tau_3 = 0.050 \text{ sec}$$

$$\tau_4 = 2.00 \text{ sec}$$

$$\tau_5 = 0.0063 \text{ sec}$$

$$\tau_6 = 0.025 \text{ sec}$$

$$\tau_7 = 0.012 \text{ sec}$$

Once the block diagram has been obtained, the next task is to replace each function block with an electrical network to simulate the transfer function as discussed earlier in this section. The equivalent networks resulting from these techniques are shown in Table 7.2.

When the simulated function blocks are interconnected, as in the block diagram in Fig. 7.5, the resulting network is an analog to the control system; this new network can be readily analyzed by ECAP. The complete analog

[2]Gordon J. Murphy, *Basic Automatic Control Theory* (Princeton, N.J.: D. Van Nostrand Co., Inc., 1957), pp. 349–51.

Table 7.2. Turbojet-engine speed-control system function-block equivalent networks.

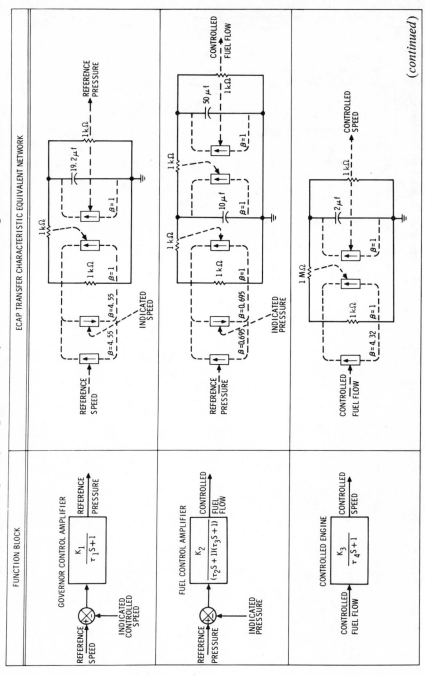

(*continued*)

Table 7.2 (continued). Turbojet-engine speed-control system function-block equivalent networks.

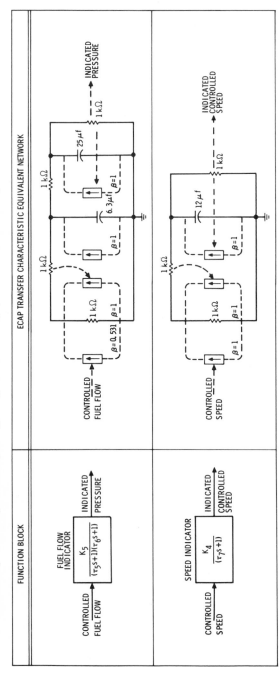

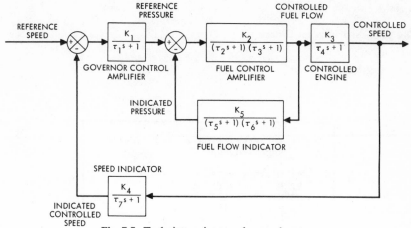

Fig. 7.5. Turbojet-engine speed-control system.

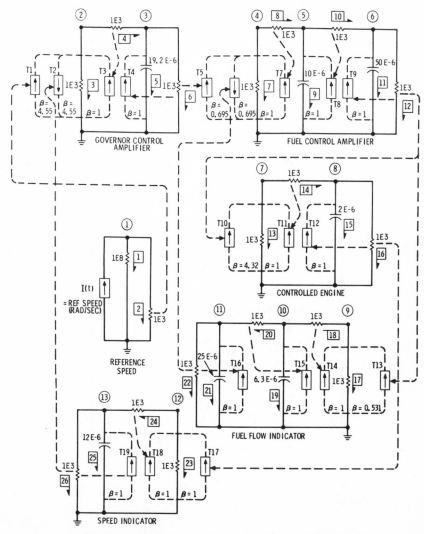

Fig. 7.6. Turbojet-engine speed-control system electrical analog network.

239

```
C      AC ANALYSIS OF TURBOJET ENGINE
C      SPEED-CONTROL SYSTEM
       AC
 B1  N(1,0),R=1E8,         I=1/0
 B2  N(1,0),R=1E3
 B3  N(2,0),R=1E3
 B4  N(2,3),R=1E3
 B5  N(3,0),C=19.2E-6
 B6  N(3,0),R=1E3
 B7  N(4,0),R=1E3
 B8  N(4,5),R=1E3
 B9  N(5,0),C=10E-6
B10  N(5,6),R=1E3
B11  N(6,0),C=50E-6
B12  N(6,0),R=1E3
B13  N(7,0),R=1E3
B14  N(7,8),R=1E6
B15  N(8,0),C=2E-6
B16  N(8,0),R=1E3
B17  N(9,0),R=1E3
B18  N(9,10),R=1E3
B19  N(10,0),C=6.3E-6
B20  N(10,11),R=1E3
B21  N(11,0),C=25E-6
B22  N(11,0),R=1E3
B23  N(12,0),R=1E3
B24  N(12,13),R=1E3
B25  N(13,0),C=12E-6
B26  N(13,0),R=1E3
 T1  B(2,3),BETA=-4.55
 T2  B(26,3),BETA=+4.55
 T3  B(4,3),BETA=-1
 T4  B(6,5),BETA=-1
 T5  B(6,7),BETA=-0.695
 T6  B(22,7),BETA=+0.695
 T7  B(8,7),BETA=-1
 T8  B(10,9),BETA=-1
 T9  B(12,11),BETA=-1
T10  B(12,13),BETA=-4.32
T11  B(14,13),BETA=-1
T12  B(16,15),BETA=-1
T13  B(12,17),BETA=-0.531
T14  B(18,17),BETA=-1
T15  B(20,19),BETA=-1
T16  B(22,21),BETA=-1
T17  B(16,23),BETA=-1
T18  B(24,23),BETA=-1
T19  B(26,25),BETA=-1
     FREQ=0.1
     PRINT,NV,CA
     EX
     MO
     FREQ=0.00159(1.095)100
     EX
```

Fig. 7.7. Input data for the turbojet-engine speed-control system closed-loop frequency response.

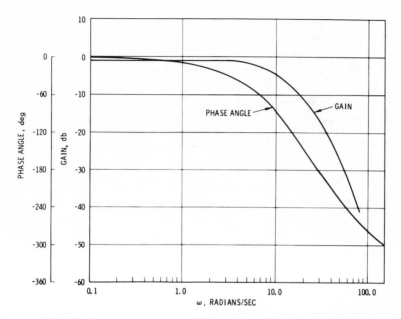

Fig. 7.8. Frequency response of the turbojet-engine speed-control system.

network in Fig. 7.6 can be used for the frequency-response and transient analyses since all *from*-branches for the dependent current sources are resistances and no integration is being performed in any transfer function. (The integration procedure is discussed in the following section.) The analysis could easily determine the effects of changes in gain values or time constants; however, this part of the study is omitted to maintain simplicity. By using the modify routine available in the AC analysis program, the effect of a change in any one, or all, of the system parameters on the closed-loop frequency response of the system can be computed.

The input data used to conduct the frequency-response analysis for this example are shown in Fig. 7.7. The closed-loop system frequency response obtained from the calculations is plotted in Fig. 7.8.

Since the transient and the frequency-response analyses can be obtained from the same network, the only changes that must be made in the input data are the command cards (AC is changed to TR), the control cards (FREQ is changed to TI, OU, FI), and the input signals. The input data for the transient response analysis appear in Fig. 7.9. The input reference-speed signal arbitrarily selected for this example and the resulting controlled engine speed are plotted in Fig. 7.10. The error in speed and the system delay are both characteristic of this type of feedback control system.

```
C       TRANSIENT RESPONSE ANALYSIS OF
C       TURBOJET ENGINE
C       SPEED-CONTROL SYSTEM
        TR
 B1 N(1,0),R=1E8,
 B2 N(1,0),R=1E3
 B3 N(2,0),R=1E3
 B4 N(2,3),R=1E3
 B5 N(3,0),C=19.2E-6
 B6 N(3,0),R=1E3
 B7 N(4,0),R=1E3
 B8 N(4,5),R=1E3
 B9 N(5,0),C=10E-6
B10 N(5,6),R=1E3
B11 N(6,0),C=50E-6
B12 N(6,0),R=1E3
B13 N(7,0),R=1E3
B14 N(7,8),R=1E6
B15 N(8,0),C=2E-6
B16 N(8,0),R=1E3
B17 N(9,0),R=1E3
B18 N(9,10),R=1E3
B19 N(10,0),C=6.3E-6
B20 N(10,11),R=1E3
 I1 (100),0,100,100,100,100,100,100,100,100,100,100,100,100,100,100,
    *90,80,70,60,50
B21 N(11,0),C=25E-6
B22 N(11,0),R=1E3
B23 N(12,0),R=1E3
B24 N(12,13),R=1E3
B25 N(13,0),C=12E-6
B26 N(13,0),R=1E3
 T1 B(2,3),BETA=-4.55
 T2 B(26,3),BETA=+4.55
 T3 B(4,3),BETA=-1
 T4 B(6,5),BETA=-1
 T5 B(6,7),BETA=-0.695
 T6 B(22,7),BETA=+0.695
 T7 B(8,7),BETA=-1
 T8 B(10,9),BETA=-1
 T9 B(12,11),BETA=-1
T10 B(12,13),BETA=-4.32
T11 B(14,13),BETA=-1
T12 B(16,15),BETA=-1
T13 B(12,17),BETA=-0.531
T14 B(18,17),BETA=-1
T15 B(20,19),BETA=-1
T16 B(22,21),BETA=-1
T17 B(16,23),BETA=-1
T18 B(24,23),BETA=-1
T19 B(26,25),BETA=-1
    TI=0.001
    OU=5
    FI=3
    PRINT,NV,CA
    EX
```

Fig. 7.9. Transient response analysis input data for the turbojet-engine speed-control system.

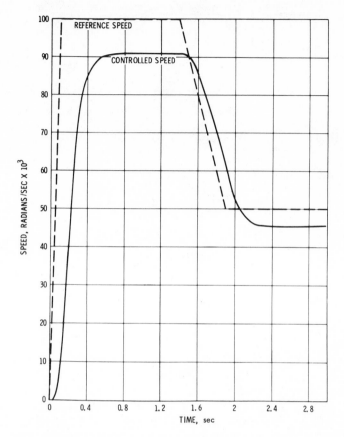

Fig. 7.10. Controlled speed response to reference speed changes of turbojet-engine speed-control system.

7.2.4   Simulation of Function Blocks Requiring Integration

Many function blocks occurring in control systems require an integration term (pole at $s = 0$) in the transfer characteristic. Some common examples of such elements are the DC motor used for position control, the galvanometer which contains a $1/s^2$ term, and the gyroscope. In the process of factoring the transfer function to find the RC networks to be used in the simulation, the integral term appears as $1/s$ or even $1/s^2$. In the AC analysis program, an integrator using a capacitance is the most convenient network to use to simulate this characteristic. This becomes apparent as development progresses.

The response of a parallel RC network (Fig. 7.11) to an arbitrary input current is a function of the time constant of the network. If $RC \gg 1$, then the

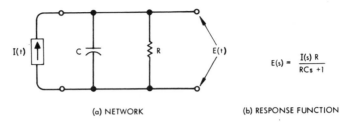

(a) NETWORK   (b) RESPONSE FUNCTION

$$E(s) = \frac{I(s)\,R}{RCs + 1}$$

Fig. 7.11. Simple RC integrator.

equation in Fig. 7.11b reduces to

$$E(s) = \frac{I(s)}{Cs} \tag{7.7}$$

Transforming

$$E(t) = \frac{1}{C}\int i\,dt \tag{7.8}$$

Arbitrarily selecting $C = 1$ further reduces the equation to

$$E(t) = \int i\,dt \tag{7.9}$$

The value selected for R is determined by the value of R_x used in the other networks for input and output resistances (see Fig. 7.4) if simplicity in the system network is to be maintained. The loading effect of R on the integration network can be eliminated ($R_{\text{eff}} \to \infty$) by returning the current flowing in the resistance to the source node m as shown in Fig. 7.12. Since the resistance presents no load to the network, the transfer function of the network becomes

$$G(s) = \frac{E_o(s)}{E_{ox}(s)} = \frac{K}{s} \tag{7.10}$$

The gain term in the integration network is determined by the current amplifi-

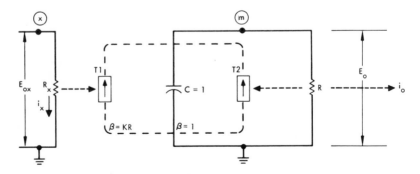

Fig. 7.12. AC analysis integration network.

cation of the dependent current source T1. The input current to the integration network equals

$$KR(i_x) = KR\left(\frac{E_{ox}}{R}\right) = KE_{ox} \qquad (7.11)$$

Hence

$$E_o = \int KE_{ox}\, dt = K \int E_{ox}\, dt \qquad (7.12)$$

The network shown in Fig. 7.12 is used in the AC analysis program because (1) only resistances can be used as *from*-branches for dependent current sources specified by BETA trans-terms, and (2) the current in the resistance is directly proportional to the integral of the input current. The input of the integrating network can also be used as a summing element, as described in Sec. 7.2.2, but remember that a current is being applied to the input of the integrator instead of developing a voltage across R.

When studying the transient response of a control system, the network used to simulate the integration term $1/s$ is different from that used in the AC analysis. The reason for this is that the basic capacitance model used in the transient analysis program is an approximation to the ideal capacitance; thus some error results from its use as an integrator element. (For a detailed discussion of the transient analysis capacitance model, refer to Appendix B. 3.) The error obtained will probably be less than 0.1 per cent, but may still be undesirable. The inductance model used in the transient analysis program is considerably more accurate than the capacitance model. Whenever possible, the inductance model should be used in the integrating network for greater model accuracy, particularly when the integration error can be multiplied.

The mathematical development of the integration network using an inductance is nearly identical to the development of that using an RC network; this development is omitted here. The integration network is shown in Fig. 7.13. The output current i_o can be expressed as

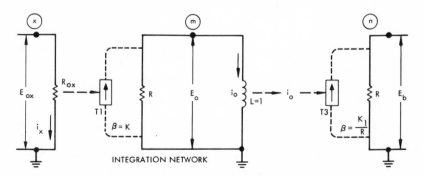

Fig. 7.13. Transient analysis integration network.

$$i_o(t) = \left(\frac{KR}{R_{ox}}\right) \int E_{ox} \, dt \tag{7.13}$$

If $R = R_{ox}$, then

$$i_o(t) = K \int E_{ox} \, dt \tag{7.14}$$

Here, as in the transfer-function simulation networks developed in Sec. 7.2.2, the input- and output-resistance values become important. Since the output current i_o in the integrator is not related to a resistance but equals $K \int e \, dt$, the value of BETA for the dependent current source at the input of the following stage must equal K_1/R. This compensates for the fact that the voltage developed at the input of the next stage is

$$E_b = \beta i_o R = \beta KR \int E_{ox} \, dt \tag{7.15}$$

Then

$$E_b = K_1 K \int E_{ox} \, dt \tag{7.16}$$

as desired. The input of this network can be used as a summing element exactly as described in Sec. 7.2.2.

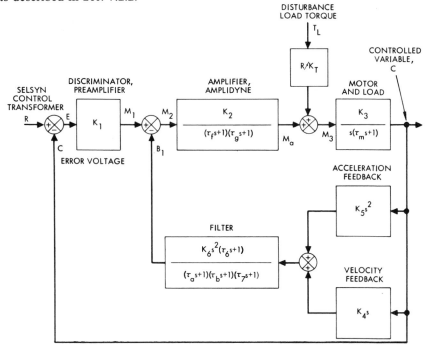

Fig. 7.14. Position control system subjected to input position and load torque disturbances.

7.2.5 Control-System Simulation Requiring Integration Networks: Example

The position control system[3] in Fig. 7.14 is an excellent example of a feedback control system which uses a wide variety of networks ranging from integration (motor plus load) to second derivatives (acceleration feedback). This control system is subjected to both input position and load disturbances to indicate the applicability of ECAP techniques in the analysis of this type of problem.

The values for the gain terms and time constants for this system are

$$K_1 = 14.8 \text{ v/radian}$$

$$K_2 = 2300 \text{ v/v}$$

$$K_3 = 0.526 \text{ radians per sec/v}$$

$$K_4 = 0.98 \text{ v/radian per sec}$$

$$K_5 = 0.01368 \text{ v/radian per sec}^2$$

$$K_6 = 0.193 \text{ sec}^2$$

$$\tau_f = 1/31.5 = 0.0317 \text{ sec}$$

$$\tau_q = 1/17.6 = 0.0569 \text{ sec}$$

$$\tau_m = 1/9.1 = 0.110 \text{ sec}$$

$$\tau_a = 1/0.986 = 1.014 \text{ sec}$$

$$\tau_b = 1/5.26 = 0.19 \text{ sec}$$

$$\tau_6 = 0.016 \text{ sec}$$

$$\tau_7 = 0.0016 \text{ sec}$$

$$K_5/K_4 = 1/71.5 = 0.0140 \text{ sec}$$

$$R/K_T = 0.855 \text{ v/lb-ft}$$

When using ECAP to aid in the analysis of physical systems, it is not desirable to attempt any function-block simulation in which the transfer characteristic involves differentiation. This is also true in analog computer simulation techniques. In this system, the velocity and acceleration feedback networks require first- and second-order derivatives, respectively.

The first step required in the analysis procedure is to remove any differentiation terms by modifying the function blocks without changing the system characteristics. This is accomplished as shown in Fig. 7.15. The motor-and-load function block is modified so that first and second derivatives of position C are available internally in the block. Using these outputs makes the acceleration and

[3]Harold Chestnut and Robert W. Mayer, *Servomechanisms and Regulating System Design* (New York: John Wiley & Sons, Inc., 1951), pp. 377–88.

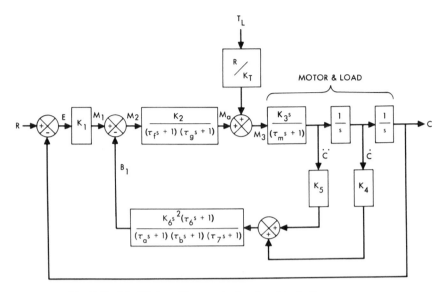

Fig. 7.15. Modified position control system block diagram removing differentiation networks from function blocks.

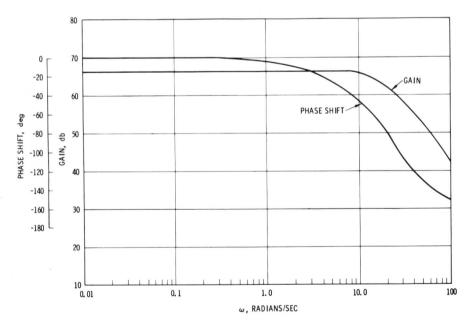

Fig. 7.16. Frequency response of the amplidyne, amplifier network (position control system).

Table 7.3. Position-control system function-block equivalent networks.

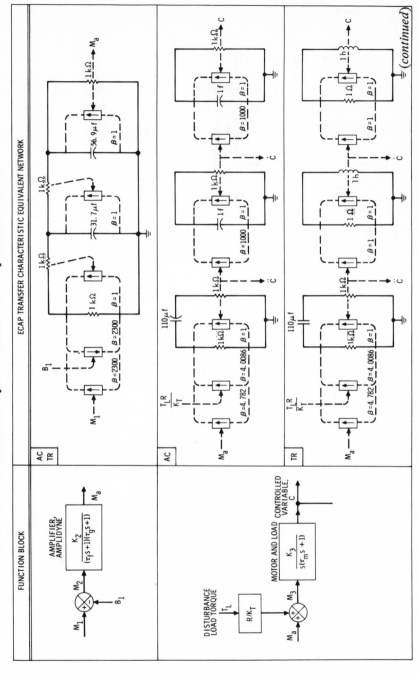

(continued)

Table 7.3 (continued). Position-control system function-block equivalent networks.

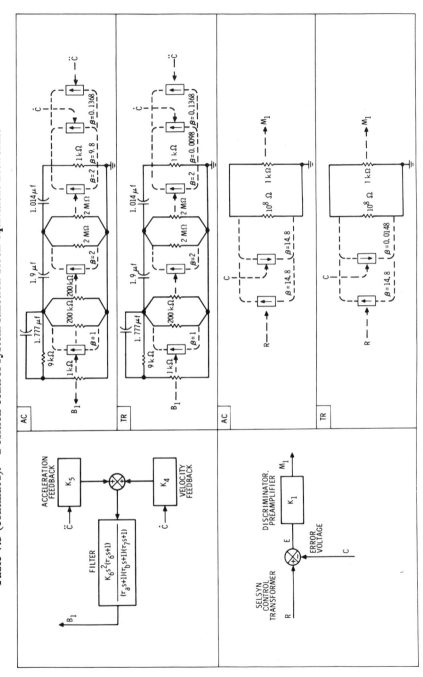

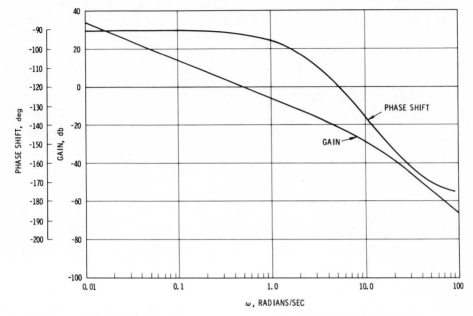

Fig. 7.17. Frequency response of the motor and load network (position control system).

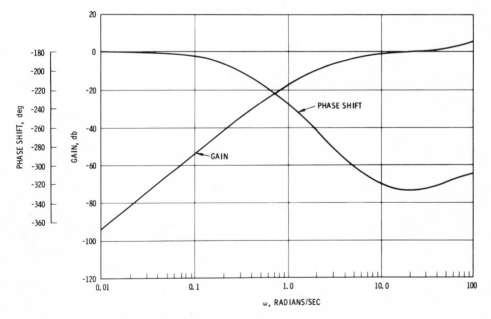

Fig. 7.18. Frequency response of filter network (position control system).

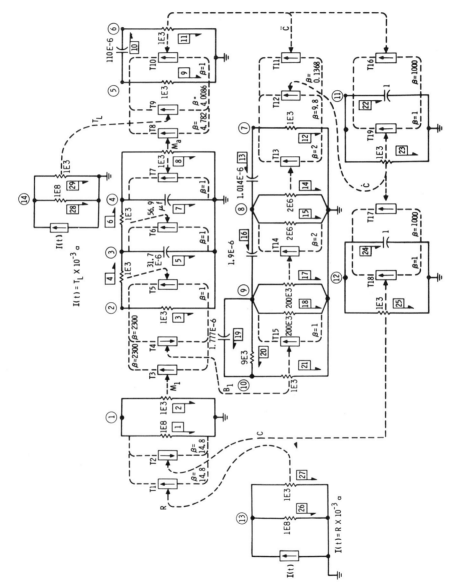

Fig. 7.19. ECAP analog network for AC analysis of the position control system.

velocity-feedback-function blocks simple amplifiers at the input of the filter network.

With the block diagram now in an acceptable form, each function block is replaced with an electrical-transfer-characteristic simulation network. ECAP networks for the AC and transient analyses are in Table 7.3 (pp. 249–50). The frequency response of major individual function blocks are shown in Figs. 7.16 (p. 248), 7.17, and 7.18.

When the networks in Table 7.3 for the AC analysis are interconnected, as shown in the modified block diagram, the resulting network is an analog to the position-control system that can easily be analyzed using ECAP. The complete analog network is in Fig. 7.19 for the AC analysis of the system. The input data used to conduct the closed-loop frequency-response analysis of the complete system appears in Fig. 7.20.

The forward-gain characteristic to position disturbance can be calculated by removing the feedback from C to R (removing the dependent current source at the input summing element) and setting $T_L = 0$. The first modify routine of the AC analysis input data computes the closed-loop response of the system to this type of disturbance.

The frequency-response of the system to a load-torque disturbance is accomplished in the second modify routine by setting the position input R equal to zero and applying an arbitrary signal to T_L.

The forward-gain characteristic with $T_L = 0$ and the position disturbance response are plotted in Figs. 7.21 and 7.22, respectively. The forward-gain characteristic with $R = 0$ and the load-torque-disturbance response are plotted in Figs. 7.23 and 7.24 (p. 256).

The analog network for the transient response analysis of the position control system is very similar to the one used for the frequency-response analysis. Differences in the two analog networks are due to the use of an inductance in the transient analysis integration networks to achieve greater accuracy than can be obtained using a capacitance. Notice that the values of BETA for the dependent current sources associated with the input and output of the integration networks have been changed to compensate for the change in integration technique. (Refer to Sec. 7.2.4 for more information.)

The complete network and input data required for the system transient analysis are shown in Figs. 7.25 (p. 257) and 7.26 (p. 258). The controlled position C is available as element current CA25 in this analysis. The reference position R and load torque T_L are equivalent to the input currents I_{26} and I_{28}, respectively. These currents are specified in milliamperes instead of amperes because of the 1000 Ω output resistances in the input networks (i.e., NV13(v) $\propto$ R(radians), and NV14(v) $\propto T_L$(ft-lb)).

This example determines the response of the system to a step reference-position change of +10 radians followed 1 sec later by a linear change in reference position from +10 radians to 0 radians over a 0.2 sec period. After the controlled position stabilizes at +10 radians, a load torque of 100 ft-lb is applied

```
C        FREQUENCY RESPONSE ANALYSIS OF A POSITION
C        CONTROL SYSTEM SUBJECTED TO INPUT AND LOAD
C        TORQUE DISTURBANCES
         AC
   B1  N(1,0),R=1E8
   B2  N(1,0),R=1E3
   B3  N(2,0),R=1E3
   B4  N(2,3),R=1E3
   B5  N(3,0),C=31.7E-6
   B6  N(3,4),R=1E3
   B7  N(4,0),C=56.9E-6
   B8  N(4,0),R=1E3
   B9  N(5,0),R=1E3
  B10  N(5,6),C=110E-6
  B11  N(6,0),R=1E3
  B12  N(7,0),R=1E3
  B13  N(7,8),C=1.014E-6
  B14  N(8,0),R=2E6
  B15  N(8,0),R=2E6
  B16  N(8,9),C=1.9E-6
  B17  N(9,0),R=200E3
  B18  N(9,0),R=200E3
  B19  N(9,10),C=1.777E-6
  B20  N(9,10),R=9E3
  B21  N(10,0),R=1E3
  B22  N(11,0),C=1
  B23  N(11,0),R=1E3
  B24  N(12,0),C=1
  B25  N(12,0),R=1E3
  B26  N(13,0),R=1E8,I=1/0
  B27  N(13,0),R=1E3
  B28  N(14,0),R=1E8,I=0/0
  B29  N(14,0),R=1E3
   T1  B(27,1),BETA=-14.8
   T2  B(25,1),BETA=+14.8
   T3  B(2,3),BETA=-2300
   T4  B(21,3),BETA=+2300
   T5  B(4,3),BETA=-1
   T6  B(6,5),BETA=-1
   T7  B(8,7),BETA=-1
   T8  B(8,9),BETA=-4.782
   T9  B(29,9),BETA=-4.0086
  T10  B(11,9),BETA=-1
  T11  B(11,12),BETA=-0.1368
  T12  P(23,12),BETA=-9.8
  T13  B(14,12),BETA=-2
  T14  B(17,15),BETA=-2
  T15  B(21,18),BETA=-1
  T16  B(11,22),BETA=-1000
  T17  B(23,24),BETA=-100C
  T18  B(25,24),BETA=-1
  T19  B(23,22),BETA=-1
         FPEQ=0.15923
         PRINT,NV,CA
         EX
         MODIFY 1
         FREQ=0.0015(1.095)100
         EX
         MODIFY 2
  B26  I=0/0
  B28  I=1/0
         FREQ=0.0015(1.095)100
         EX
```

Fig. 7.20. Input data for the position control system closed-loop response analysis.

254

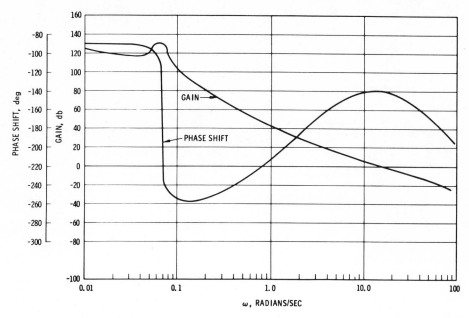

Fig. 7.21. Forward gain characteristic ($T_L = 0$) to a position disturbance for the position control system.

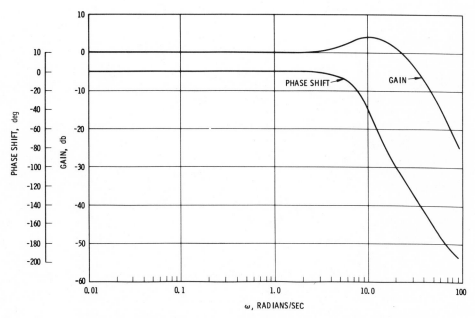

Fig. 7.22. Frequency response of the position control system to a position disturbance.

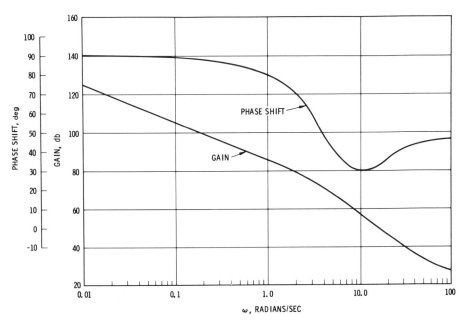

Fig. 7.23. Forward gain characteristic ($R = 0$) to a load torque disturbance for position control system.

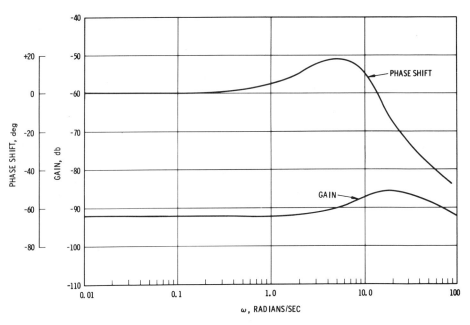

Fig. 7.24. Frequency response of position control system to a load torque disturbance.

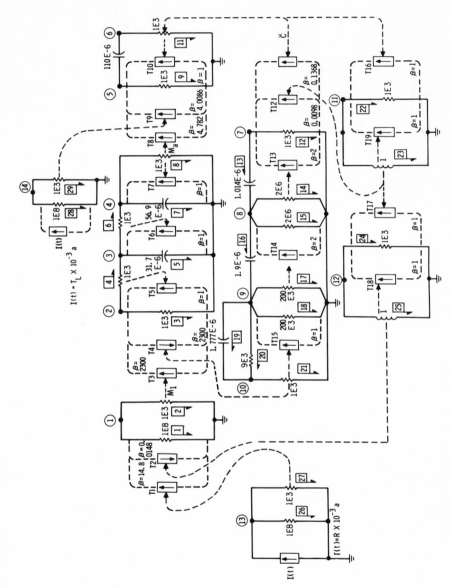

Fig. 7.25. ECAP analog network for the transient analysis of the position control system.

```
C       TRANSIENT RESPONSE ANALYSIS OF A POSITION
C       CONTROL SYSTEM SUBJECTED TO AN INPUT
C       POSITION AND LOAD TORQUE
C       DISTURBANCE
        TRANSIENT
   B1   N(1,0),R=1E8
   B2   N(1,0),R=1E3
   B3   N(2,0),R=1E3
   B4   N(2,3),R=1E3
   B5   N(3,0),C=31.7E-6
   B6   N(3,4),R=1E3
   B7   N(4,0),C=56.9E-6
   B8   N(4,0),R=1E3
   B9   N(5,0),R=1E3
  B10   N(5,6),C=110E-6
  B11   N(6,0),R=1E3
  B12   N(7,0),R=1E3
  B13   N(7,8),C=1.014E-6
  B14   N(8,0),R=2E6
  B15   N(8,0),R=2E6
  B16   N(8,9),C=1.9E-6
  B17   N(9,0),R=200E3
  B18   N(9,0),R=200E3
  B19   N(9,10),C=1.777E-6
  B20   N(9,10),R=9E3
  B21   N(10,0),R=1E3
  B22   N(11,0),R=1E3
  B23   N(11,0),L=1
  B24   N(12,0),R=1
  B25   N(12,0),L=1
  B26   N(13,0),R=1E8,I=1/0
  I26   (1000),10E-3,10E-3,10E-3,10E-3,10E-3,10E-3,0
  B27   N(13,0),R=1E3
  B28   N(14,0),R=1E8,I=0/0
  I28   (1000),0,0,0,.1,.1,0
  B29   N(14,0),R=1E3
  B30   N(15,0),R=1E4
  B31   N(15,0),R=1
   T1   B(27,1),BETA=-14.8
   T2   B(25,1),BETA=0.0148
   T3   B(2,3),BETA=-2300
   T4   B(21,3),BETA=+2300
   T5   B(4,3),BETA=-1
   T6   B(6,5),BETA=-1
   T7   B(8,7),BETA=-1
   T8   B(8,9),BETA=-4.782
   T9   B(29,9),BETA=-4.0086
  T10   B(11,9),BETA=-1
  T11   B(11,12),BETA=-0.1368
  T12   B(23,12),BETA=-.0098
  T13   B(14,12),BETA=-2
  T14   B(17,15),BETA=-2
  T15   B(21,18),BETA=-1
  T16   B(11,22),BETA=-1
  T17   B(23,24),BETA=-1
  T18   B(25,24),BETA=-1
  T19   B(23,22),BETA=-1
  T20   B(25,30),BETA=-1
        TI=0.0002
        OU=20
        FI=2
        PRINT,NV,CA
        EX
```

Fig. 7.26. Transient response analysis input data for the position control system.

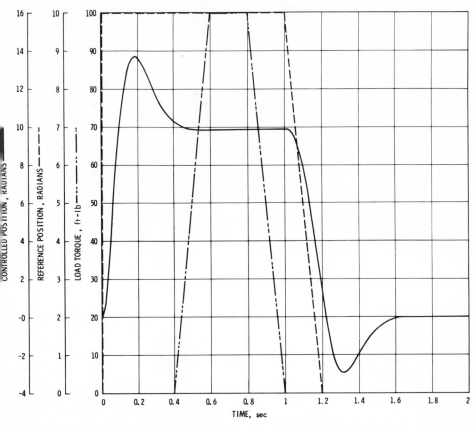

Fig. 7.27. Controlled position response to reference position changes and load torque disturbances.

to the system as in Fig. 7.27. This figure also contains the reference-position signal and the controlled-position transient response of the system. The responses of the major function blocks to the disturbing signals are plotted in Fig. 7.28.

Solutions to problems of this type are generally obtained using an analog computer or a digital simulation program. However, if the user is familiar with ECAP techniques, application of this program to the simulation of control systems can be both convenient and timesaving.

7.3 HEAT-TRANSFER ANALYSIS

ECAP can be used effectively in the analysis of heat-transfer problems since the form of the general differential equations representing heat flow in a thermal system are identical to the equations representing current flow in an electrical

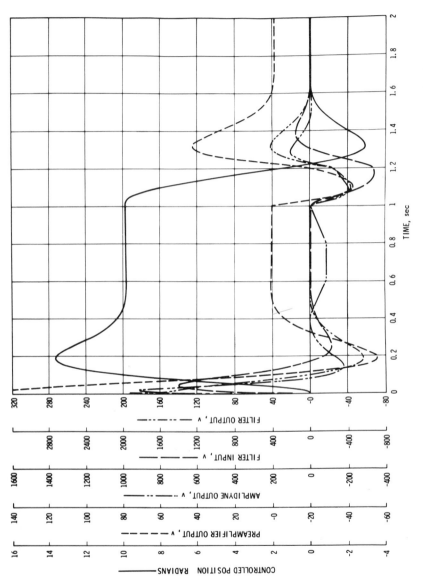

Fig. 7.28. Response of function blocks of the position control system to reference position and load torque disturbances (disturbances shown in Fig. 7.27).

system. Because of this similarity in the basic relationships defining the two systems, it is possible to construct an equivalent electrical network which is an exact analogy of the thermal system.

7.3.1   Electrical Analog (Conduction)

Heat flow between any two points in a medium is directly proportional to the temperature difference between the points and the thermal conductance of the medium. Mathematically

$$Q = G_T(T_a - T_b) \tag{7.17}$$

where Q = heat flow
 $T_a - T_b$ = temperature difference between the two points in the medium
 $G_T = kA/d$ = thermal conductance of the medium
where k = thermal conductivity
 A = cross-sectional area perpendicular to the direction of the heat flow
 d = effective length parallel to the heat flow

Equation (7.17) is similar to Equation (7.18), which relates current and voltage in the electrical system.

$$i = G_E(V_a - V_b) \tag{7.18}$$

$$G_E = \frac{A}{\rho l}$$

where G_E = electrical conductance

The temperature rise in a medium is proportional to the amount of energy stored and the thermal capacitance of the storing medium. Mathematically

$$\Delta T = \frac{1}{C_T} \int_t^{t+\Delta t} Q \, dt \tag{7.19}$$

or

$$Q = C_T \frac{dT}{dt} \tag{7.20}$$

where $C_T = \rho c v$ = thermal capacitance of the medium
 ρ = density
 c = specific heat
 v = volume

This is analogous to

$$\Delta V = \frac{1}{C_E} \int_t^{t+\Delta t} i \, dt \tag{7.21}$$

where C_E = electrical capacitance

Kirchhoff's first and second laws for the electrical system can be easily redefined for the thermal system as follows:

Table 7.4. Relationships for ideal, linear, two-terminal network elements.*

QUANTITY	PROTOTYPE	ELECTRICAL	TRANSLATIONAL	ROTATIONAL	FLUID-FLOW	THERMAL
FLOW	FLOW, f	CURRENT, i	FORCE, f	TORQUE, τ	VOLUME FLOW, q	HEAT FLOW, Q
POTENTIAL	POTENTIAL, p	VOLTAGE, e	VELOCITY, v	ANGULAR VEL., ω	PRESSURE, p	TEMPERATURE, T
FLOW EQUATION	$f=Yp$	$i=Ye$	$f=Yv$	$\tau=Y\omega$	$q=Yp$	$Q=YT$
INSTANTANEOUS POWER (WATTS)	$P=fp$	$P=ie$	$P=fv$	$P=\tau\omega$	$P=qp$	$P=Q$
OTHER VARIABLES		CHARGE: $q=\int i\,dt$	DISPLACEMENT $x=\int v\,dt$	ANGLE, $\theta=\int \omega\,dt$		
INTEGRATING ELEMENT (AS AN ADMITTANCE) AND FLOW EQUATION	$f=A\int p\,dt$	$\Gamma=1/L$; $i=\dfrac{1}{L}\int e\,dt$; L = INDUCTANCE	$f=K\int v\,dt$; K = SPRING CONSTANT	$\tau=K\int \omega\,dt$; K = SPRING CONSTANT	$Y=\dfrac{1}{I}$; $q=\dfrac{1}{I}\int p\,dt$; I = INERTANCE	
i-d ADMITTANCE	$Y=A\int dt$	$\dfrac{1}{L}\int dt$	$K\int dt$	$K\int dt$	$\dfrac{1}{I}\int dt$	
i-d IMPEDANCE	$\dfrac{1}{A}\dfrac{d}{dt}$	$L\dfrac{d}{dt}$	$\dfrac{1}{K}\dfrac{d}{dt}$	$\dfrac{1}{K}\dfrac{d}{dt}$	$I\dfrac{d}{dt}$	
STORED ENERGY (JOULES)	$\dfrac{f^2}{2A}$	$\dfrac{Li^2}{2}$	$\dfrac{f^2}{2K}$ OR $\dfrac{Kx^2}{2}$	$\dfrac{\tau^2}{2K}$ OR $\dfrac{K\theta^2}{2}$	$\dfrac{Iq^2}{2}$	
INITIAL CONDITION	$f(0)$	$i(0)$	$f(0)$ OR $x(0)$	$\tau(0)$ OR $\theta(0)$	$q(0)$	
PROPORTIONAL ELEMENT (AS AN ADMITTANCE) AND FLOW EQUATION	$f=Bp$	$G=1/R$; $i=Ge=e/R$; R = RESISTANCE	$f=Bv$; B = VISCOUS FRICTION	$\tau=D\omega$; D = VISCOUS FRICTION	$q=p/R$; R = FLUID RESISTANCE	$G_A=1/R$; $Q=GT=T/R$; R = THERMAL RESISTANCE
i-d ADMITTANCE	B	G OR $1/R$	B	D	$1/R$	G OR $1/R$
i-d IMPEDANCE	$1/B$	R	$1/B$	$1/D$	R	R
POWER (WATTS)	$fp=f^2/B=p^2B$	$ie=i^2R=e^2/R$	$fv=f^2/B=v^2B$	$\omega=\tau^2/B=\omega^2D$	$qp=q^2R=p^2/R$	$Q=T/R=GT$
DIFFERENTIATING ELEMENT (AS AN ADMITTANCE) AND FLOW EQUATION	$f=C\dfrac{dv}{dt}$	$i=C\dfrac{de}{dt}$; C = CAPACITANCE	$f=M\dfrac{dv}{dt}$; M = MASS	$\tau=J\dfrac{d\omega}{dt}$; J = MOMENT OF INERTIA	$q=C\dfrac{dp}{dt}$; C = FLUID CAPACITANCE	$Q=C\dfrac{dt}{dt}$; C = THERMAL CAPACITANCE
i-d ADMITTANCE	$C\dfrac{d}{dt}$	$C\dfrac{d}{dt}$	$M\dfrac{d}{dt}$	$J\dfrac{d}{dt}$	$C\dfrac{d}{dt}$	$C\dfrac{d}{dt}$
i-d IMPEDANCE	$\dfrac{1}{C}\int dt$	$\dfrac{1}{C}\int dt$	$\dfrac{1}{M}\int dt$	$\dfrac{1}{J}\int dt$	$\dfrac{1}{C}\int dt$	$\dfrac{1}{C}\int dt$
STORED ENERGY (JOULES)	$\dfrac{Cp^2}{2}$	$\dfrac{Ce^2}{2}$	$\dfrac{Mv^2}{2}$	$\dfrac{J\omega^2}{2}$	$\dfrac{Cp^2}{2}$	$C\,T$
INITIAL CONDITION	$p(0)$	$e(0)$ OR $q(0)$	$v(0)$	$\omega(0)$	$p(0)$	$T(0)$

*R.S. Sanford, *Physical Networks* (Englewood Cliffs, N.J.: Prentice-Hall, Inc., 1965), p. 113.

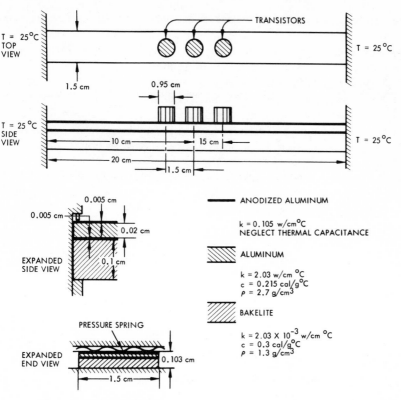

Fig. 7.29. Printed circuit board containing three transistor heat sources.

Kirchhoff's first law. The algebraic sum of the heat flow at a common point, or junction, equals zero.

Kirchhoff's second law. The algebraic sum of the temperature drops around a closed path equals zero.

The parameters and elements of interest in the analysis of thermal networks are summarized in Table 7.4.

To visualize the application of the electric analogy of the analysis of conductive heat-transfer problems, we will analyze the thermal network illustrated in Fig. 7.29.

7.3.2 Thermal Analysis of Printed Circuit Board: Example

A 1.5 cm by 20 cm printed circuit board containing three transistors and an anodized aluminum heat sink is mounted between two supports as shown in Fig. 7.29. The supports are maintained at a constant temperature of +25°C.

Each transistor dissipates 0.5 w peak power during the first half-cycle and 0.0 w during the second half-cycle. The objective of the analysis is to determine the maximum operating junction temperature of each of the three transistors on the board. The maximum allowable junction temperature is +200°C.

The first part of the problem is to construct an electrical network analogous to the thermal circuit being investigated. For this example, the printed circuit board will be divided arbitrarily into seven sections, as shown in Fig. 7.30a. The greater the number of sections used in the analog network, the greater the accuracy of the solution. Each printed circuit board section is modeled by an electrical network as in Fig. 7.30b. The subscripts on the R and C elements indicate the material type (i.e., subscripts A, M, and B specify anodized aluminum, aluminum, and bakelite, respectively). Using the thermal coefficients listed in Fig. 7.29 and the equations

$$R = \frac{d}{kA} \frac{o_C}{w} \tag{7.22}$$

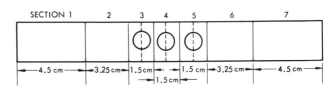

(a) SUBDIVISION OF PRINTED CIRCUIT BOARD

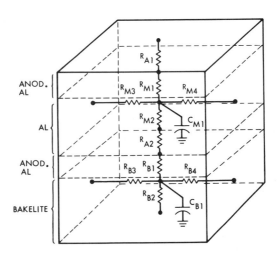

(b) ELECTRICAL ANALOG NETWORK ONE SECTION OF PRINTED CIRCUIT BOARD

Fig. 7.30. Subdivision of printed circuit board for the analysis of thermal response of transistors.

where $d =$ path length (cm)
 $k =$ thermal conductivity (w/cm°C)
 $A =$ cross-sectional area of element (cm²)
and

$$C = 4.186 \, \rho c v \text{ joules/°C} \qquad (7.23)$$

where $\rho =$ density of element (g/cm³)
 $c =$ specific heat of element (cal/g°C)
 $v =$ volume (cm³)

the values of the elements in each section can be computed. Calculated values are tabulated in Table 7.5.

**Table 7.5.   Element values for printed circuit board
thermal analysis.**

Element	Units	Section 1, 7	Section 2, 6	Section 3, 4, 5
R_{A1}	°C/w	—	—	2.11×10^{-2}
R_{M1}	°C/w	—	—	2.19×10^{-3}
R_{M2}	°C/w	0.73×10^{-3}	1.01×10^{-3}	2.19×10^{-3}
R_{M3}	°C/w	37	26.65	12.3
R_{M4}	°C/w	37	26.65	12.3
R_{A2}	°C/w	0.705×10^{-2}	0.977×10^{-2}	2.11×10^{-2}
R_{B1}	°C/w	3.65	5.05	10.95
R_{B2}	°C/w	—	—	—
R_{B3}	°C/w	14.75×10^3	10.7×10^3	4.92×10^3
R_{B4}	°C/w	14.75×10^3	10.7×10^3	4.92×10^3
C_{M1}	joules/°C	0.328	0.2365	0.1092
C_{B1}	joules/°C	1.10	0.795	0.367

Thermal contact resistance between the transistor case and the anodized aluminum heat sink is assumed to be 10°C/w. Contact resistance between the heat sink and the pressure spring is assumed to be 100°C/w. A 10°C/w thermal resistance is used between the bakelite board and the frame. Thermal parameters for the transistor are

$$C_{\text{junction}} = 0.0025 \text{ joules/°C}$$

$$C_{\text{case}} = 0.04 \text{ joules/°C}$$

$$R_{\text{junction-case}} = 50°\text{C/w}$$

The effects of convection are neglected. The effects of radiation are also neglected to simplify analysis. The error in the junction temperature resulting from neglected radiant heat flow will raise the computed junction temperature slightly, yielding what might be considered a worst case condition. The electrical analog

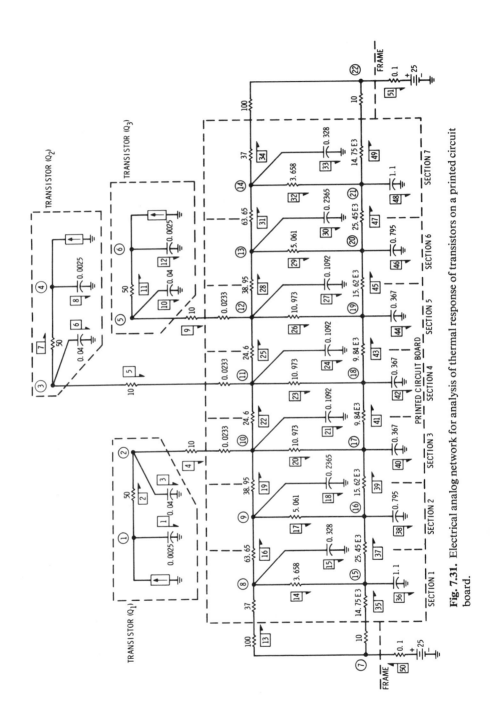

Fig. 7.31. Electrical analog network for analysis of thermal response of transistors on a printed circuit board.

```
C       THERMAL ANALYSIS OF PRINTED CIRCUIT STRIP
C       ASSUMING STABLE AT 25 C AT T=0
C       FIRST SOLUTION DUTY CYCLE  50 PERCENT
        TR
   B1  N(1,0),C=.0025,EO=-25
   B2  N(1,2),R=50
   B3  N(2,0),C=.04,EO=-25
   B4  N(2,10),R=10.0233
   B5  N(3,11),P=10.0233
   B6  N(3,0),C=.04,EO=-25
   B7  N(3,4),R=50
   B8  N(4,0),C=0.0025,EO=-25
   B9  N(5,12),R=10.0233
  B10  N(5,0),C=.04,EO=-25
  B11  N(5,6),R=50
  B12  N(6,0),C=0.0025,EO=-25
  B13  N(7,8),R=137
  B14  N(8,15),R=3.658
  B15  N(8,0),C=0.328,EO=-25
  B16  N(8,9),R=63.65
  B17  N(9,16),R=5.061
  B18  N(9,0),C=0.2365,EO=-25
  B19  N(9,10),R=38.95
  B20  N(10,17),R=10.973
  B21  N(10,0),C=0.1092,EO=-25
  B22  N(10,11),R=24.6
  B23  N(11,18),R=10.973
  B24  N(11,0),C=0.1092,EO=-25
  B25  N(11,12),R=24.6
  B26  N(12,19),R=10.973
  B27  N(12,0),C=0.1092,EO=-25
  B28  N(12,13),R=38.95
  B29  N(13,20),R=5.061
  B30  N(13,0),C=0.2365,EO=-25
  B31  N(13,14),R=63.65
  B32  N(14,21),R=3.658
  B33  N(14,0),C=0.328,EO=-25
  B34  N(14,22),R=137
  B35  N(7,15),R=14.76E3
  B36  N(15,0),C=1.1,EO=-25
  B37  N(15,16),R=25.45E3
  B38  N(16,0),C=0.795,EO=-25
  B39  N(16,17),R=15.62E3
  B40  N(17,0),C=0.367,EO=-25
  B41  N(17,18),R=9.84E3
  B42  N(18,0),C=0.367,EO=-25
  B43  N(18,19),R=9.84E3
  B44  N(19,0),C=0.367,EO=-25
  B45  N(19,20),R=15.62E3
  B46  N(20,0),C=0.795,EO=-25
  B47  N(20,21),R=25.45E3
  B48  N(21,0),C=1.1,EO=-25
  B49  N(21,22),R=14.76E3
  B50  N(0,7),R=0.1,E=25
  B51  N(22,0),R=0.1,E=-25
   I1  (50),0,.5,.5,.5,.5,0,0,0,0,.5,.5,.5,.5,0,0,0,0,.5,.5,.5,.5,
       *0,0,0,0,.5,.5,.5,.5,0,0,0,0,.5,.5,.5,.5,0,0,0,0,.5,.5,.5,.5,
       *0,0,0,0,.5,.5,.5,.5,0,0,0,0,.5,.5,.5,.5
   I8  (50),0,.5,.5,.5,.5,0,0,0,0,.5,.5,.5,.5,0,0,0,0,.5,.5,.5,.5,
       *0,0,0,0,.5,.5,.5,.5,0,0,0,0,.5,.5,.5,.5,0,0,0,0,.5,.5,.5,.5,
       *0,0,0,0,.5,.5,.5,.5,0,0,0,0,.5,.5,.5,.5
  I12  (50),0,.5,.5,.5,.5,0,0,0,0,.5,.5,.5,.5,0,0,0,0,.5,.5,.5,.5,
       *0,0,0,0,.5,.5,.5,.5,0,0,0,0,.5,.5,.5,.5,0,0,0,0,.5,.5,.5,.5,
       *0,0,0,0,.5,.5,.5,.5,0,0,0,0,.5,.5,.5,.5
        TI=0.5
        OU=4
        FI=1000
        PRINT,NV,CA
        EX
```

Fig. 7.32. Transient analysis input data for investigation of thermal response of transistors on a printed circuit board.

network for the thermal circuit is shown in Fig. 7.31. Using this network gives a good approximation of the thermal characteristics of the printed circuit board and allows computation of the junction temperature response of each transistor.

Input data for the analysis in which each transistor is conducting simultaneously on a 50 per cent duty cycle are shown in Fig. 7.32. Thermal response for each transistor junction for this condition is plotted in Fig. 7.33. The maximum junction temperature of each transistor is definitely less than the maximum allowable temperature of $+200°C$.

Engineers are usually concerned about the thermal response of the system if a failure occurs that will cause the transistors to achieve a 100 per cent duty cycle, and, if such a failure does occur, the time that will elapse before transistor junction temperatures exceed the critical temperature of $+200°C$. To calculate this information, the driving functions I_1, I_8, and I_{12} are modified and the analysis is repeated. Input data for the analysis assuming a 100 per cent duty cycle

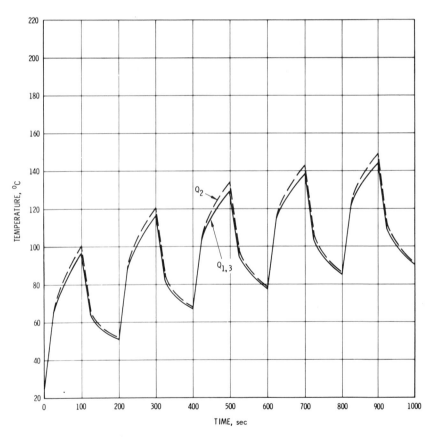

Fig. 7.33. Thermal response of transistor junction temperatures with 0.5 w transistor power dissipation and 50 per cent duty cycle.

```
C       THERMAL ANALYSIS OF PRINTED CIRCUIT STRIP
C       ASSUMING STABLE AT 25 C AT T=0
C       SECOND SOLUTION DUTY CYCLE 100 PERCENT
        TR
  B1  N(1,0),C=.0025,EO=-25
  B2  N(1,2),R=50
  B3  N(2,0),C=.04,EO=-25
  B4  N(2,10),R=10.0233
  B5  N(3,11),R=10.0233
  B6  N(3,0),C=.04,EO=-25
  B7  N(3,4),R=50
  B8  N(4,0),C=0.0025,EO=-25
  B9  N(5,12),R=10.0233
 B10  N(5,0),C=.04,EO=-25
 B11  N(5,6),R=50
 B12  N(6,0),C=0.0025,EO=-25
 B13  N(7,8),R=137
 B14  N(8,15),R=3.658
 B15  N(8,0),C=0.328,EO=-25
 B16  N(8,9),R=63.65
 B17  N(9,16),R=5.061
 B18  N(9,0),C=0.2365,EO=-25
 B19  N(9,10),R=38.95
 B20  N(10,17),R=10.973
 B21  N(10,0),C=0.1092,EO=-25
 B22  N(10,11),R=24.6
 B23  N(11,18),R=10.973
 B24  N(11,0),C=0.1092,EO=-25
 B25  N(11,12),R=24.6
 B26  N(12,19),R=10.973
 B27  N(12,0),C=0.1092,EO=-25
 B28  N(12,13),R=38.95
 B29  N(13,20),R=5.061
 B30  N(13,0),C=0.2365,EO=-25
 B31  N(13,14),R=63.65
 B32  N(14,21),R=3.658
 B33  N(14,0),C=0.328,EO=-25
 B34  N(14,22),R=137
 B35  N(7,15),R=14.76E3
 B36  N(15,0),C=1.1,EO=-25
 B37  N(15,16),R=25.45E3
 B38  N(16,0),C=0.795,EO=-25
 B39  N(16,17),R=15.62E3
 B40  N(17,0),C=0.367,EO=-25
 B41  N(17,18),R=9.84E3
 B42  N(18,0),C=0.367,EO=-25
 B43  N(18,19),R=9.84E3
 B44  N(19,0),C=0.367,EO=-25
 B45  N(19,20),R=15.62E3
 B46  N(20,0),C=0.795,EO=-25
 B47  N(20,21),R=25.45E3
 B48  N(21,0),C=1.1,EO=-25
 B49  N(21,22),R=14.76E3
 B50  N(0,7),R=0.1,E=25
 B51  N(22,0),R=0.1,E=-25
  I1  (50),0,.5
  I8  (50),0,.5
 I12  (50),0,.5
        TI=0.5
        OU=4
        FI=1000
        PRINT,NV,CA
        EX
```

Fig. 7.34. Transient analysis input data for investigation of thermal response of transistors on a printed circuit board with a 100 per cent duty cycle.

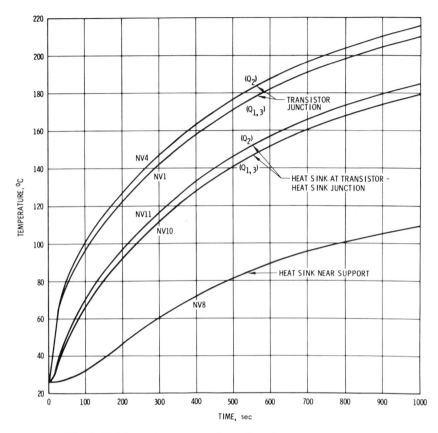

Fig. 7.35. Thermal response of transistor junction temperatures with 0.5 w transistor power dissipation and a 100 per cent duty cycle.

are shown in Fig. 7.34, and the resulting thermal response is plotted in Fig. 7.35. The center transistor Q_2 does exceed $+200°C$ within 750 sec after the application of power. If transistors in the system had been operating at about $+150°C$ initially, the time to failure would have been much shorter.

7.3.3 Electrical Analog (Convection)

Equivalent circuits for thermal conduction can also be applied to problems in which cooling by convection is important. To illustrate this point, consider the relationship for heat transfer by convection, which is

$$Q = Ah(T_2 - T_1) = \frac{(T_2 - T_1)}{1/Ah} = \frac{\Delta T}{R_t} \tag{7.24}$$

then

$$R_t = 1/Ah \tag{7.25}$$

where Q = heat flow
$\quad\ \ T_2$ = fluid temperature
$\quad\ \ T_1$ = surface temperature
$\quad\ \ A$ = cross-sectional area of surface exposed to convection
$\quad\ \ h$ = heat-transfer coefficient

The heat-transfer coefficient h is a function of many variables, such as shape, roughness, and dimensions of the surface; velocity and direction of fluid flow; and temperature, density, viscosity, thermal conductivity, and specific heat of the fluid. Generally speaking, however, the heat-transfer coefficient is not too difficult to obtain and the techniques used in the last example can be used to study systems in which heat is transferred by both conduction and convection. The references at the end of the chapter are recommended for additional information on this subject.

7.3.4   Electrical Analog (Radiation)

In the analysis of thermal problems involving heat transmission by radiation, the effects of reflectivities and emissivities of all surfaces in the system and the geometrical relationships between them must be accounted for in construction of the electrical analog network. The electrical analogy to radiant heat transfer, originally described in 1956 by A. K. Oppenheim,[4] is an excellent link between radiant heat-transfer problems and ECAP techniques.

Basic mathematical relationships used in constructing the analog are as follows: For black-body radiation, radiosity J is equal to total emissive power E. Thus, for a black surface

$$J = E = \sigma T^4 \tag{7.26}$$

where σ = Stefan-Boltzman constant ($0.1713 \times 10^{-8}(\text{BTU/ft}^2\text{-hr}°\text{R}^4)$)
$\quad\ \ T$ = temperature (°R)

Between any two black surfaces A_i and A_k, the net heat transfer is

$$Q_{ik} = \sigma A_i F_{ik}(T_i^4 - T_k^4) = -\sigma A_k F_{ki}(T_k^4 - T_i^4) = -Q_{ki} \tag{7.27}$$

where F_{ik} and F_{ki} are geometric factors. Comparing this equation with

$$i_{ik} = G_{ik}(V_i - V_k)$$

gives the analogy between radiation heat flow and electric current, in which $E/\sigma \propto V$ and $G_{ik} \propto A_i F_{ik} = A_k F_{ki}$.

For any surface A_i within an enclosure consisting of n surfaces, the total net rate of radiation heat transfer from A_i is

[4]"Radiation Analysis by the Network Method," *Transactions of the American Society of Mechanical Engineers,* **78**, No. 725 (1956).

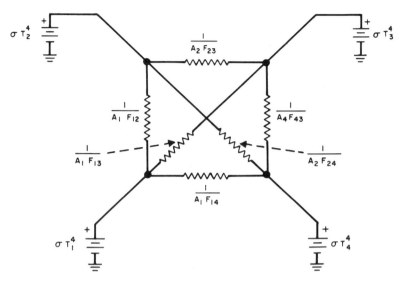

Fig. 7.36. Network for enclosures consisting of four radiating, black surfaces.

$$Q_i(\text{net}) = \sum_{k=1}^{n} \sigma A_i F_{ik}(T_i^4 - T_k^4) \tag{7.28}$$

To illustrate the principle, an analog network for an enclosure of four black surfaces is shown in Fig. 7.36. In this figure, the net rate of radiation heat transfer from surface 1 of the enclosure is

$$Q_1(\text{net}) = \sigma A_1 F_{12}(T_1^4 - T_2^4) + \sigma A_1 A_{13}(T_1^4 - T_3^4) + \sigma A_1 F_{14}(T_1^4 - T_4^4) \tag{7.29}$$

or, in the electrical analog network

$$i_1(\text{net}) = G_{12}V_{12} + G_{13}V_{13} + G_{14}V_{14} \tag{7.30}$$

Now, consider the same development with gray instead of black surfaces. In this case, the radiosity J, defined as the radiant energy leaving the surface, is

$$J = \epsilon E + \gamma G = \epsilon E + (1 - \epsilon)G \tag{7.31}$$

where ϵ = surface emissivity
γ = reflectivity
G = irradiation
E = total emissive power of a black surface at the same temperature

For an opaque body $\epsilon + \gamma = 1$. The rate of direct radiant heat transfer between any two opaque, diffuse surfaces A_i and A_k is

$$Q_{ik} = A_i F_{ik}(J_i - J_k) \tag{7.32}$$

If bodies i and k are black, then the net rate of radiation energy leaving a gray surface A is

$$Q_{\text{net}} = A(J - G) \tag{7.33}$$

Substituting Eq. (7.31) into Eq. (7.33)

$$Q_{\text{net}} = A\left[\frac{\epsilon}{\gamma} E - \frac{1 - \gamma}{\gamma} J\right] = A \frac{\epsilon}{(1 - \epsilon)}(E - J) \tag{7.34}$$

The electrical network analogy of radiant heat transfer between gray surfaces in an enclosure is based upon Eqs. (7.32) and (7.34). If J_i and J_k are the voltages of two nodes in the electrical network shown in Fig. 7.37, and the resis-

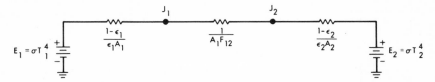

Fig. 7.37. Network for enclosures consisting of two radiating, opaque gray surfaces.

tance between these nodes equals $1/A_i F_{ik}$, the current flowing between the nodes is analogous to the rate of heat flow Q_{ik}. (See Eq. (7.32).) If E and J of Eq. (7.34) are the potentials of two nodes and the resistance between the nodes equals $(1 - \epsilon)/\epsilon A$, then the current flowing in the network is analogous to Q_{net}. The gray surface may be considered to have a potential J with respect to the potential $E = \sigma T^4$ of a black surface and to have a resistance $(1 - \epsilon)/\epsilon A$.

Figure 7.38 shows the extension of the electrical network analogy for an enclosure consisting of four black radiating surfaces (Fig. 7.36) to an enclosure with

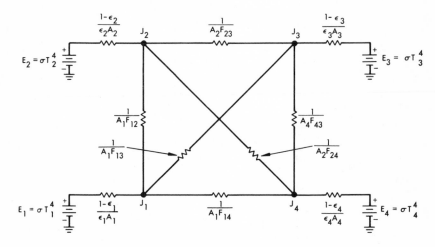

Fig. 7.38. Network for enclosures consisting of four radiating, opaque gray surfaces.

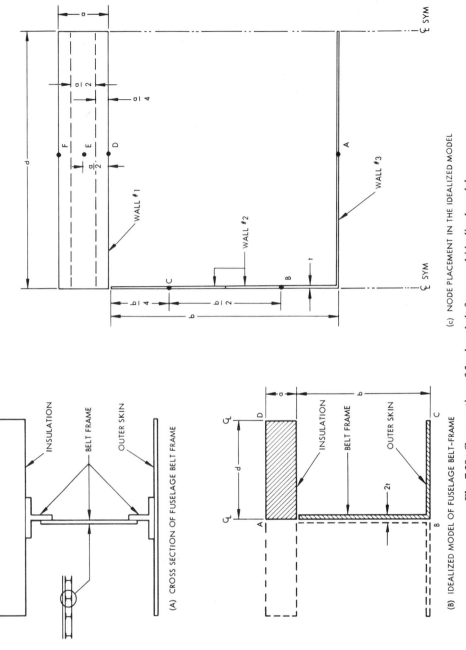

(A) CROSS SECTION OF FUSELAGE BELT FRAME

INSULATION

BELT FRAME

OUTER SKIN

(B) IDEALIZED MODEL OF FUSELAGE BELT-FRAME

INSULATION

BELT FRAME

OUTER SKIN

(c) NODE PLACEMENT IN THE IDEALIZED MODEL

WALL #1

WALL #2

WALL #3

Fig. 7.39. Cross section of fuselage belt frame and idealized model.

opaque gray surfaces. Notice that if any surface A_i is black instead of gray, the resistance $(1 - \epsilon_i)/A_i\epsilon_i$ equals zero and the potential of the floating node J_i equals σT_i^4.

The references at the end of the chapter present more information on the electrical analogy application to thermal radiation problems.

7.3.5 Thermal Analysis of a Fuselage Belt Frame: Example

This example, furnished by H. N. Tyson, Jr., of IBM Corp., illustrates the application of ECAP to an important problem in the aircraft and missile industry. Aerodynamic heating of the outer skin of an aircraft moving at high velocity often results in exceedingly high skin temperatures. The temperature may become so high that a significant quantity of heat is transferred to the interior surfaces by thermal radiation. In this example, heat transmission by thermal radiation and conduction in a typical portion of a fuselage belt frame shown in Fig. 7.39a will be investigated. Two assumptions are made concerning the thermal characteristics of the structure. First, the length (depth) of the fuselage belt frame in the direction normal to the cross section drawn in Fig. 7.39a is large compared with the other dimensions. Second, the temperature gradient in the direction normal to the cross section of the frame is negligible. This assumption allows the geometric view factors to be calculated for a structure of infinite depth. Using these assumptions, we can perform a thermal analysis of this structure for an arbitrary structure section of unit depth.

The first step in the analysis is the simplification of the structure to the idealized model of unit depth shown in Figs. 7.39b and 7.39c. In Fig. 7.39c, the structure is reduced to a set of six nodes, which will represent the frame in the analysis. The thermal and physical parameters necessary in the analysis are tabulated in Table 7.6. Assuming the depth of the structure to be 1 in., we can calculate the values for the equivalent thermal resistances and capacitances associated with the heat conduction in the structure by using the relationships presented in Sec. 7.3.1.

The next step in construction of the analog network is calculation of the equivalent resistance values for the radiation network. Before proceeding with this, the view factors F_{ij} must be calculated for the enclosure. The geometric

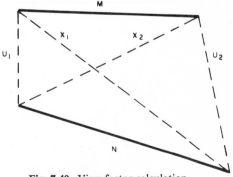

Fig. 7.40. View factor calculation.

view factors F_{MN} and F_{NM} are associated with two planar surfaces M and N (Fig. 7.40), which are exchanging radiant energy. Assuming that the dimension normal to M and N is large compared with M and N and that the temperature gradient in that direction is small, we can compute from the relationship

$$MF_{MN} = NF_{NM} = \frac{(x_1 + x_2) - (u_1 + u_2)}{2} \qquad (7.35)$$

Table 7.6. Thermal and physical parameters of belt frame.

Wall (node)	Length, in.	Thickness, in.	Density, lb/ft³	Conductivity, Btu-ft/hr-ft²-°R	Specific heat, Btu/lb-°R	Emissivity
A	5.0	—	270	6.0	0.126	0.35
B	2.25	0.0625	270	6.0	0.126	0.35
C	2.25	0.0625	270	6.0	0.126	0.35
D	5.0	1.0	1.5	0.029	0.23	0.15

where x_1 and x_2 are the lengths of line segments connecting the diagonally opposite end points of M and N. The terms u_1 and u_2 are the length of line segments connecting the directly opposite end points of M and N. In computing view factors for the surfaces in the enclosure, we must account for the radiation heat transfer between the other surfaces in the enclosure, as indicated in Fig. 7.41, which shows the geometrical relationships between the surfaces. For example, the view factor F_{AD} is defined as

$$F_{AD} = f_{AD} + f_{A'D}$$

$$= \frac{l_{1,4} + l_{2,5} - l_{1,5} - l_{2,4}}{2l_{1,2}} + \frac{l_{1,5} + l_{2',4} - l_{1,4} - l_{2',5}}{2l_{1,2'}}$$

$$F_{AD} = \frac{l_{2',4} - l_{4,2}}{2l_{1,2}} \qquad (7.36)$$

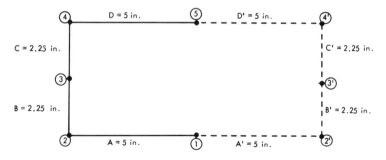

Fig. 7.41. Fuselage belt frame enclosure for view factor calculations.

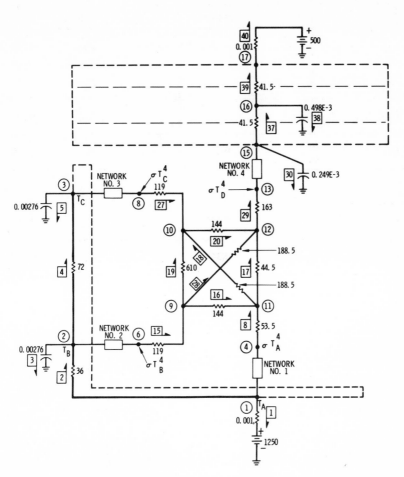

Fig. 7.42. Electric analog network for simulation of heat transfer in the fuselage belt frame.

The view factors and the equations used to compute them are listed in Table 7.7.

Using information now available, resistance values can be calculated for the radiation network used in the electrical analog network representing the thermal properties of the belt frame in Fig. 7.42. In this network, the potentials σT^4 at nodes 4, 6, 8, and 13 still have not been generated. The required network characteristic for the generation of σT^4 appears in Fig. 7.43 superimposed on the piecewise linear approximation to the characteristic. To obtain σT^4 at the nodes mentioned, the network current (ma) is transferred to a 1000 Ω resistance via a dependent current source as in Fig. 7.44. To prevent loading of the network generating σT^4 the currents CA8, CA15, CA27, and CA29 are returned to their respective nodes. The effects of the radiation network are then coupled to the radiating surfaces, nodes 2, 3, and 15, with dependent current sources T3, T7, and

Table 7.7. View factors for belt-frame analysis.

F_{ij}	Equation	Value	F_{ij}	Equation	Value
F_{AB}	$\dfrac{l_{32} + l_{22'} - l_{32'}}{2l_{21}}$	0.200	F_{DA}	F_{AD}	0.647
F_{AC}	$\dfrac{l_{43} + l_{32'} - l_{42'}}{2l_{21}}$	0.153	F_{DB}	F_{AC}	0.153
F_{AD}	$\dfrac{l_{42'} - l_{42}}{2l_{21}}$	0.647	F_{DC}	F_{AB}	0.200
F_{BA}	$F_{AB}\dfrac{l_{21}}{l_{32}}$	0.444	F_{CA}	F_{BD}	0.340
F_{BB}	$\dfrac{l_{23'} - l_{33'}}{2l_{32}}$	0.111	F_{CB}	F_{BC}	0.105
F_{BC}	$\dfrac{l_{42'} + l_{33'} - 2l_{32'}}{2l_{32}}$	0.105	F_{CC}	F_{BB}	0.111
F_{BD}	$F_{AC}\dfrac{l_{21}}{l_{32}}$	0.340	F_{CD}	F_{BA}	0.444

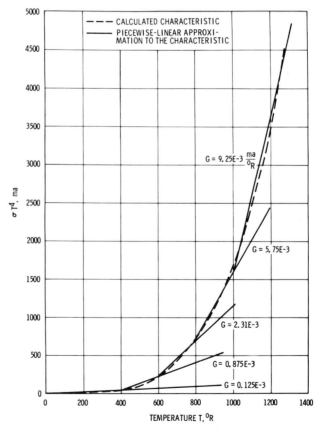

Fig. 7.43. Nonlinear characteristic for fourth-power networks in the fuselage belt frame analysis.

278

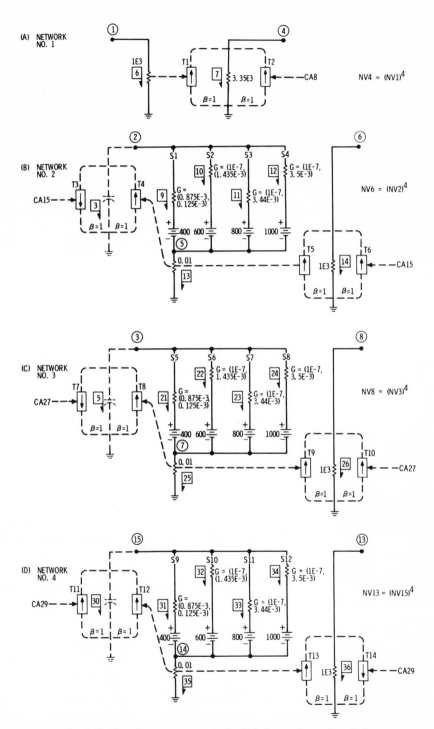

Fig. 7.44. Fourth-power networks for belt frame thermal analysis.

```
C        THERMAL TRANSIENT RESPONSE OF THE FUSELAGE
C        BELT FRAME INCLUDING RADIATION EFFECTS
         TR
  B1   N(1,0),R=0.001,E=-1250
  B2   N(1,2),R=36
  B3   N(2,0),C=0.00276,EO=-500
  B4   N(2,3),R=72
  B5   N(3,0),C=0.00276,EO=-500
  B6   N(1,0),R=1E3
  B7   N(4,0),P=3.35E3
  B8   N(4,11),P=53.5
  B9   N(2,5),G=(.875E-3,.125E-3),E=-400
 B10   N(2,5),G=(1E-7,1.435E-3),E=-600
 B11   N(2,5),G=(1E-7,3.44E-3),E=-800
 B12   N(2,5),G=(1E-7,3.5E-3),E=-1000
 B13   N(5,0),R=0.01
 B14   N(6,0),R=1E3
 B15   N(6,9),R=119
 B16   N(9,11),P=144
 B17   N(11,12),R=44.5
 B18   N(11,10),R=188.5
 B19   N(9,10),R=610
 B20   N(10,12),R=144
 B21   N(3,7),G=(.875E-3,.125E-3),E=-400
 B22   N(3,7),G=(1E-7,1.435E-3),E=-600
 B23   N(3,7),G=(1E-7,3.44E-3),E=-800
 B24   N(3,7),G=(1E-7,3.5E-3),E=-1000
 B25   N(7,0),R=0.01
 B26   N(8,0),R=1E3
 B27   N(8,10),P=119
 B28   N(9,12),R=188.5
 B29   N(12,13),R=163
 B30   N(15,0),C=.249E-3,EO=-500
 B31   N(15,14),G=(.875E-3,.125E-3),E=-400
 B32   N(15,14),G=(1E-7,1.435E-3),E=-600
 B33   N(15,14),G=(1E-7,3.44E-3),E=-800
 B34   N(15,14),G=(1E-7,3.5E-3),E=-1000
 B35   N(14,0),R=.01
 B36   N(13,0),R=1E3
 B37   N(15,16),R=41.5
 B38   N(16,0),C=.498E-3,EO=-500
 B39   N(16,17),R=41.5
 B40   N(17,0),R=.001,E=-500
  T1   B(6,7),BETA=-1
  T2   B(8,7),BETA=-1
  T3   B(15,3),BETA=1
  T4   B(13,3),BETA=-1
  T5   B(13,14),BETA=-1
  T6   B(15,14),BETA=-1
  T7   B(27,5),BETA=1
  T8   B(25,5),BETA=-1
  T9   B(25,26),BETA=-1
 T10   B(27,26),BETA=-1
 T11   B(29,30),BETA=-1
 T12   B(35,30),BETA=-1
 T13   B(35,36),BETA=-1
 T14   B(29,36),BETA=1
  S1   B=9,(9),ON
  S2   B=10,(10),OFF
  S3   B=11,(11),OFF
  S4   B=12,(12),OFF
  S5   B=21,(21),ON
  S6   B=22,(22),OFF
  S7   B=23,(23),OFF
  S8   B=24,(24),OFF
  S9   B=31,(31),ON
 S10   B=32,(32),OFF
 S11   B=33,(33),OFF
 S12   B=34,(34),OFF
       TI=1E-3
       OU=1
       FI=0.5
       1ERROR=0.1
       PRINT,NV,CA
       EX
```

Fig. 7.45. Transient thermal response input data for the fuselage belt frame analysis.

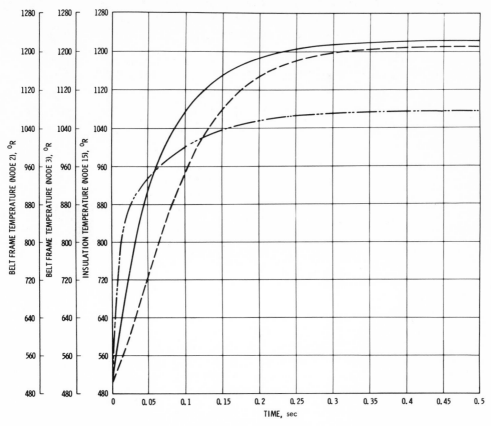

Fig. 7.46. Thermal response of the belt frame and the interior insulation of the fuselage belt frame structure.

T11, respectively. This is not done in network 1 since the node is at a fixed temperature and the radiation network has no effect on it.

Finally, the σT^4 networks are added to the belt-frame analog network in Fig. 7.42, and the analysis input data in Fig. 7.45 are prepared. The results of the analysis are plotted in Fig. 7.46.

7.4 MECHANICAL AND HYDRAULIC SYSTEMS ANALYSIS

This section explains the techniques that can be used to solve problems in a general class of physical systems which includes translational-mechanical, rotational-mechanical, and hydraulic systems. The discussion is limited to the translational-mechanical system since the other systems are analyzed using identical methods. Tables 7.4 and 7.8 relate the mechanical and hydraulic physical elements to the analogous electrical elements.

Table 7.8. Analogs. Any pair of even or pair of odd rows (1 and 3, 2 and 6, etc.) are topological analogs. Any pair of even and odd rows (1 and 2, 2 and 5, etc.) are dual analogs.*

	ROW NUMBER	ANALOGOUS VARIABLES	ANALOGOUS ELEMENTS	EXAMPLES OF ANALOGOUS NETWORKS AND EQUATIONS
FLUID FLOW (HYDRAULIC)	8	VOLUME FLOW q	(Z) I R $1/C$	$\left(R + I\frac{d}{dt} + \frac{1}{C}\int dt\right)q = P$
	7	PRESSURE p	(Y) C $1/R$ $1/I$	$\left(\frac{1}{R} + C\frac{d}{dt} + \frac{1}{I}\int dt\right)p = Q$
ROTATIONAL-MECHANICAL	6	TORQUE τ	(Z) $1/K$ $1/D$ $1/J$	$\left(\frac{1}{D} + \frac{1}{K}\frac{d}{dt} + \frac{1}{J}\int dt\right)\tau = \Omega$
	5	ANGULAR VELOCITY ω	(Y) J D K	$\left(D + J\frac{d}{dt} + K\int dt\right)\omega = T$
TRANSLATIONAL-MECHANICAL	4	FORCE f	(Z) $1/K$ $1/B$ $1/M$	$\left(\frac{1}{B} + \frac{1}{K}\frac{d}{dt} + \frac{1}{M}\int dt\right)f = V$
	3	VELOCITY v	(Y) M B K	$\left(B + M\frac{d}{dt} + K\int dt\right)v = F$
ELECTRICAL	2	CURRENT i	(Z) L R $1/C$	$\left(R + L\frac{d}{dt} + \frac{1}{C}\int dt\right)i = E$
	1	VOLTAGE e	(Y) C $G = 1/R$ $1/L$	$\left(\frac{1}{R} + C\frac{d}{dt} + \frac{1}{L}\int dt\right)e = I$

*R. S. Sanford, *Physical Networks* (Englewood Cliffs, N. J.: Prentice-Hall, Inc. ,1965), p. 124.

7.4.1   Electrical-Mechanical System Analogs

The three types of ideal, linear, two-terminal, passive elements in the majority of physical systems are: the proportional element, in which flow is directly proportional to the potential (electrical resistance); the integrating element, in which flow is proportional to the integral of the potential across it (electrical inductance); and the differentiating element, in which flow is proportional to the derivative of the potential across the element (electrical capacitance). Although most physical systems are not composed of ideal elements, nonlinearities generally can be reasonably approximated. Many nonlinear elements can be adequately modeled using the techniques described in chapters 4 and 5.

In any event, we must always bear in mind the limitations inherent in any analog approach. In this type of analysis, as well as in a circuit analysis, we are not working with the actual physical system but with equations defining a model of the system. Thus, the accuracy of the solution for the system response depends almost entirely upon the degree to which the analog model simulates the actual system.

Since there are many books which deal solely with mechanical network analysis and the analogies between mechanical and electrical networks, we will limit ourselves to a very brief summary of the relationships between the ideal network elements for the different systems. A summary of the characteristics of the ideal, linear, two-terminal network elements is presented in Table 7.4. This table also provides a good comparison of the elements in the electrical, mechanical, hydraulic, and thermal systems, which may be of great use in formulating the analog network.

Frequently, the analog network resulting from direct modeling of a physical problem is not in a desirable form. Consider the mechanical system defined by

$$\frac{d^2x}{dt^2} + \frac{dx}{dt} + x = F$$

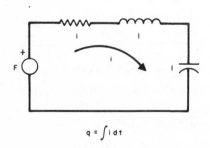

$$q = \int i \, dt$$

(a) SERIES ANALOG NETWORK

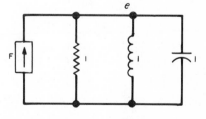

(b) PARALLEL ANALOG NETWORK

Fig. 7.47. Electrical analog networks for mechanical system defined by

$$\frac{d^2x}{dt^2} + \frac{dx}{dt} + x = F$$

This system is equivalent to the electrical network shown in Fig. 7.47a, in which $x \propto q$ in the electrical analog network. The network in Fig. 7.47b is also analogous to the same mechanical system. The second network is frequently referred to as the *dual* of the first network. In the second network $x \propto e$. Typical dual analogs from the electrical, mechanical, and hydraulic networks are illustrated in Table 7.8 to compare the configurations available. It is obviously very convenient when attempting to formulate an analog network to first write the equations of the system if possible, then select the electrical network which best fits the requirements.

In Secs. 7.4.2 and 7.4.3, analyses of two translational-mechanical networks are presented as examples. The first is the familiar problem of a weight suspended on a spring, with damping present in the form of viscous friction. The second example analyzes the motion in a simplified automobile suspension system in which the suspension spring constant is nonlinear.

7.4.2 Displacement of a Weight Suspended on a Spring: Example

Let us study the frequency-response and the transient response (displacement) of a weight suspended on a spring (Fig. 7.48) subjected to an arbitrary driving force $F(t)$.

The displacement y of the weight w in the one-dimensional translational-mechanical system is defined by the equation

$$\frac{w}{g}\frac{d^2y}{dt^2} + c_M \frac{dy}{dt} + ky = F(t) \tag{7.37}$$

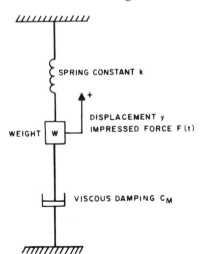

Fig. 7.48. Weight suspended on a spring with viscous damping present.

By defining the following system constants

$$w = 100 \text{ lb}$$

$$g = 32.2 \text{ ft/sec}^2$$

$$k = 300 \text{ lb/ft}$$

$$c_M = 0.5c_c = 30.52 \text{ lb-sec/ft}$$

where c_c = critical damping coefficient $2\sqrt{kw/g} = 61.04$ lb-sec/ft, the equation of motion is

$$3.105\frac{d^2y}{dt^2} + 30.52\frac{dy}{dt} + 300y = F(t) \tag{7.38}$$

The equation for an analogous electrical network is

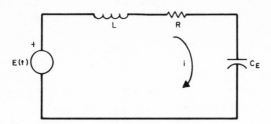

Fig. 7.49. Series analog electrical network.

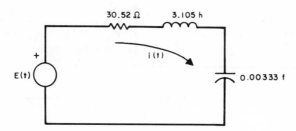

Fig. 7.50. Electric network analog to the mechanical system defined by Eq. (7.38).

$$L\frac{d^2q}{dt^2} + R\frac{dq}{dt} + \frac{q}{C_E} = E(t) \qquad (7.39)$$

From this, the corresponding electrical network is shown in Fig. 7.49, where $q = \int i\,dt$. Then the corresponding electrical network is that in Fig. 7.50.

To obtain the analog to displacement, we must find $q = \int i\,dt$. This can be easily done by using a crude RC integration network as in the ECAP equivalent circuit for the solution of the problem in Fig. 7.51.

The arbitrary driving function $E(t)$ in Fig. 7.51 is replaced by a sinusoidal

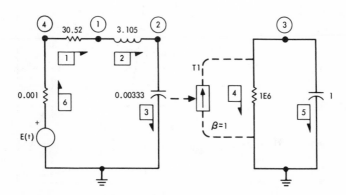

Fig. 7.51. ECAP equivalent network for AC and transient analysis of a weight suspended on a spring.

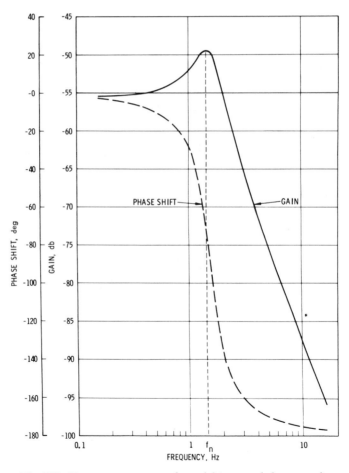

```
AC ANALYSIS OF ANALOGY PROBLEM
B1 N(4,1),R=30.52
B2 N(1,2),L=3.105
B3 N(2,0),C=0.00333
B4 N(3,0),R=1E6
B5 N(3,0),C=1
B6 N(0,4),R=.001,E=10/0
T1 B(1,4),BETA=-1
   FREQ=1
   PRINT,NV,CA
   EX
   MO
   FREQ=0.1563(1.047)15.63
   EX
   MO
B1 R=15.26
   FREQ=0.1563(1.047)15.63
   EX
   MO
B1 R=61.04
   FREQ=0.1563(1.047)15.63
   EX
```

Fig. 7.52. Frequency response analysis input data for the weight on a spring analog example.

Fig. 7.53. Frequency response of a weight suspended on a spring ($c_M = 0.25\ c_c$).

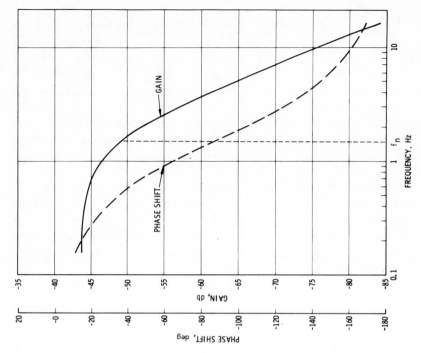

Fig. 7.55. Frequency response of a weight suspended on a spring $(c_M = 1.0c_c)$.

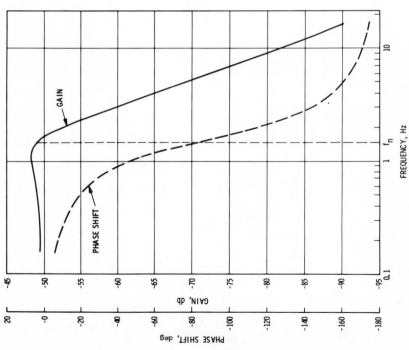

Fig. 7.54. Frequency response of a weight suspended on a spring $(c_M = 0.5c_c)$.

source in branch 6 for the AC analysis of the network. The AC analysis input data (Fig. 7.52) include requests for the calculated frequency-response of the system when the viscous friction c_M is equal to $0.25c_c$, $0.5c_c$, and $1.0c_c$. The frequency and phase response for the three test conditions are plotted in Figs. 7.53, 7.54, and 7.55, respectively.

To study the transient response of the system, an instantaneous downward (negative) force of 20 lb is applied to the weight. This is analogous to the driving voltage of -20 v in branch 6 in the transient analysis input data shown in Fig. 7.56. The analysis is conducted for damping coefficient values of $0.25c_c$, $0.5c_c$, and $1.0c_c$. Results of the transient analyses are plotted in Fig. 7.57. The steady-state displacement and the overshoot agree with the values obtained by classical techniques. The step function was used as a driving force to make the results relatively simple to verify. However, the usefulness of the ECAP analogy technique becomes much more apparent as the complexity of the system and of the driving function increases.

```
        TR ANALYSIS OF ANALOGY PROBLEM
   B1  N(4,1),R=30.52
   B2  N(1,2),L=3.105
   B3  N(2,0),C=0.00333
   B4  N(3,0),R=1E5
   B5  N(3,0),C=1
   B6  N(0,4),R=.001,E=-20
   T1  B(1,4),BETA=-1
        TI=0.02
        OU=1
        FI=2.0
        PRINT,NV,CA
        EX
        TR ANALYSIS OF ANALOGY PROBLEM
   B1  N(4,1),R=15.26
   B2  N(1,2),L=3.105
   B3  N(2,0),C=0.00333
   B4  N(3,0),R=1E5
   B5  N(3,0),C=1
   B6  N(0,4),R=0.001,E=-20
   T1  B(1,4),BETA=-1
        TI=0.02
        OU=1
        FI=2
        PRINT,NV,CA
        EX
        TR ANALYSIS OF ANALOGY PROBLEM
   B1  N(4,1),R=61.04
   B2  N(1,2),L=3.105
   B3  N(2,0),C=0.00333
   B4  N(3,0),R=1E5
   B5  N(3,0),C=1
   B6  N(0,4),R=0.001,E=-20
   T1  B(1,4),BETA=-1
        TI=0.02
        OU=1
        FI=2.0
        PRINT,NV,CA
        EX
```

Fig. 7.56. Transient analysis input data for the weight-on-spring analog example.

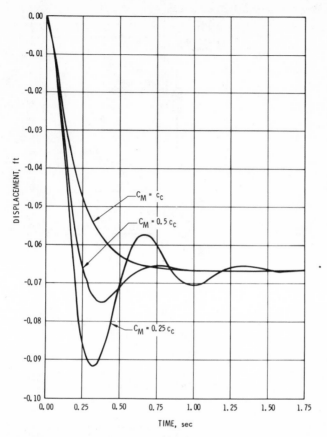

Fig. 7.57. Transient response of the weight suspended on a spring in the presence of various degrees of damping subjected to a step function driving force.

7.4.3 Simulation of a Mechanical System Containing Nonlinear Elements (Automobile Suspension System): Example

The analysis of the simplified automobile suspension system[5] shown in Fig. 7.58 is an excellent example of the application of ECAP simulation techniques to mechanical problems which contain nonlinear elements.

The system consists of wheel and body masses, a tire with an assumed linear spring rate, a linear viscous shock absorber, and a nonlinear suspension spring. The automobile moves at a constant velocity in the direction indicated in the figure. The road surface is defined by

[5]*DYANA Programming Manual I*, General Motors Research Laboratories publication GMR 229 (n. p.: General Motors Corp., 1959), pp. 64ff.

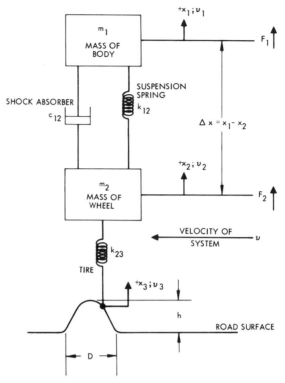

Fig. 7.58. Simplified automobile suspension system.

$$x_3 = \frac{h}{2}\left(1 - \cos\frac{2\pi}{T}t\right) \tag{7.40}$$

where $T = D/v$ for any time $0 \le t \le T$ sec. For values of $t \ge T$, x_3 is equal to 0. For this example, the velocity of the system and the road parameters are

$$v = 44 \text{ ft/sec (30 mph)}$$

$$D = 4.4 \text{ ft}$$

$$T = 0.1 \text{ sec}$$

$$h = 3 \text{ in.}$$

The remaining system parameters, with the exception of the nonlinear suspension spring constant, are

$$m_1 = 3.4 \text{ lb-sec}^2/\text{in.}$$

$$m_2 = 0.15 \text{ lb-sec}^2/\text{in.}$$

$$c_{12} = 5 \text{ lb-sec/in.}$$

$$k_{23} = 1000 \text{ lb/in.}$$

The nonlinear suspension spring constant k_{12} is defined by

$$k_{12} = \frac{A_0 P_0}{(x_1 - x_2)}\left[1.0 - \left(\frac{V_0}{V_0 + A_0(x_1 - x_2)}\right)^\gamma\right] \tag{7.41}$$

where $P_0 = 100$ lb/in.^{2} (initial pressure of the cylindrical air chamber)
$V_0 = 235$ in.^{3} (initial volume of the cylindrical air chamber)
$A_0 = 19.635$ in.^{2} (cross-sectional area of the cylindrical air chamber)
$\gamma = 1.4$ (specific heat ratio)

We will compute the vertical motion of the front end of the automobile body x_1 and the wheel displacement x_2 when the tire strikes the road bump, as defined by Eq. (7.40).

The electrical network analogous to the suspension system can be developed from the equations describing the system. The equations are

$$m_1\left(\frac{dv_1}{dt}\right) + c_{12}(v_1 - v_2) + k_{12}\int(v_1 - v_2)dt = 0 \tag{7.42}$$

$$m_2\left(\frac{dv_2}{dt}\right) - c_{12}(v_1 - v_2) - k_{12}\int(v_1 - v_2)dt + k_{23}(v_2 - v_3)dt = 0 \tag{7.43}$$

The equivalent electrical equations are

$$C_1\left(\frac{de_1}{dt}\right) + \frac{1}{R_{12}}(e_1 - e_2) + \frac{1}{L_{12}}\int(e_1 - e_2)dt = 0 \tag{7.44}$$

$$C_2\left(\frac{de_2}{dt}\right) - \frac{1}{R_{12}}(e_1 - e_2) - \frac{1}{L_{12}}\int(e_1 - e_2)dt + \frac{1}{L_{23}}(e_2 - e_3)dt = 0 \tag{7.45}$$

By comparing Eqs. (7.42) through (7.45), we see that

$$m \text{ lb-sec}^2/\text{in.} \propto C\text{ f}$$

$$c \text{ lb-sec}^2/\text{in.} \propto \frac{1}{R}\text{ }\mho$$

$$k \text{ lb/in.} \propto \frac{1}{L}\text{ h}$$

$$v \text{ in./sec} \propto e\text{ v}$$

Mechanical and analogous electrical networks derived from the equations are shown in Fig. 7.59. The simple electrical network is all that would be required for solution of the motion problem if all elements were linear. However, k_{12} must be modeled as a nonlinear element before analysis can begin. The nonlinear suspension spring constant k_{12} is plotted as a function of the spring displacement $(x_1 - x_2)$ in Fig. 7.60. The spring displacement $(x_1 - x_2)$ equals $\int(v_1 - v_2)dt$. Thus the current $i_{L_{12}}$, which is analogous to $\int(v_1 - v_2)dt$, must be a nonlinear function of $\int(e_1 - e_2)\,dt$. Notice that $x_1 = x_2 = 0$ at $t = 0$ (i.e., the steady-state value of $\Delta x = (x_1 - x_2)$ is equal to 0, which is consistent with the definition of k_{12}). Thus, a negative value for Δx implies the suspension spring is

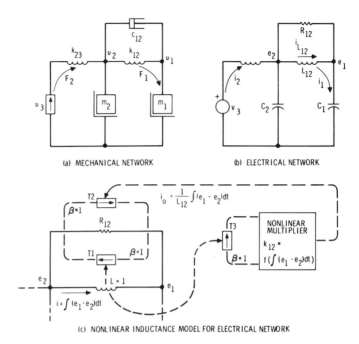

(a) MECHANICAL NETWORK   (b) ELECTRICAL NETWORK

(c) NONLINEAR INDUCTANCE MODEL FOR ELECTRICAL NETWORK

Fig. 7.59. Physical networks defining the simplified automobile suspension system.

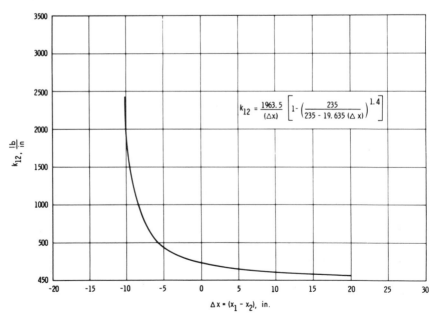

$$k_{12} = \frac{1963.5}{(\Delta x)} \left[1 - \left(\frac{235}{235 - 19.635\,(\Delta x)} \right)^{1.4} \right]$$

Fig. 7.60. Nonlinear suspension spring constant k_{12} for suspension system simulation example.

compressed. Inductance L_{12} of Fig. 7.59b can be replaced with the network in Fig. 7.59c. The current flowing in the 1 h inductance is analogous to $(x_1 - x_2)$. This current is transferred to the nonlinear multiplier network via dependent current source T3. The output of the multiplier network, which is equal to $(1/L_{12}) \int (e_1 - e_2)dt$, is returned to the analog network via T2. The current flowing in L is returned to the network via dependent current source T1 to prevent any loading of the network due to L. Hence, the net current flowing between nodes 2 and 1 is

$$i_1 = G_{12}(e_2 - e_1) + \frac{1}{L_{12}} \int (e_2 - e_1)dt$$

The nonlinear multiplier network (see Fig. 7.62) is constructed from a piecewise linear approximation of the displacement vs. force characteristic—specified by k_{12}—which is shown in Fig. 7.61. If the displacement $(x_1 - x_2)$ is analogous to the input current, the force will be analogous to the voltage in a nonlinear resistance network. A 1 Ω resistance R_{12} is placed across the multiplier network to transform the force from a voltage to an equivalent current since the output of the network must be a current whose value is equal to the value of the force. The current in R_{12} is returned to the multiplier network to prevent loading via the dependent current source T5.

The piecewise linear characteristic in Fig. 7.61 assumes that the off-state conductance of the multiplier network is equal to 0. This is not possible in a

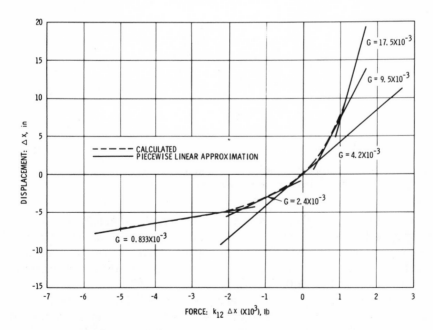

Fig. 7.61. Nonlinear displacement-force characteristic for k_{12} (automobile suspension system simulation).

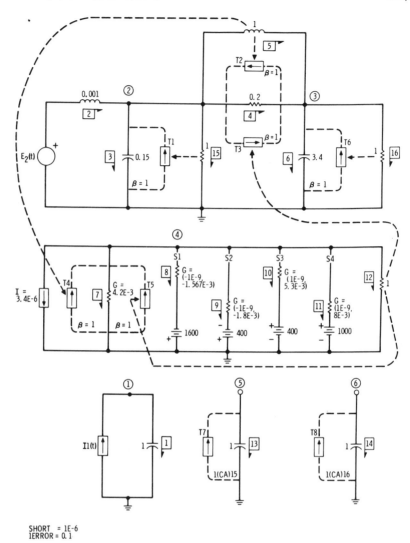

Fig. 7.62. ECAP equivalent network for response of the wheel and body of the simplified automobile suspension system.

practical problem, and the resulting error will cause the characteristic to intercept the displacement axis at a value not equal to 0. To prevent the initial condition error ($i_{L_{12}} \neq 0$ at $t = 0$), an independent current source I_7 is added to cancel the intercept error created by the bias sources in the switch branches B8 through B11 at NV4 = 0.

The complete analog network is shown in Fig. 7.62, including the network simulating the nonlinear spring constant k_{12}. One-ohm resistances are used at nodes 2 and 3 to transform the voltages (velocities) at these nodes to currents

```
C       TRANSIENT ANALYSIS OF NONLINEAR AUTOMOBILE
C       SUSPENSION SYSTEM USING ELECTRICAL
C       NETWORK ANALOGS FOR SOLUTION
        TRANSIENT ANALYSIS
   B1 N(1,0),C=1
   I1 (1),0,8.2,16.35,24.4,32.2,39.8,47.1,54,60.6,66.7,72.2,
     *77.2,81.6,85.4,88.6,91.1,92.8,93.9,94.2,93.9,92.8,91.1,
     *88.6,85.4,81.6,77.2,72.2,66.7,60.6,54,47.1,39.8,32.2,
     *24.4,16.35,8.2,0,-8.2,-16.35,-24.4,-32.2,-39.8,-47.1,
     *-54,-60.6,-66.7,-72.2,-77.2,-81.6,-85.4,-88.6,-91.1,
     *-92.8,-93.9,-94.2,-93.9,-92.8,-91.1,-88.6,-85.4,-81.6,
     *-77.2,-72.2,-66.7,-60.6,-54,-47.1,-39.8,-32.2,-24.4,
     *-16.35,-8.2,0
   B2 N(0,2),L=0.001
   E2 (1),0,8.2,16.35,24.4,32.2,39.8,47.1,54,60.6,66.7,72.2,
     *77.2,81.6,85.4,88.6,91.1,92.8,93.9,94.2,93.9,92.8,91.1,
     *88.6,85.4,81.6,77.2,72.2,66.7,60.6,54,47.1,39.8,32.2,
     *24.4,16.35,8.2,0,-8.2,-16.35,-24.4,-32.2,-39.8,-47.1,
     *-54,-60.6,-66.7,-72.2,-77.2,-81.6,-85.4,-88.6,-91.1,
     *-92.8,-93.9,-94.2,-93.9,-92.8,-91.1,-88.6,-85.4,-81.6,
     *-77.2,-72.2,-66.7,-60.6,-54,-47.1,-39.8,-32.2,-24.4,
     *-16.35,-8.2,0
   B3 N(2,0),C=0.15
   B4 N(2,3),R=0.2
   B5 N(2,3),L=1
   B6 N(3,0),C=3.4
   B7 N(4,0),G=4.2E-3,I=-3.40E-6
   B8 N(4,0),G=(-1E-9,-1.567E-3),E=1600
   B9 N(4,0),G=(-1E-9,-1.8E-3),E=400
  B10 N(4,0),G=(1E-9,5.3E-3),E=-400
  B11 N(4,0),G=(1E-9,8E-3),E=-1000
  B12 N(4,0),R=1
  B13 N(5,0),C=1
  B14 N(6,0),C=1
  B15 N(2,0),R=1
  B16 N(3,0),R=1
   T1 B(15,3),BETA=-1
   T2 B(5,4),BETA=-1
   T3 B(12,4),BETA=-1
   T4 B(5,7),BETA=1
   T5 B(12,7),BETA=-1
   T6 B(16,6),BETA=-1
   T7 B(15,13),BETA=-1
   T8 B(16,14),BETA=-1
   S1 B=8,(8),OFF
   S2 B=9,(9),OFF
   S3 B=10,(10),OFF
   S4 B=11,(11),OFF
      1ERROR=0.1
      TI=0.0013889
      OU=1
      FI=0.6
      SHORT=1E-6
      PRINT,NV,CA
      EXECUTE
```

Fig. 7.63. Transient analysis input data for the nonlinear automobile suspension system analog example.

CA15 and CA16, respectively. These currents are integrated by the 1 f capacitances in B13 and B14 to provide the displacements x_2 and x_1, respectively.

The driving function is

$$E_2 = \frac{d(x_3)}{dt} = \frac{h\pi}{T} \sin \frac{2\pi}{T} t = 94.2 \sin 20 \, \pi t \qquad (7.46)$$

when $t \leq T$, and $E_2 = 0$

when $t > T$. To obtain a graph of the road bump, the driving function was duplicated and integrated by the network at node 1.

The time step selected for the analysis is TI $= 0.0013889$ sec (corresponding to five-degree steps in the sinusoidal shape of the road bump). The normal equivalent resistance for the capacitance is then (see Appendix B.3)

$$\frac{\Delta t}{C_1} = \frac{0.0013889}{3.4} = 0.000408 \, \Omega$$

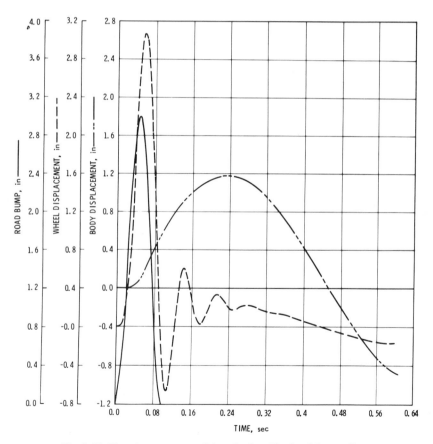

Fig. 7.64. Transient response of the wheel and body of the nonlinear automobile suspension system analog.

The value for SHORT must be reduced to a value less than $1 \times 10^{-6} \, \Omega$ to prevent errors in the results immediately following a switch actuation.

Input data for the transient analysis of the nonlinear suspension system appear in Fig. 7.63, followed by results of the analysis plotted in Fig. 7.64. Results obtained for the motion of the automobile body in Fig. 7.64 differ by approximately 5 per cent from those obtained by other solution techniques. This error can be attributed to the coarse, nonlinear, spring-constant approximation used in the analysis. The error could be significantly reduced by refining the nonlinear spring-constant approximation.

7.4.4   Simulation of Mechanical and Hydraulic Systems using ECAP Analog Computer Techniques

It is possible to solve the analysis problems described in Secs. 7.4.2 and 7.4.3 by obtaining directly a solution to the integral-differential equations which define the system. This is the basic approach used in the analog computer and in digital-simulation programs such as MIMIC, DSL-90, and CSMP. The digital-simulation programs are more accurate in their calculations and generally easier to use because they can operate on nonlinear equations and generate functions that are difficult to derive on the analog computer. Also, scaling and element degradation are not problems with the simulation programs. Analog computers, however, allow the user to adjust parameters manually during the course of the operation to study the effects on the response of the system of varying any given parameter without the time loss derived from making changes on a coding sheet and the delay in processing a digital-computer problem. The advent in the digital-computer systems of graphic terminals, which allow direct man-machine interaction during the problem solution, lessens these delay problems.

If one of the digital-simulation programs is not available, and the problem to be solved does not lie beyond the capabilities of ECAP, then it may be preferable to use ECAP to solve the problem rather than an analog computer. Notice that replacement of the other tools by ECAP is not advocated, but ECAP's use as a supplement is suggested when it is more convenient or advantageous to do so. With these thoughts in mind, the use of ECAP in the solution of integral-differential equations will now be demonstrated.

In Sec. 7.2.4, networks were developed which make it possible to model transfer characteristics requiring integration. The motor-and-load transient analysis equivalent network for the position control system in Table 7.3 (Sec. 7.2.5) uses two of these networks to generate $\ddot{C}$, $\dot{C}$, and C. The same networks can be used as integrators when using ECAP as a pseudo-analog computer. Basic function blocks that will be used in the pseudo-analog computer are shown in Table 7.9. Using the basic function blocks or modifications of the basic

Table 7.9. ECAP pseudo-analog computer elements.

blocks, we can solve many problems involving integral-differential equations with constant or variable coefficients. As examples of the application of these techniques, the problems in Secs. 7.4.2 and 7.4.3 will be solved using the pseudo-analog computer techniques discussed in Sec. 7.4.5.

7.4.5 Application of ECAP Pseudo-Analog Computer Techniques to Mechanical and Hydraulic Systems: Examples

WEIGHT SUSPENDED ON A SPRING

In this example, equations defining the mechanical system in Sec. 7.4.2 will be solved for a damping coefficient $c_M = 0.5\, c_c$.

The equation defining the system is

$$3.105 \frac{d^2y}{dt^2} + 30.52 \frac{dy}{dt} + 300y = 20u(t) \tag{7.47}$$

Solving for d^2y/dt^2,

$$\frac{d^2y}{dt^2} = -9.82901 \frac{dy}{dt} - 96.6151y + 6.4412u(t) \tag{7.48}$$

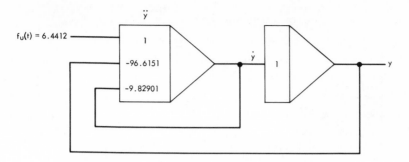

Fig. 7.65. Analog computer block diagram for the solution of

$$3.105 \frac{d^2y}{dt^2} + 30.52 \frac{dy}{dt} + 300y = 20u(t)$$

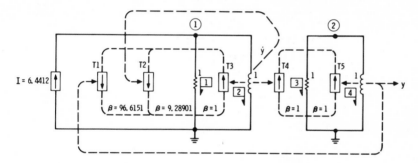

Fig. 7.66. ECAP equivalent network for the solution of

$$3.105 \frac{d^2y}{dt^2} + 30.52 \frac{dy}{dt} + 300y = 20u(t)$$

The analog block diagram representing this system is shown in Fig. 7.65. The corresponding ECAP equivalent network appears in Fig. 7.66. The displacement y in feet is equal in value to the element current in branch 4 (CA4) in amperes. NV1 and NV2 correspond to d^2y/dt^2 and dy/dt, respectively. Input data for the transient solution of the problem are listed in Fig. 7.67. The resulting calculated displacement as a function of time is identical to that obtained in Sec. 7.4.2 and plotted in Fig. 7.57.

NONLINEAR AUTOMOBILE SUSPENSION SYSTEM

The equations defining the automobile suspension system are

$$3.4 \frac{d^2x_1}{dt^2} + 5 \frac{d(x_1 - x_2)}{dt} + k_{12}(x_1 - x_2) = 0 \qquad (7.49)$$

and

$$0.15 \frac{d^2x_2}{dt^2} + 5 \frac{d(x_2 - x_1)}{dt} + k_{12}(x_2 - x_1) + 1000(x_2 - x_3) = 0 \qquad (7.50)$$

```
C        TRANSIENT ANALYSIS OF THE MECHANICAL ANALOGY
C        PROBLEM USING THE INDUCTOR INTEGRATOR MODEL
C        EQUATION
C             3.105Y''+30.52Y'+300Y=EU(T)=20U(T)
         TR
   B1  N(1,0),R=1,I=6.4412
   B2  N(1,0),L=1
   B3  N(2,0),R=1
   B4  N(2,0),L=1
   B5  N(3,0),R=1E4
   B6  N(3,0),R=1
   T1  B(4,1),BETA=96.6151
   T2  B(2,1),BETA=9.82901
   T3  B(2,1),BETA=-1
   T4  B(2,3),BETA=-1
   T5  B(4,3),BETA=-1
   T6  B(4,5),BETA=-1
       TI=.001
       OU=5
       FI=2.5
       PRINT,NV,CA
       EX
```

Fig. 7.67. Transient analysis input data for the solution of

$$3.105\frac{d^2y}{dt^2} + 30.52\frac{dy}{dt} + 300y = 20u(t)$$

using ECAP pseudo-analog computer techniques.

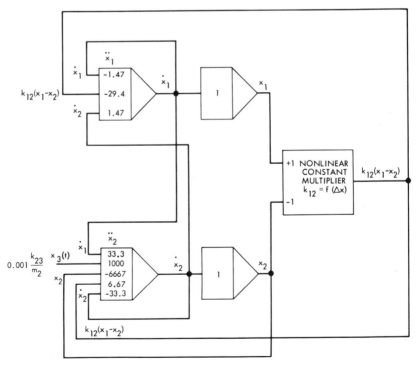

Fig. 7.68. Analog computer block diagram for solution of the differential equations defining the displacement of the wheel and body of the nonlinear automobile suspension system.

where
$$k_{12} = \frac{1963.5}{(x_1 - x_2)}\left[1.0 - \left(\frac{235}{235 + 19.635(x_1 - x_2)}\right)^{1.4}\right] \qquad (7.51)$$

The analog block diagram representing this system is shown in Fig. 7.68. The diagram contains a function block to provide a nonlinear constant-multiplier network whose output is a function of $(x_1 - x_2)$. The nonlinear constant-multiplier can be of the same form as the nonlinear network (node 4 to ground) in Fig. 7.62 since the output current $i = k_{12}(x_1 - x_2)$ in both the electrical analogy solution and the ECAP analog-computer simulation network. Details of the nonlinear network development are discussed in Sec. 7.4.3.

In this analog-simulation network (shown in Fig. 7.69), care must be exercised in selecting the BETA values used since each input, with the exception of those associated with the output of the nonlinear constant-multiplier, has an

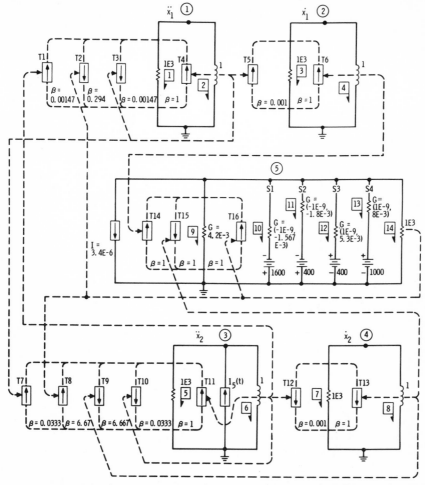

Fig. 7.69. ECAP equivalent network for solution of the displacement of wheel and body of the nonlinear automobile suspension system.

```
C      ECAP ANALOG COMPUTER SIMULATION TO DETERMINE
C      THE TRANSIENT RESPONSE OF A
C      SIMPLIFIED AUTOMOBILE SUSPENSION SYSTEM
C      CONTAINING A NONLINEAR SPRING
       TR
   B1  N(1,0),R=1E3
   B2  N(1,0),L=1
   B3  N(2,0),R=1E3
   B4  N(2,0),L=1
   B5  N(3,0),R=1E3
   I5  (1),0,.038,.152,.341,.603,.937,1.340,1.808,2.340,2.929,3.572,
      *4.264,5.0,5.774,6.58,7.412,8.264,9.1284,10.0,10.87,11.74,12.59,
      *13.42,14.23,15.0,15.74,16.43,17.07,17.66,18.19,18.66,19.06,19.40,
      *19.66,19.85,19.96,20.0,19.96,19.85,19.66,19.40,19.06,18.66,18.19,
      *17.66,17.07,16.43,15.74,15.0,14.23,13.42,12.59,11.74,10.87,10.0,
      *    9.1284,8.264,7.412,6.58,5.774,5.0,4.264,3.572,2.929,2.340,
      *1.808,1.340,.937,.603,.341,.152,.038,0
   B6  N(3,0),L=1
   B7  N(4,0),R=1E3
   B8  N(4,0),L=1
   B9  N(5,0),G=4.2E-3,I=-3.4E-6
  B10  N(5,0),G=(-1E-9,-1.567E-3),E=1600
  B11  N(5,0),G=(-1E-9,-1.8E-3),E=400
  B12  N(5,0),G=(1E-9,5.3E-3),E=-400
  B13  N(5,0),G=(1E-9,8E-3),E=-1000
  B14  N(5,0),R=1E3
   T1  B(6,1),BETA=-0.00147
   T2  B(14,1),BETA=0.294
   T3  B(2,1),BETA=0.00147
   T4  B(2,1),BETA=-1
   T5  B(2,3),BETA=-0.001
   T6  B(4,3),BETA=-1
   T7  B(2,5),BETA=-0.0333
   T8  B(14,5),BETA=-6.667
   T9  B(8,5),BETA=+6.667
  T10  B(6,5),BETA=0.0333
  T11  B(6,5),BETA=-1
  T12  B(6,7),BETA=-0.001
  T13  B(8,7),BETA=-1
  T14  B(4,9),BETA=-1
  T15  B(8,9),BETA=1
  T16  B(14,9),BETA=-1
   S1  B=10,(10),OFF
   S2  B=11,(11),OFF
   S3  B=12,(12),OFF
   S4  B=13,(13),OFF
       TI=0.001388
       OU=1
       FI=0.6
       1ERROR=0.1
       PRINT,NV,CA
       EXECUTE
```

Fig. 7.70. Transient analysis input data for solution of the displacement of wheel and body of the nonlinear automobile suspension system.

automatic gain of 1000 due to the 1000 Ω resistance at the input of each integration network (i.e., $E_{iy} = R_{iy}i_{ox} = 1000i_{ox}$). The network construction is quite straightforward with the exception of this detail.

The displacements $x_1(t)$ and $x_2(t)$ in inches are numerically equal to element currents CA4 and CA8, respectively, in Fig. 7.69. Input data for the transient response analysis derived from the analog-simulation network are listed in Fig. 7.70. The displacements calculated are (as in Sec. 7.4.3) within 5 per cent of the values obtained using the digital-simulation programs.

7.5 SUMMARY

In this chapter we have illustrated ECAP solutions to physical problems which are not directly related to electronic circuits but which are indirectly related through the similarity of their equations to the electrical network equations. This material was included to illustrate the enlarged capabilities of the program to the solution of physical problems of a nonelectrical nature.

The topics discussed were control system analysis, heat transfer, mechanical and hydraulic systems, and the use of ECAP as a pseudo-analog computer in the solution of certain integral-differential equations.

For problems of the type presented in this chapter, the use of one of the digital-simulation programs might be more effective. However, effective methods of analyzing physical systems by using ECAP techniques have been developed and are valuable. For additional information on the topics discussed and on related areas, see the reference material listed at the end of this chapter.

PROBLEMS

7.1. Construct an equivalent circuit for a complex network whose transfer function is

$$\frac{E_o(s)}{E_i(s)} = G(s) = \frac{100(s + 5)(s + 20)}{s(s + 1)(s + 10)(s^2 + 30s + 100)}$$

The input resistance equals 1000 Ω and is independent of frequency. Determine the frequency-response of the equivalent circuit to verify the model.

7.2. Construct an equivalent circuit for a complex network whose transfer function is

$$\frac{E_o(s)}{E_i(s)} = G(s) = \frac{100(s + 2 \times 10^3)(s^2 + 400s + 2 \times 10^4)}{(s + 10^3)(s + 10^5)(s^2 + 500s + 2 \times 10^4)}$$

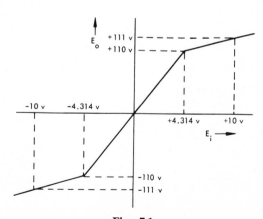

Fig. p7.1

303

and the input impedance is

$$\frac{E_i(s)}{I_i(s)} = Z_i(s) = \frac{10^6}{s + 2 \times 10^4}\ \Omega$$

Verify the equivalent circuit using the AC analysis program.

7.3. Derive a network to simulate the nonlinear amplifier characteristic shown in Fig. p7.1.

7.4. Determine the frequency-response (gain and phase) of the functions C/R and M_1/E for the feedback control system in Fig. p7.2.[6]

[6]Abraham, R. G., and A. F. Harsch, "Digital Computers in Control-System Design," *Electro-Technology*, 77, No. 6 (June 6, 1966), 77.

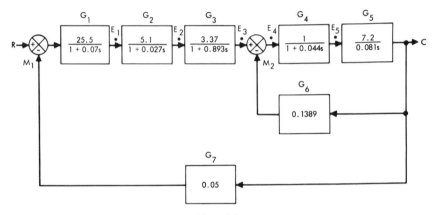

Fig. p7.2

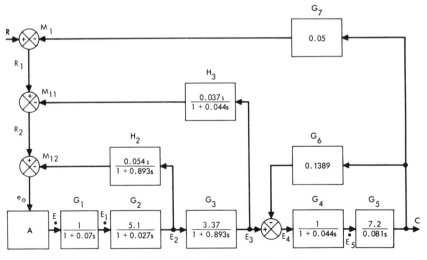

AMPLIFIER BLOCK A IS SHOWN IN FIG. p7.1

Fig. p7.3

7.5. Determine the transient response of the output C to a step change in R from 0 v to $+88$ v for the feedback control system in Fig. p7.2. Explain the results obtained using the information calculated in prob. 7.4.

7.6. Determine the frequency-response (gain and phase) for the functions C/R, M_1/R_1, E_3/R_1, and E_2/R_2 for the feedback control system in Fig. p7.3.[7] The transfer function $G(s) = E(s)/e_o(s) = 25.5$ for the amplifier block A.

7.7. Determine the transient response over a 1.0 sec period of the output C and the error signal e_o to a step change in R from 0 v to $+88$ v for the feedback control system in Fig. p7.3 assuming the amplifier block A is as described in prob. 7.6.

7.8. Determine the transient response over a 1.0 sec period of the output C and the error signal e_o to a step change in R from 0 v to $+88$ v for the nonlinear feedback control system in Fig. p7.3. The characteristic of the amplifier block A is shown in Fig. p7.1. Compare the results obtained for C and e_o in this problem with those obtained in prob. 7.7 and describe the advantages and disadvantages of the nonlinear amplifier A.

7.9. A 10 w transmitter mounted inside a spacecraft to a metal disk on the spacecraft surface is to be tested in a large vacuum chamber which is maintained at a constant temperature of 0°C. The surface of the metal disk is covered with a dull finish which has an emissivity of 0.85. The configuration, thermal parameters, and dimensions of the transmitter and disk are shown in Fig. p7.4.

[7]Abraham and Harsch, "Digital Computers in Control-System Design," p. 79.

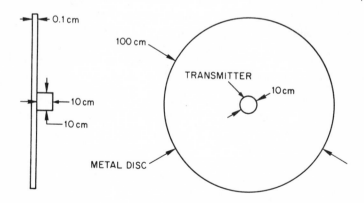

METAL DISC	TRANSMITTER
$k = 2.03$ w/cm.°C	$k = 2.1$ w/cm.°C
$c = 0.215$ cal/g.°C	$c = 0.3$ cal/g.°C
$\rho = 2.7$ g/cm^3	$\rho = 2.1$ g/cm^3

Fig. p7.4

The edge of the metal disk is insulated from the spacecraft. The geometric view factor between the disk and the chamber wall can be assumed equal to unity.

Determine the temperature of the transmitter and the temperature gradient of the metal disk as a function of time assuming the transmitter power is applied at $t = 0$ sec.

7.10. A railroad car coasting at 2 m/sec has a mass of 10,000 kg and a viscous friction of 10 newtons/m/sec. It couples with a similar car at $t = 0$ through a spring of force constant $k = 100,000$ newtons/m. Find the mechanical diagram and the analogous electrical network that can be used to find the motion of the cars after impact.

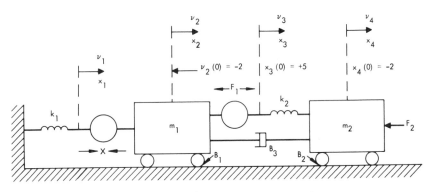

Fig. p7.5

7.11. Find the mechanical and electrical network diagrams for the system in Fig. p7.5. References are taken at zero spring displacement, and initial conditions are as shown. (If not shown, they are to be assumed zero.) The arrows on the sources indicate F_1 is initially an expansion force tending to move nodes 2 and 3 apart and X is a "tension" displacement source tending to pull nodes 1 and 2 together. Force F_2 tends to move m_2 to the left and is pushing against "ground."

7.12. In Fig. p7.6 mass m_2 is stretched to the right 1.2 m from its rest posi-

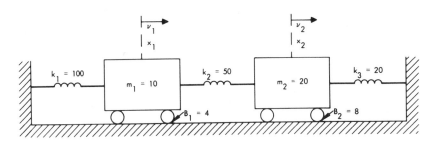

Fig. p7.6

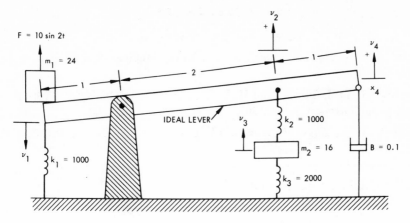

F = 10 sin 2t

m_1 = 24

v_2

v_4

IDEAL LEVER

v_3

k_2 = 1000

x_4

v_1

k_1 = 1000

m_2 = 16

B = 0.1

k_3 = 2000

Fig. p7.7

tion, and then, after steady-state conditions have been reached, it is released. Draw the network showing initial conditions and determine the displacements x_1 and x_2 as a function of time over a 5 sec period.

7.13. Find the equivalent mechanical and electrical networks for the lever system in Fig. p7.7. Assume that the lever is massless with no friction at the pivot and that the angular displacement is small. Determine the displacement x_4 for ω values from 2 to 20 using an adequate number of frequency steps to describe a smooth curve.

7.14. Solve the equation $(dy/dt) + y = u(t)$ assuming $y(0) = 0$ and using an ECAP integration network containing an inductance. Compare the ECAP output results with published tables of e^{-t} to determine the accuracy of the integration network.

7.15. Solve the equation $(dy/dt) + y = u(t)$ assuming $y(0) = 0$ using an ECAP integration network containing a capacitance. Compare the ECAP output results with published tables of e^{-t} to determine the accuracy of the integration network. Also compare the calculated $y(t)$ for this problem with the results of prob. 7.14 to determine which network yields the greater accuracy.

7.16. Find the solution to the equation $d^2y/dt^2 - 4(dy/dt) + 4y = 0$ for which $y = 3$ and $dy/dt = 4$ at $t = 0$ for $0 \le t \le 2$.

7.17. Repeat prob. 7.12 using pseudo-analog computer techniques.

7.18. Using the techniques described in this chapter derive a single network to generate $I_1(t) = 10 \sin (2\pi \times 10^3)t$ and $I_2(t) = 10 \cos (2\pi \times 10^3)t$. Verify. (*Hint:* $(d^2y/dt^2) + \omega^2 y = 0$.)

7.19. Derive a network to generate a 1000 Hz sine wave modulated by a 60 Hz sine wave. The amplitude of the 1000 Hz signal should be 10 v p-p. The modulating signal amplitude should be 2 v p-p.

REFERENCES

GENERAL PHYSICAL NETWORK ANALYSIS

Karplus, W. J., *Analog Simulation*. New York: McGraw-Hill Book Company, 1958.

Murphy, G., D. J. Shippy, and H. L. Luo, *Engineering Analogies*. Ames, Iowa: Iowa State University Press, 1963.

Sanford, R. S., *Physical Networks*. Englewood Cliffs, N. J.: Prentice-Hall, Inc., 1965.

Thorn, D. C., *An Introduction to Generalized Circuits*. Belmont, Calif.: Wadsworth Publishing Co., Inc., 1965.

CONTROL-SYSTEM ANALYSIS

Chestnut, H., and R. W. Mayer, *Servomechanisms and Regulating System Design*. New York: John Wiley & Sons, Inc., 1951.

Jensen, R. W., "ECAP and Extensions," *Computer Aids for Large Circuit Arrays* (*Conference Proceedings*, University of Wisconsin, Madison, Wisc., September 27, 1967).

Jensen, R. W., "Control System Analysis with a Circuit Analysis Program," *Electro-Technology* (November 1967).

Kuo, B. C., *Automatic Control Systems*. Englewood Cliffs, N. J.: Prentice-Hall, Inc., 1962.

Murphy, G. J., *Basic Automatic Control Theory*. Princeton, N. J.: D. Van Nostrand Co., Inc., 1957.

HEAT-TRANSFER ANALYSIS

Hsu, S. T., *Engineering Heat Transfer*. Princeton, N. J.: D. Van Nostrand Co., Inc., 1957.

Jakob, M., *Heat Transfer* (2 vols.). New York: John Wiley & Sons, Inc., 1949, 1957.

Jakob, M., and G. A. Hawkins, *Elements of Heat Transfer*. New York: John Wiley & Sons, Inc., 1957.

McAdams, W. H., *Heat Transmission* (3rd ed.). New York: McGraw-Hill Book Company, 1954.

Oppenheim, A. K., "Radiation Analysis by the Network Method," *American Society of Mechanical Engineers Paper* 54-A-75, December 1954.

Oppenheim, A. K., "Radiation Analysis by the Network Method," *Transactions of the American Society of Mechanical Engineers* 78, No. 725 (1956).

STRUCTURE ANALYSIS

MacNeal, R. H., *Electric Circuit Analogies for Elastic Structures*. New York: John Wiley & Sons, Inc., 1962.

ANALOG COMPUTERS

Johnson, C. L., *Analog Computer Techniques*. New York: McGraw-Hill Book Company, 1956.

Karplus, W. J., *Analog Simulation*. New York: McGraw-Hill Book Company, 1958.

Karplus, W. J., and W. W. Soroka, *Analog Methods* (2nd ed.). New York: McGraw-Hill Book Company, 1959.

Levine, L., *Methods for Solving Engineering Problems Using Analog Computers*. New York: McGraw-Hill Book Company, 1964.

APPENDIX A

TRANSISTOR
EQUIVALENT
CIRCUITS

A.1. SMALL-SIGNAL TRANSISTOR EQUIVALENT CIRCUITS

The hybrid-π transistor equivalent circuit is probably the most useful of the small-signal transistor models available to the ECAP user. It is valid from very low frequencies to the cutoff frequency of the transistor. The model's low frequency limit can be extended to DC with the addition of an independent voltage source to simulate the base-emitter voltage of the transistor. The parameters of the model are not difficult to measure and are usually derivable from the transistor data sheet. A very important feature of the hybrid-π model is that the elements used are standard ECAP elements. The hybrid-π equivalent circuit is shown in Fig. A.1.

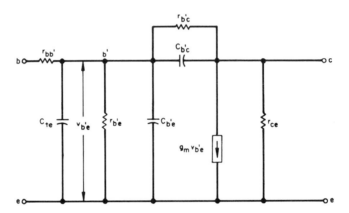

Fig. A.1. Hybrid-π transistor model.

The *base spreading resistance* r'_{bb} is the resistance between the base lead and the base-emitter junction. The value of the resistance is dependent upon base resistivity and device geometry, and is generally between 5 and 300 Ω. This resistance is relatively constant (± 10 per cent) for any specific transistor type and is not very sensitive to the operating point.

Two elements $r_{b'e}$ and $C_{b'e}$ are functions of the emitter resistance r_e, which in turn is related to the emitter current as

$$r_e = \frac{kT}{qI_e} \tag{A.1}$$

where k = Boltzmann's constant = 1.38×10^{-23} joules/°K

T = absolute temperature in °K

q = electron charge = 1.6×10^{-19} coulombs

This equation simplifies to $r_e \simeq 26/I_e$ in ohms, where I_e is in milliamperes at +25°C. The *base-emitter resistance* $r_{b'e}$ can be expressed as

$$r_{b'e} = r_e(1 + h_{fe}) \simeq r_e h_{fe} \tag{A.2}$$

where h_{fe} is the low frequency, common-emitter current gain.

The *emitter junction capacitance* $C_{b'e}$ accounts for the decrease in gain and the increase in phase shift with frequency in a transistor. In terms of the gain-bandwidth product f_T, which is usually specified by the manufacturer,

$$C_{b'e} = \frac{1}{2\pi f_T r_e} \qquad \text{(A.3)}$$

The value of f is a function of the operating point of the transistor as indicated in Fig. A.2 for a typical transistor.

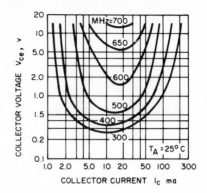

Fig. A.2 Contours of constant gain bandwidth product (f_T).

The *emitter transition capacitance* C_{te}, also called the depletion layer capacitance, can be neglected if the emitter current is large enough that $C_{b'e} \gg C_{te}$.

The *feedback resistance* $r_{b'c}$ is not important in the model at higher frequencies, where

$$r_{b'c} \gg \frac{1}{\omega C_{b'c}} \qquad \text{(A.4)}$$

At lower frequencies, ignoring $r_{b'c}$ is analogous to setting the h parameter $h_r = 0$ (i.e., ignoring the feedback term). The term $r_{b'c}$ can be expressed as

$$r_{b'c} = \frac{r_e(1 + h_{fe})}{h_{re}} = \frac{r_{b'e}}{h_{re}} \qquad \text{(A.5)}$$

The reverse biased *collector-base junction capacitance* $C_{b'c}$ is usually referred to as C_{ob} on the transistor data sheets. The value of the capacitance is related to the reverse bias voltage V by

$$C_{b'c} = A\left(\frac{k}{V}\right)^n \qquad \text{(A.6)}$$

where A = a constant related to the junction area

k = a constant related to the junction properties

n = a constant depending on the junction type

Figure A.3 illustrates the effect of the junction bias voltage on C_{te} and C_{ob}.

The *common emitter output resistance* r_{ce} is defined as

$$r_{ce} = \frac{1}{h_{oe} - h_{re}/r_e} \simeq \frac{1}{h_{oe}} \qquad \text{(A.7)}$$

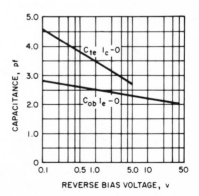

Fig. A.3. Emitter transition and output capacitance versus reverse bias voltage.

This parameter is usually very large, and thus it can very often be neglected.

The last parameter of the hybrid-π model is the *transconductance g_m*, which is

$$g_m = \frac{h_{fe}}{r_e(1 + h_{fe})} \simeq \frac{1}{r_e} \tag{A.8}$$

For use in ECAP, it is sometimes advantageous to convert the GM dependent current source to a BETA dependent current source and to let

$$i_c = h_{fe}i_{r_{b'e}} = \frac{h_{fe}v_{b'e}}{r_e(1 + h_{fe})} = g_m v_{b'e} \tag{A.9}$$

Thus, either GM or BETA sources can be used with this model.

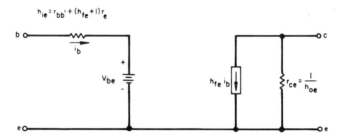

Fig. A.4. DC transistor model.

When considering a DC problem, the hybrid-π model reduces to the circuit shown in Fig. A.4 assuming $r_{b'c} \gg r_{b'e}$. The addition of the *bias voltage V_{be}* accounts for the base-emitter voltage drop.

REFERENCES

Joyce, M., and K. Clark, *Transistor Circuit Analysis*. Reading, Mass.: Addison-Wesley Publishing Co., Inc., 1961.

Mammano, R. A., "Try the Hybrid-Pi," *Electronic Design* (Sept. 13, 1966), pp. 66–68.

Phillips, A., *Transistor Engineering*. New York: McGraw-Hill Book Company, Inc., 1962.

A.2. *LARGE-SIGNAL TRANSISTOR EQUIVALENT CIRCUIT*

The charge-control transistor model is developed in several steps. First, a simplified equivalent circuit, applicable in the cutoff and active regions, is developed without parasitic effects. Second, the parasitic elements are added to define a model, which is useful if not operated near saturation. Finally, the equivalent circuit is expanded so that the model is valid in the cutoff, active, partial saturation, and saturation regions of operation.

The transistor models discussed in most of the literature are limited to applications in which the signal level is small enough that the transistor output voltage does not vary appreciably from the DC operating point. This limitation makes it possible to assume that the transistor model parameters are constant. This is not true in many applications since

$$\beta = \beta_0 f(I_c); \qquad h_{ie} \simeq r_{bb'} + \frac{(\beta + 1)26\,\Omega}{I_e(\text{ma})}; \qquad V_{be} = f(I_c, I_b); \qquad \text{etc.}$$

A transistor switch is usually assumed to operate only in a cutoff region of operation or in a fully saturated region in which the device can be specified in terms of forced β, $V_{be(\text{sat})}$, $V_{ce(\text{sat})}$, and occasionally C_{ob} and f_T.

In reality, transistor performance under large-signal conditions can be arbitrarily divided into four separate regions of operation:

1. *Cutoff Region.* The emitter-base and base-collector junctions are both reverse biased.

2. *Normal Active Region.* The emitter-base junction is forward biased, while the base-collector junction is reverse biased.

3. *Partial Saturation Region.* Both junctions of the transistor are forward biased in this region, yet the base-collector junction is still in a low-conductance state. It might be stated that the current in the base-emitter junction producing the base control charge is greater than the current producing the stored charge in the base-collector junction.

4. *Saturation Region.* Both junctions are forward biased, and the voltage across the base-collector junction is such that the current flow in the forward direction is of the same order of magnitude as the base current.

To avoid confusion from this point on, the discussion is limited to *NPN* transistors, although the analysis is also valid for *PNP* transistors providing the proper current direction changes are made.

It is useful at this point to introduce a charge distribution diagram[1]

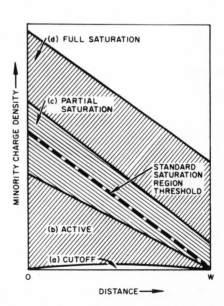

Fig. A.5. Charge distribution diagram in the base region of an *NPN* transistor.

[1]H. H. El-Sherif, "Transistor Large Signal Analysis," *IBM Technical Report, TR22–150,* November 20, 1964.

(Fig. A.5) which shows the approximate minority charge distribution in the base region for the four regions of operation. It has been shown that the total charge in the base region (which changes parabolically with distance) can be approximated as shown in the figure during both turn-on and turn-off times for the majority of cases. The only difference between this diagram and the common three-region charge diagram is the partial saturation region. The saturation region threshold, indicated as a dashed line in Fig. A.5, separates the active and saturation regions in the three-region diagram.

CUTOFF REGION

In this region an equilibrium concentration of holes and electrons in the base region exists, and the base charge Q_B can be assumed to be zero.

NORMAL ACTIVE REGION

It has been demonstrated that the base charge Q_B of electrons is

$$Q_B = \frac{W^2 I_e}{2D_n} \tag{A.10}$$

where W = base width between depletion layers, cm
  I_e = emitter current, ma
  D_n = electron diffusion constant, cm²/sec
  The basic relationship used in the development of the charge-control transistor theory is that current is defined as charge per unit time, or

$$\Delta I = \frac{\Delta Q}{\tau} \tag{A.11}$$

where τ is a constant that relates the current to the charge.
  The term $W^2/2D_n$ is defined as the emitter time constant

$$T_E = \frac{\Delta Q_B}{\Delta I_e} = \frac{W^2}{2D_n} \simeq \frac{1.22}{\omega_n} \tag{A.12}$$

where ω_n is the alpha cutoff frequency of the transistor.
  The base time constant T_B and the collector time constant T_C are derived from the emitter time constant:

$$T_B = \frac{T_E}{(1 - \alpha_n)} \simeq \frac{1.22}{\omega_n(1 - \alpha_n)} = \begin{matrix} \text{effective lifetime of holes or} \\ \text{electrons in the base region} \end{matrix} \tag{A.13}$$

$$T_C = \frac{T_B}{\beta} = \frac{1.22}{\alpha_n \omega_n} \tag{A.14}$$

The equivalent circuit of a transistor operating in the active region is shown in Fig. A.6.
  The current I_{be} is divided into two components, Q_B/T_B and $d(Q_B)/dt$. The

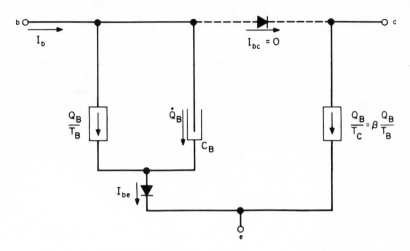

Fig. A.6. Simplified large-signal equivalent circuit of a transistor operating in the active region.

first component Q_B/T_B is the base current necessary to maintain the collector current I_c and accounts for the charge that is recombined in the base region. The second component $d(Q_B)/dt$ accounts for the base control charge built up in the base region and stored in the base storage region.

A diode, rather than an emitter resistance, is shown in the equivalent circuit since it has to represent the actual behavior of the emitter junction, in that for positive bias, the emitter current is approximately an exponential function of the emitter-base voltage, and for negative bias, the current is small and constant. The collector-emitter voltage is the difference between the base-collector and the base-emitter diode voltage drops.

The characteristic of the base-emitter diode is a function of the collector current. To differentiate between the base-emitter and base-collector diode characteristics, it is appropriate to introduce a symbol for a diode whose forward-voltage characteristic is a function of collector current. The symbol

 $\longleftarrow I_c$

will be used throughout the discussion.

In the normal active region, $I_{bc} \simeq 0$ and the base-collector diode can be omitted. The diode will be added to the transistor model when the partial saturation and saturation characteristics of the transistor are studied.

Recombination is not the only imperfection to appear in the practical transistor, as indicated in the complete large-signal active region transistor equivalent circuit of Fig. A.7. The imperfections are as follows:

    1. Collector leakage current

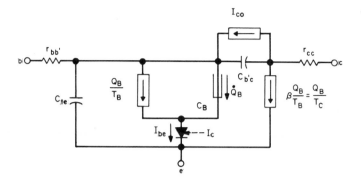

Fig. A.7. Large-signal equivalent circuit for the normal active region including parasitic elements.

2. Base resistance
3. Junction capacitance

PARTIAL SATURATION REGION

Figure A.8 shows the large-signal equivalent circuit for the partial saturation and total saturation regions.

The following equations define the current and charge relationships of the transistor (neglecting I_{co} and the junction capacitances).

$$I_{be} = \frac{Q_B}{T_B} + \dot{Q}_B + \frac{Q_S}{T_S} + \dot{Q}_S \qquad (A.15)$$

$$I_{bc} = \frac{Q_S}{T_S} + \dot{Q}_S \qquad (A.16)$$

where I_{be} = base-emitter diode current
        I_{bc} = base-collector diode current
        Q_B = base control charge

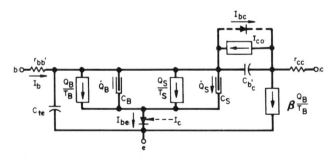

Fig. A.8. Large-signal equivalent circuit for the partial saturation and total saturation regions.

$T_B =$ base control charge time constant

$Q_S =$ base stored charge

$T_S =$ excess stored charge recombination time constant

The current I_{bc} does not contribute to the collector current but contributes to the stored charge in the base region. The collector current is related to the base control charge by the equation

$$I_c = \beta \frac{Q_B}{T_B} \tag{A.17}$$

A transistor is defined as operating in the saturation region when the base-collector diode is forward biased. The transistor is not considered to be totally saturated until the collector-emitter voltage is much less than the base-emitter voltage. (In silicon devices $V_{be(sat)}$ is usually greater than 0.8 v and $V_{ce(sat)}$ is less than 0.5 v.)

In the partial saturation region, the collector current is increasing as more base control charge Q_B is stored. As the transistor approaches total saturation, the percentage of the charge contributing to base control decreases and more of the charge Q_S is stored as excess in the base region.

An arbitrary line dividing the partial and total saturation regions might be drawn where the current in the base-emitter junction producing the base control charge is equal to the base-collector current I_{bc}. Mathematically

$$\dot{Q}_B = I_{bc} \tag{A.18}$$

Therefore

$$I_{be} - \frac{Q_B}{T_B} = 2I_{bc} \tag{A.19}$$

or

$$\dot{Q}_B = \frac{Q_S}{T_S} + \dot{Q}_S \tag{A.20}$$

TOTAL SATURATION

Both junctions are forward biased in this region and

$$\frac{Q_S}{T_S} + \dot{Q}_S > \dot{Q}_B \tag{A.21}$$

The excess current $(I_{be} - Q_B/T_B)$ contributes little to an increase of the base control charge Q_B, but contributes primarily to the buildup of storage charge Q_S.

APPENDIX B

PROGRAM
SOLUTION
TECHNIQUES

In all ECAP analysis links, the basic solution technique centers on solving the circuit nodal equations for the node voltages. The other circuit values, such as element currents and branch voltages, are then determined in accordance with their definitions in Sec. 2.3. In this appendix only those techniques and underlying assumptions that are of practical significance to the ECAP user are discussed in detail.

B.1. DC ANALYSIS

This section presents derivations of the general circuit equations and of the explicit formulas used for the partial derivative calculations and discusses the techniques and assumptions used in the worst case and standard deviation calculations.

B.1.1   Derivation of Circuit Equations[1]

The topological properties of an electrical circuit can be represented by a *graph*. A graph simply illustrates the arrangement (nodal connections) of the branches of a circuit. A DC circuit and its corresponding graph are illustrated in Figs. B.1 and B.2, respectively. The DC circuit contains six branches (B1 through

[1]*The 1620 Electronic Circuit Analysis Program (ECAP) (1620-EE-02X) User's Manual*, (White Plains, N.Y.: International Business Machines Corp., 1965), pp. 160–63.

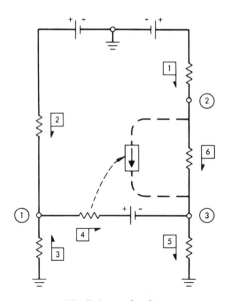

Fig. B.1. DC circuit.

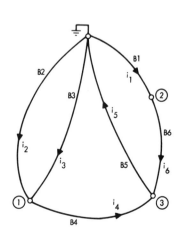

Fig. B.2. Graph.

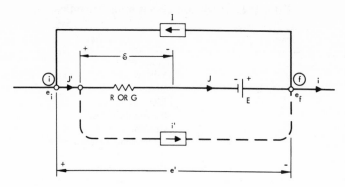

Fig. B.3. Standard circuit branch.

B6). The structure of each of these branches is consistent with the standard circuit branch, as defined in Sec. 2.3. The standard circuit branch for DC analyses is redrawn in Fig. B.3, using terminology appropriate for the developments in this section.

The minimum configuration of the standard circuit branch is just the resistance alone, although, as shown, it can also contain a voltage source, a current source, and/or one or more dependent current sources. In the DC circuit shown in Fig. B.1, branches B3 and B5 contain resistances only, whereas B1, B2, and B4 each consist of a resistance and a voltage source. Branch 6 contains a resistance and a dependent current source. The standard circuit branch symbols (Fig. B.3) are defined in Table B.1.

The topology of a graph can be conveniently represented by a matrix containing only the elements $+1$, -1, and 0. For instance, the matrix corresponding to the graph of Fig. B.2 is

$$\mathbf{A} = \begin{bmatrix} 0 & -1 & 0 \\ -1 & 0 & 0 \\ -1 & 0 & 0 \\ 1 & 0 & -1 \\ 0 & 0 & 1 \\ 0 & 1 & -1 \end{bmatrix} \tag{B.1}$$

where the value of each element a_{ij} of the matrix is determined by the rules given in Table B.2. The subscripts i and j of each term a_{ij} are the corresponding branch number and node number, respectively. Consequently, the matrix $\mathbf{A}$ generally has the dimensions $m \times n$, where m is the number of branches in the graph and n is the number of nodes excluding the ground node. Node 0 (the ground node) is excluded since the corresponding column contains only redundant information.

Table B.1. Standard circuit branch notation.

Symbol	Definition
e	Node voltage
E	Voltage source
e'	Branch voltage
$\mathscr{E}$	Element voltage
J	Resistance current
I	Current source
i'	Dependent current source
i	Branch current
J'	Element current

Table B.2. Connectivity values of a_{ij}.

a_{ij}	Connectivity: branch i, node j
$+1$	Node j is the initial node of branch i
-1	Node j is the final node of branch i
0	Branch i is not connected to node j

The application of Kirchhoff's current law, which states that the algebraic sum of the currents flowing into a node from all the branches connected to it is zero, to the graph of Fig. B.2 yields the following expressions

$$-i_2 \ -i_3 \ +i_4 \qquad\qquad = 0$$
$$-i_1 \qquad\qquad\qquad +i_6 = 0$$
$$-i_4 \ +i_5 \ -i_6 = 0$$

In matrix form this is

$$\begin{bmatrix} 0 & -1 & -1 & +1 & 0 & 0 \\ -1 & 0 & 0 & 0 & 0 & +1 \\ 0 & 0 & 0 & -1 & +1 & -1 \end{bmatrix} \begin{bmatrix} i_1 \\ i_2 \\ i_3 \\ i_4 \\ i_5 \\ i_6 \end{bmatrix} = 0$$

or, in general

$$\mathbf{A}_t i = 0 \tag{B.2}$$

where $\mathbf{A}_t$ is the transpose of the matrix $\mathbf{A}$ and i is the branch current vector.

Next, let e_1, e_2, and e_3 represent node voltages at nodes 1, 2, and 3, respectively. Then, the branch voltages can be written as

$$e'_1 = \quad\quad - e_2$$

$$e'_2 = - e_1$$

$$e'_3 = - e_1$$

$$e'_4 = \quad e_1 \quad\quad\quad - e_3$$

$$e'_5 = \quad\quad\quad\quad\quad e_3$$

$$e'_6 = \quad\quad\quad\quad e_2 \quad - e_3$$

Again, in matrix form this is

$$
\begin{bmatrix} e'_1 \\ e'_2 \\ e'_3 \\ e'_4 \\ e'_5 \\ e'_6 \end{bmatrix} =
\begin{bmatrix} 0 & -1 & 0 \\ -1 & 0 & 0 \\ -1 & 0 & 0 \\ 1 & 0 & -1 \\ 0 & 0 & 1 \\ 0 & 1 & -1 \end{bmatrix}
\begin{bmatrix} e_1 \\ e_2 \\ e_3 \end{bmatrix}
$$

or

$$e' = Ae \tag{B.3}$$

where e' is the branch voltage vector and e is the node voltage vector.

From the standard circuit branch (Fig. B.3 and Table B.1), the element voltage is related to the branch voltage vector by the expression

$$\mathscr{E} = E + e' \tag{B.4}$$

The branch current is

$$i = J' - I \tag{B.5}$$

Each of the terms in Eqs. (B.4) and (B.5) are vectors of order m, where m is the number of branches. Each element of the vector J' is related to the voltage vector $\mathscr{E}$ by the expression

$$J'_i = \sum_{j=1}^{m} Y_{ij}\mathscr{E}_j$$

where Y_{ii} is the conductance of the ith branch, and Y_{ij} is the transconductance (if any) from the jth to the ith branch and in general does not equal Y_{ji}. Consequently, the vector J' can be expressed in terms of a matrix Y and the vector $\mathscr{E}$ by the equation

$$J' = Y\mathscr{E} \tag{B.6}$$

This is a generalized statement of Ohm's law.

To obtain a solution for e, Eqs. (B.2), (B.4), (B.5), and (B.6) can be manipulated to show that

$$A_t Y e' = A_t (I - YE) \tag{B.7}$$

Finally, from Eqs. (B.3) and (B.7),

$$(\mathbf{A}_t\mathbf{YA})e = \mathbf{A}_t(\mathbf{I} - \mathbf{YE})\tag{B.8}$$

and the solution for e is

$$e = (\mathbf{A}_t\mathbf{YA})^{-1}\mathbf{A}_t(\mathbf{I} - \mathbf{YE})\tag{B.9}$$

The principal terms in Eqs. (B.8) and (B.9) are identified in Table B.3. The evaluation of Eq. (B.9) for a DC circuit yields the nominal solution (if the matrix $\mathbf{Y}$ and the vectors $\mathbf{I}$ and $\mathbf{E}$ reflect the nominal values of the parameters).

Table B.3.   Nodal equation terminology.

Term	Identification
e	Node voltage vector
$\mathbf{A}_t\mathbf{YA}$	Nodal conductance matrix
$\mathbf{A}_t(\mathbf{I} - \mathbf{YE})$	Equivalent current vector
$(\mathbf{A}_t\mathbf{YA})^{-1}$	Nodal impedance matrix

B.1.2   Calculation of the Partial Derivatives[2]

Exact formulas are used in the ECAP DC analysis program for the calculation of the partial derivatives of the node voltages e with respect to each of the circuit parameters. The development of these formulas follows.

The derivative of the node voltages e, as presented in Eq. (B.9), with respect to the resistance in the ith branch is

$$\frac{\partial e}{\partial R_i} = \frac{\partial(\mathbf{A}_t\mathbf{YA})^{-1}}{\partial R_i}\mathbf{A}_t(\mathbf{I} - \mathbf{YE}) + (\mathbf{A}_t\mathbf{YA})^{-1}\mathbf{A}_t\frac{\partial(\mathbf{I} - \mathbf{YE})}{\partial R_i}$$

Realizing that $\dfrac{\partial(\mathbf{A}_t\mathbf{YA})^{-1}}{\partial R_i} = -(\mathbf{A}_t\mathbf{YA})^{-1}\mathbf{A}_t\dfrac{\partial\mathbf{Y}}{\partial R_i}\mathbf{A}(\mathbf{A}_t\mathbf{YA})^{-1}$

and $\dfrac{\partial\mathbf{E}}{\partial R_i} = \dfrac{\partial\mathbf{I}}{\partial R_i} = 0$

and recalling that $e = (\mathbf{A}_t\mathbf{YA})^{-1}\mathbf{A}_t(\mathbf{I} - \mathbf{YE})$

and $e' = \mathbf{A}e$

we can reduce the above expression for $\partial e/\partial R_i$ to

$$\frac{\partial e}{\partial R_i} = -(\mathbf{A}_t\mathbf{YA})^{-1}\mathbf{A}_t\frac{\partial\mathbf{Y}}{\partial R_i}(\mathbf{E} + e')\tag{B.10}$$

In a similar manner, the differentiations of Eq. (B.9) with respect to the other

[2]*The 1620 Electronic Circuit Analysis Program (ECAP) (1620-EE-02X) User's Manual*, (White Plains, N.Y.: International Business Machines Corp., 1965), pp. 163–65.

circuit parameters yield

$$\frac{\partial e}{\partial \beta_{ji}} = -(\mathbf{A}_t \mathbf{Y} \mathbf{A})^{-1} \mathbf{A}_t \frac{\partial \mathbf{Y}}{\partial \beta_{ji}} (E + e') \tag{B.11}$$

$$\frac{\partial e}{\partial E_i} = -(\mathbf{A}_t \mathbf{Y} \mathbf{A})^{-1} \mathbf{A}_t \mathbf{Y} \frac{\partial E}{\partial E_i} \tag{B.12}$$

$$\frac{\partial e}{\partial I_i} = (\mathbf{A}_t \mathbf{Y} \mathbf{A})^{-1} \mathbf{A}_t \frac{\partial I}{\partial I_i} \tag{B.13}$$

Next, ϵ_i and $\bar{\epsilon}_j$ are defined, respectively, as column and row vectors of dimension m.

$$\epsilon_i = \begin{bmatrix} 0 \\ \cdot \\ \cdot \\ \cdot \\ 0 \\ 1 \\ 0 \\ \cdot \\ \cdot \\ \cdot \\ 0 \end{bmatrix} \quad ; \quad \bar{\epsilon}_j = [0 \; \ldots \; 0 \; 0 \; 1 \; \ldots \; 0]$$

The vector ϵ_i has a 1 in the ith location and zeros elsewhere. The ith location represents the ith row of a matrix. Similarly, $\bar{\epsilon}_j$ has a 1 in the jth location (representing the jth column of a matrix) and zeros elsewhere. Then, for example, the product $\epsilon_i \bar{\epsilon}_j$ is a matrix with a 1 in the ijth location and zeros elsewhere.
    With this notation

$$\frac{\partial \mathbf{Y}}{\partial R_i} = -\frac{1}{R_i^2} \epsilon_i \bar{\epsilon}_i - \sum_{j \neq i} \frac{\beta_{ji}}{R_i^2} \epsilon_j \bar{\epsilon}_i$$

or

$$\frac{\partial \mathbf{Y}}{\partial R_i} = -\frac{1}{R_i^2} \sum_{j=1}^{m} \beta_{ji} \epsilon_j \bar{\epsilon}_i, \quad \text{where} \quad \beta_{ii} = 1$$

where i is the column in the $\mathbf{Y}$ matrix corresponding to the ith branch, and j is the row corresponding to the jth branch. The term

$$\frac{\beta_{ji}}{R_i^2} \epsilon_j \bar{\epsilon}_i$$

is present only if the ith branch is the controlling (or *from*-) branch for a dependent current source assigned to the jth branch. If the ith branch is the *from*-branch for dependent current sources assigned to other branches as well, then additional similar terms must appear in the above expression. In a similar manner

$$\frac{\partial \mathbf{Y}}{\partial \beta_{ji}} = \frac{1}{R_j} \boldsymbol{\epsilon}_i \bar{\boldsymbol{\epsilon}}_j$$

$$\frac{\partial \mathbf{E}}{\partial E_i} = \boldsymbol{\epsilon}_i$$

$$\frac{\partial \mathbf{I}}{\partial I_i} = \boldsymbol{\epsilon}_i$$

Thus, the formulas for the partial derivatives become

$$\frac{\partial e}{\partial R_i} = (\mathbf{A}_t \mathbf{YA})^{-1} \mathbf{A}_t \left(\boldsymbol{\epsilon}_i + \sum_{j \neq i} \boldsymbol{\epsilon}_j \beta_{ji} \right) \frac{1}{R_i^2} (E_i + e_i') \tag{B.14}$$

and

$$\frac{\partial e}{\partial \beta_{ji}} = -(\mathbf{A}_t \mathbf{YA})^{-1} \mathbf{A}_t \boldsymbol{\epsilon}_i \frac{1}{R_j} (E_j + e_j') \tag{B.15}$$

$$\frac{\partial e}{\partial E_i} = -(\mathbf{A}_t \mathbf{YA})^{-1} \mathbf{A}_t \mathbf{Y} \boldsymbol{\epsilon}_i \tag{B.16}$$

$$\frac{\partial e}{\partial I_i} = (\mathbf{A}_t \mathbf{YA})^{-1} \mathbf{A}_t \boldsymbol{\epsilon}_i \tag{B.17}$$

In ECAP, the partial derivatives $(\partial e / \partial \mathrm{GM}_{ji})$ are actually computed, instead of $(\partial e / \partial \beta_{ji})$, where the quantities GM_{ji} and β_{ji} are related by the expression

$$\mathrm{GM}_{ji} = \frac{\beta_{ji}}{R_i}$$

The resistance R_i is contained by the controlling (or *from-*) branch of the dependent current source assigned to the jth branch. Hence, the partial derivatives are related as follows:

$$\frac{\partial e}{\partial \mathrm{GM}_{ji}} = \frac{\partial e}{\partial \beta_{ji}} \frac{\partial \beta_{ji}}{\partial \mathrm{GM}_{ji}} + \frac{\partial e}{\partial R_i} \frac{\partial R_i}{\partial \mathrm{GM}_{ji}} = \frac{\partial e}{\partial \beta_{ji}} R_i$$

where the quantity $(\partial R_i / \partial \mathrm{GM}_{ji})$ is zero since R_i is an independent variable.

B.1.3   Worst Case Node Voltages

The ECAP user should understand the techniques used in the worst case analysis solutions: Suppose a wo card such as

WO,3,8,17

is included in a DC problem. After the normal node voltage solutions have been obtained, the partial derivatives of all node voltages with respect to all circuit parameters have been calculated, and the partial signs have been stored, the minimum node voltage at node 3 (i.e., the first node voltage to be worst-cased) is calculated as follows. Parameters which yielded positive partial derivatives for

e_3 are set to their minimum values; those which resulted in negative partials are set to their maximum values. Node voltage 3 is then calculated in the normal manner, but since the selected extreme values are used, the result is $e_{3\ min}$. The partial derivatives for $e_{3\ min}$ are then calculated using the extreme parameter values as set for the $e_{3\ min}$ solution. The sign of each of the new partials is compared with the sign of the same partial calculated when nominal values were used. In each case in which the signs differ, a message is printed accordingly. (See Sec. 2.6.) A similar procedure is followed to determine the maximum value $e_{3\ max}$. The minimum and maximum values for the other requested node voltages (e_8 and e_{17} in this example) are then calculated in the same manner.

The validity of ECAP's worst case analysis solutions rests solely on the relations of the node voltages and the circuit parameters. That is, the node voltages must be monotonic functions of every circuit parameter over all allowed parameter ranges. Obviously, circuits with smaller parameter variations are less likely to violate this requirement.

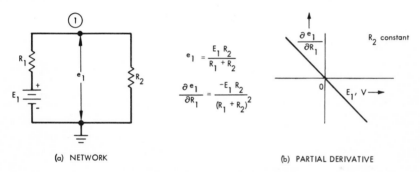

(a) NETWORK   (b) PARTIAL DERIVATIVE

Fig. B.4. Effect of E_1 on $\partial e_1/\partial R_1$.

Consider Fig. B.4, which shows how $\partial e_1/\partial R_1$ can change sign as a function of the independent voltage source variation in violation of the condition required for the true worst case. Indeed, if E_1 had minimum and maximum values of opposite signs, R_1 would not be properly set to calculate both worst case e_1 values. For such a situation a partial sign-change message would be printed and remedial action could be taken. (See Sec. 2.6.)

When no partial sign-change message occurs, a possibility of a double sign change, such as in Fig. B.5, exists. This condition would not result in a true worst case, but its occurrence is rare in most practical circuits.

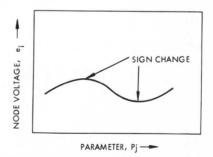

Fig. B.5. Double sign change causing no partial derivative sign-change message.

In spite of these possible trouble areas, the worst case analysis feature, when used with an understanding of its techniques and an awareness of its possible pitfalls, is quite powerful.

B.1.4   Standard Deviation Calculation

The purpose of this section is to point out the assumptions necessary to permit ECAP's standard deviation (ST) calculation to represent the true standard deviation σ. This will be done by a derivation of the ST equation.

From basic probability theory[3] the variance of the variable y is

$$\sigma^2(y) = \sum_i a_i^2 \sigma^2(x_i)$$

where $\sigma(y)$ is defined as the standard deviation of y when

$$y = \sum_i a_i x_i$$

and where the $\{x_i\}$ are statistically independent random variables with standard deviations $\{\sigma(x_i)\}$ and the $\{a_i\}$ are constants. If it is assumed that a node voltage e_k is a linear function of the circuit variables $\{P_j\}$ such that

$$e_k = \sum_j k_j P_j$$

where the $\{P_j\}$ are statistically independent random variables with standard deviations $\{\sigma(P_j)\}$ and the $\{k_j\}$ are constants, then

$$\sigma^2(e_k) = \sum_j k_j^2 \sigma^2(P_j)$$

Realizing that

$$k_j = \frac{\partial}{\partial P_j}\left[\sum_j k_j P_j\right] = \frac{\partial e_k}{\partial P_j}$$

then

$$\sigma^2(e_k) = \sum_j \left(\frac{\partial e_k}{\partial P_j}\right)^2 \sigma^2(P_j)$$

Further, assuming that

$$\sigma(P_j) = \frac{P_{j\,\text{max}} - P_{j\,\text{min}}}{6}$$

where $P_{j\,\text{max}}$ and $P_{j\,\text{min}}$ are, respectively, the specified maximum and minimum values for P_j, gives an expression for the standard deviation of e_k which can be written exclusively as a function of terms available in ECAP (which have previously been described in this text). That is, the standard deviation of node voltage e_k is calculated by ECAP as

[3]William Feller, *Introduction to Probability Theory and Its Applications* (New York: John Wiley and Sons, Inc., 1950).

$$ \mathrm{ST}(e_k) = \sqrt{ \sum_j \left(\frac{\partial e_k}{\partial P_j} \right)^2 \left(\frac{P_{j\,\max} - P_{j\,\min}}{6} \right)^2 } \tag{B.18} $$

By realizing that the circuit conditions implied in the development of the ST equation generally will not be obtained (for example, e_k realistically cannot be expected to be linearly related to all its variables), the circuit analyst should view cautiously any attempt to equate $\mathrm{ST}(e_k)$ to $\sigma(e_k)$. More wisely, he might view the ST values as easily determined approximations to $\sigma(e_k)$.

As an example of an ST calculation, the following equation is used by ECAP in determining $\mathrm{ST}(e_1)$ for the circuit of Fig. B.6:

$$ \mathrm{ST}(e_1) = \left[\left(\frac{\partial e_1}{\partial R_1} \right)^2 \left(\frac{R_{1\,\max} - R_{1\,\min}}{6} \right)^2 \right. $$

$$ + \left(\frac{\partial e_1}{\partial R_2} \right)^2 \left(\frac{R_{2\,\max} - R_{2\,\min}}{6} \right)^2 $$

$$ \left. + \left(\frac{\partial e_1}{\partial E_1} \right)^2 \left(\frac{E_{2\,\max} - E_{2\,\min}}{6} \right)^2 \right]^{1/2} $$

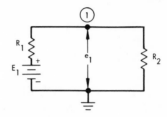

Fig. B.6. Standard deviation calculation example circuit.

B.2. AC ANALYSIS

The AC analysis program sets up a real and an imaginary nodal equation set, as is consistent with the requirements of the phasor operations. The two equation sets are solved together using a modified Crout[4] solution technique, which does not require the inversion of the nodal admittance matrix.

B.3. TRANSIENT ANALYSIS

The technique used in the transient analysis program to solve algebraic equations developed at each time interval is similar to the Crout procedure used in the AC analysis program, and is of no special user interest. The development of the algebraic equations to replace the integral-differential equations associated with a transient analysis solution is important, however, and the technique will be explained in detail in this section.

B.3.1 Capacitance Model

The model used in the ECAP transient analysis program at time t_k to permit an algebraic expression to represent the voltage-current relationship of a capaci-

[4]P. D. Crout, "A Short Method for Evaluating Determinants and Solving Systems of Linear Equations with Real or Complex Coefficients," *AIEE Transactions*, Vol. 60 (1941), pp. 1235–41.

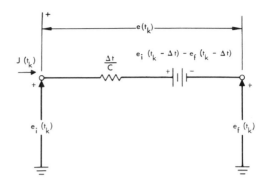

Fig. B.7. ECAP transient analysis capacitance model.

tance is shown in Fig. B.7, where $J(t_k)$ is the current through the capacitance C and $e(t_k)$ is the voltage across it at time t_k. Its justification follows.

The equation relating a capacitance's current and voltage is

$$J(t) = C\frac{de(t)}{dt} \qquad (B.19)$$

The slope of the $e(t)$ curve at time t_k is

$$\frac{J(t_k)}{C} = \left.\frac{de(t)}{dt}\right|_{t=t_k}$$

as shown graphically in Fig. B.8. This exact derivative $de(t)/dt\,|_{t=t_k}$ is approximated by an algebraic term $[\Delta e(t_k)/\Delta t]$, and therefore $J(t_k)/C$ can be approximated as $[\Delta e(t_k)/\Delta t]$, as shown in Fig. B.9.

Therefore at $t = t_k$,

$$J(t_k) \simeq \frac{C}{\Delta t}[e(t_k) - e(t_k - \Delta t)]$$

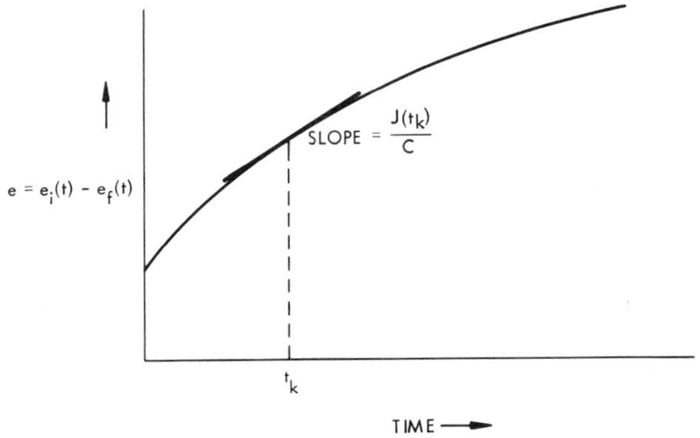

Fig. B.8. Capacitance voltage versus time.

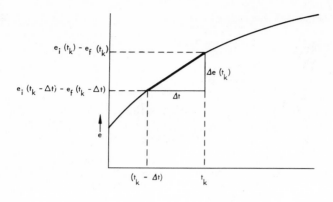

Fig. B.9. Approximation to slope $= J/C$.

From Fig. B.10, which is of the same form as Fig. B.7, we can write

$$e(t) = e_i(t) - E - e_f(t)$$

and $\quad \Delta e(t) = [e_i(t) - E - e_f(t)] - [e_i(t - \Delta t) - E - e_f(t - \Delta t)]$

Thus the voltage-current expression $\{J(t_k) \simeq [C\Delta e(t_k)]/\Delta t\}$ can now be written as

$$J(t_k) \simeq \frac{C}{\Delta t}\{[e_i(t_k) - e_f(t_k)] - [e_i(t_k - \Delta t) - e_f(t_k - \Delta t)]\} \qquad \text{(B.20)}$$

and approximates Eq. (B.19) at $t = t_k$. Since this equation is of the same form as

$$I = \frac{1}{R}[(E_1 - E_2) - (E_3)]$$

which represents the circuit in Fig. B.11, it should be clear that Eq. (B.20) is indeed the mathematical representation of the circuit in Fig. B.7, and therefore Fig. B.7 is a schematic approximation of Eq. (B.19).

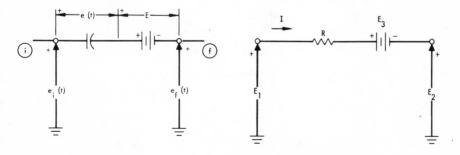

Fig. B.10. Capacitance with series voltage source.

Fig. B.11. Simplified approximation to capacitance at $t = t_k$.

333

B.3.2   Inductance Model

The model used in the ECAP transient analysis program at time t_k to permit an algebraic expression to represent the voltage-current relationship of an inductance is shown in Fig. B.12, where $J(t_k)$ is the current through the inductance L and $e(t_k)$ is the voltage across it at time t_k. Its justification follows.

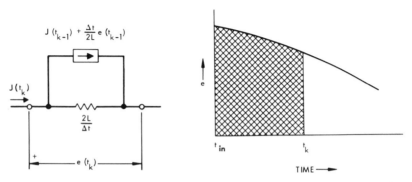

Fig. B.12. ECAP transient analysis inductance model.

Fig. B.13. Inductance voltage versus time.

The equation relating an inductance's current and voltage is

$$J(t) = \frac{1}{L} \int e(t)\, dt$$

and at t_k

$$J(t_k) = \frac{1}{L} \int_{t_{in}}^{t_k} e(t)\, dt + J_{t_{in}} \tag{B.21}$$

That is, the area under the $e(t)$ curve from the initial time t_{in} to some time t_k is

$$L[J(t_k) - J_{t_{in}}] = \int_{t_{in}}^{t_k} e(t)\, dt$$

as shown in Fig. B.13.

This exact integral $\int_{t_{in}}^{t_k} e(t)\, dt$ can be approximated by an algebraic series

$$\frac{\Delta t}{2} \cdot e(t_{in}) + 2\frac{\Delta t}{2}[e(t_1) + e(t_2) + \cdots + e(t_{k-1})] + \frac{\Delta t}{2} e(t_k)$$

$$= \frac{\Delta t}{2} \cdot e(t_{in}) + 2 \cdot \sum_{i=1}^{k-1} \left[\frac{\Delta t}{2} e(t_i) \right] + \frac{\Delta t}{2} e(t_k)$$

as shown in Fig. B.14. Therefore

$$J(t_k) \simeq J_{t_{in}} + \frac{1}{L}\left\{ \frac{\Delta t}{2} e(t_{in}) + 2\sum_{i=1}^{k-1}\left[\frac{\Delta t}{2} e(t_i) \right] + \frac{\Delta t}{2} e(t_k) \right\}$$

$$\simeq J(t_{k-1}) + \frac{\Delta t}{2L} e(t_{k-1}) + \frac{\Delta t}{2L} e(t_k) \tag{B.22}$$

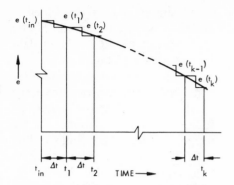

Fig. B.14. Approximation to inductance voltage.

Fig. B.15. Simplified inductance approximation at $t = t_k$.

and approximates Eq. (B.21) at $t = t_k$. Since Eq. (B.22) is of the same form as $I = I_B + Y \cdot E_1 + Y \cdot E_2$ which represents the circuit in Fig. B.15, it should be clear that Eq. (B.22) is indeed the mathematical representation of the circuit in Fig. B.12, and therefore Fig. B.12 is a schematic approximation of Eq. (B.21).

B.3.3 Capacitance and Inductance Models: Example

An example will now be presented to illustrate the use of the capacitance and inductance models and to provide insight into ECAP's transient analysis program solution techniques.

Figure B.16 presents three simple circuits and their corresponding nodal equations. Parts (a) and (b) contain capacitance and inductance networks, respectively; part (c) contains a switch network which has the sole purpose of permitting

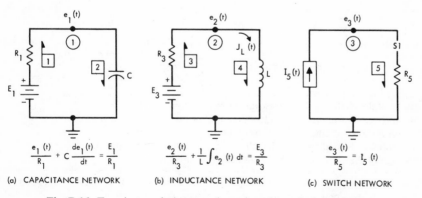

(a) CAPACITANCE NETWORK   (b) INDUCTANCE NETWORK   (c) SWITCH NETWORK

$$\frac{e_1(t)}{R_1} + C\frac{de_1(t)}{dt} = \frac{E_1}{R_1}$$

$$\frac{e_2(t)}{R_3} + \frac{1}{L}\int e_2(t)\,dt = \frac{E_3}{R_3}$$

$$\frac{e_3(t)}{R_5} = I_5(t)$$

Fig. B.16. Transient analysis networks and mathematical description.

```
        TR  L,C MODELS
    C
        B1  N(0,1),R=4.5,E=10
        B2  N(1,0),C=2
    C
        B3  N(0,2),R=1,E=20
        B4  N(2,0),L=9.5
    C
        B5  N(3,0),R=(2,5)
        I5  (3),-2.4,+0.6
        S1  B=5,(5),OFF
    C
        2ERROR=.2
        OPEN=5E3
        INITIAL TIME=3
        TI=1
        FI=8
        PR,NV,CA
        EX
```

Fig. B.17. Input data for L and C models example.

the illustration of the C and L models during a switch actuation. Figure B.17 presents the ECAP input data for the circuits of Fig. B.16.

Figure B.18 presents the algebraic nodal equations, resulting from the input data of Fig. B.17, used by ECAP as approximations to the integral-differential equations of Fig. B.16. The circuits which these equations represent are also shown.

Table B.4 indicates the values used for all variables for every solution of

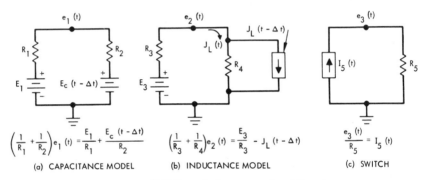

Fig. B.18. ECAP equivalent networks at time t.

Table B.4.   Transient analysis solution procedure for networks in Fig. B.18.

Time	n	$\dfrac{TI}{2^n}$	Switch considerations S	L	P	Δt	R_2	E_c	R_4	J_L
in = 3	\	\	—	\	Yes	\	SHORT	E_c(in)	OPEN	J_L(in)
4	\	\	—	\	Yes	1	$\dfrac{\Delta t}{C}$	$e_1(3)$	$\dfrac{2L}{\Delta t}$	$J_L(3) + \dfrac{\Delta t}{2L} \cdot e_2(3)$

Column S shows the sign of the switch-current CA5.
Column L shows whether $TI/2^n$ is less than ($<$) or greater than ($>$) 2ERROR·TI.
Column P shows whether or not the solution results are printed.
The symbol \ is used to show that the term is undefined or irrelevant.
*The sign of the switch-current is not checked at this time.

Note: $e_1(t)$ and $e_2(t)$ are the node voltages at time t at nodes 1 and 2, respectively; $J_L(t)$ is the element current CA4 in the inductor at time t.

**Table B.4 (continued).   Transient analysis solution
procedure for networks in Fig. B.18.**

Time	n	$\frac{\text{TI}}{2^n}$	S	L	P	Δt	R_2	E_c	R_4	J_L
				Switch considerations				Model values		
5	\	\	—	\	Yes	1	$\frac{\Delta t}{C}$	$e_1(4)$	$\frac{2L}{\Delta t}$	$J_L(4) + \frac{\Delta t}{2L} \cdot e_2(4)$
6	0	1	+	>	No	1	$\frac{\Delta t}{C}$	$e_1(5)$	$\frac{2L}{\Delta t}$	$J_L(5) + \frac{\Delta t}{2L} \cdot e_2(5)$
5.5	1	0.5	+	>	No	0.5	$\frac{\Delta t}{C}$	$e_1(5)$	$\frac{2L}{\Delta t}$	$J_L(5) + \frac{\Delta t}{2L} \cdot e_2(5)$
5.25	2	0.25	—	>	No	0.25	$\frac{\Delta t}{C}$	$e_1(5)$	$\frac{2L}{\Delta t}$	$J_L(5) + \frac{\Delta t}{2L} \cdot e_2(5)$
5.375	3	0.125	*	<	Yes	0.375	$\frac{\Delta t}{C}$	$e_1(5)$	$\frac{2L}{\Delta t}$	$J_L(5) + \frac{\Delta t}{2L} \cdot e_2(5)$
5.375+	\	\	*	\	Yes	\	SHORT	$e_1(5.375)$	OPEN	$J_L(5.375)$
6	\	\	+	\	Yes	0.625	$\frac{\Delta t}{C}$	$e_1(5.375+)$	$\frac{2L}{\Delta t}$	$J_L(5.375+)$ $+ \frac{\Delta t}{2L} \cdot e_2(5.375+)$
7	\	\	+	\	Yes	1	$\frac{\Delta t}{C}$	$e_1(6)$	$\frac{2L}{\Delta t}$	$J_L(6) + \frac{\Delta t}{2L} \cdot e_2(6)$
8	\	\	+	\	Yes	1	$\frac{\Delta t}{C}$	$e_1(7)$	$\frac{2L}{\Delta t}$	$J_L(7) + \frac{\Delta t}{2L} \cdot e_2(7)$

Column S shows the sign of the switch-current CA5.
Column L shows whether $\text{TI}/2^n$ is less than ($<$) or greater than ($>$) 2ERROR·TI.
Colume P shows whether or not the solution results are printed.
The symbol \ is used to show that the term is undefined or irrelevant.
*The sign of the switch-current is not checked at this time.

Note: $e_1(t)$ and $e_2(t)$ are the node voltages at time t at nodes 1 and 2, respectively; $J_L(t)$ is the element current CA4 in the inductor at time t.

the equations of Fig. B.18 in the range from $t = 3$ to $t = 8$. In addition, the step-by-step presentation provides insight into the transient analysis program solution procedure. Note that the number of times that TI is halved while the program searches for the switch actuation time is determined by comparing $\text{TI}/2^n$ and 2ERROR·TI. That is, the program considers that the switch actuation time has been found when $\text{TI}/2^n \leq$ 2ERROR·TI. This procedure then results in a final $n \geq 3.32 \log (2\text{ERROR})^{-1}$, or $n = 3$ for this example.

The program operation to find the switch actuation time is further described by Fig. B.19.

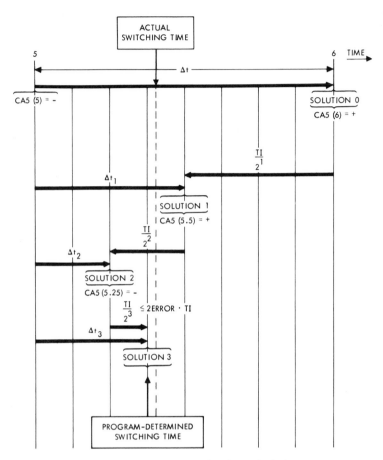

Fig. B.19. Switch actuation time calculation.

APPENDIX C

SUMMARY

OF ECAP

CAPABILITIES

Table C.1. Standard branch input parameters.

Standard branch parameter		Analysis		
		DC	AC	Transient
	R	*	*	*
	G	*	*	*
	C	—	*	*
Self and mutual	L	—	*	*
Fixed $\{$	E	*	—	*
	I	*	—	*
Sinusoidal $\{$	$E(=\|E\|/\underline{\theta_E})$	—	*	—
	$I(=\|I\|/\underline{\theta_I})$	—	*	—
Time-dependent $\{$	E	—	—	*
	I	—	—	*
Initial conditions $\{$	EO (or $E0$)	—	—	*
	IO (or $I0$)	—	—	*
	BETA	*	See Sec. 3.7	*
	GM	*	*	See Sec. 4.12

*Parameter may be used.
—Parameter may not be used.

Table C.2. Directly available output data.

Output terms	Description of output	Analysis		
		DC	AC	Transient
NV	Node voltage	*	*	*
CA	Element current	*	*	*
CV	Element voltage	*	*	*
BA	Branch current	*	*	*
BV	Branch voltage	*	*	*
BP	Element power dissipation	*	*	*
SE	Sensitivity coefficient and partial derivative	*	—	—
WO	Worst case analysis	*	—	—
ST	Standard deviation	*	—	—
MI	Miscellaneous: nodal admittance matrix, equivalent current vector, and (for DC analyses only) nodal impedance matrix	*	*	*

*Output may be requested.
—Output may not be requested.

Table C.3. Assumed parameter values (when not explicitly specified otherwise).

Term	Assumed value		
	DC	AC	Transient
E	0	0/0	0
I	0	0/0	0
EO (or E0)	—	—	0
IO (or I0)	—	—	0
INITIAL TIME	—	—	0
OUTPUT INTERVAL	—	—	1
SHORT	—	—	$0.01\ \Omega$*
OPEN	—	—	$10\ M\Omega$*
1ERROR	0.001 a*	0.001 a*	0.001 a*
2ERROR	—	—	0.001

*Based on first set in Table C.4.

Table C.4. ECAP assumed units.

Parameter	Typically assumed units	
R	Ohms	Kilohms
G	Mhos	Millimhos
C	Farads	Picofarads
L	Henries	Microhenries
E	Volts	Volts
I	Amperes	Milliamperes
BETA	(Unitless)	(Unitless)
GM	Mhos	Millimhos
Time	Seconds	Nanoseconds
Frequency	Hertz (cycles per second)	Gigahertz
Phase angle	Degrees	Degrees

Table C.5. IBM 360 program capacities.

Item	Capacity		
	DC	AC	Transient
Nodes (other than node 0)	50	50	50
B cards	200	200	200
T cards	200	200	200
M cards*	—	25	25
s cards	—	—	200
Fixed sources {Voltage	200	200	200
{Current	200	200	200
Time-dependent sources {Voltage	—	—	5
{Current	—	—	5

*Only 50 B cards may specify inductances in a problem using M cards.

Table C.5 (continued). IBM 360 program capacities.

Item		Capacity		
		DC	AC	Transient
Sources per branch	Voltage	1	1	1
	Current	1	1	1
Modify routines		No limit	No limit	—
Parameters per modify routine		50	50	—
Parameter iterations per modify routine		1	1	—
Branches affected by switches		—	—	200
Comment cards		No limit	No limit	No limit

Table C.6.   Important precautions.

1. Every node must have a path to ground through the circuit's branches.
2. The smallest impedance at any node should not be more than about six or seven orders of magnitude smaller than the next smallest impedance at that node.
3. An AC analysis must specify FREQUENCY.
4. A transient analysis must specify a TIME STEP and FINAL TIME (unless an EQUILIBRIUM solution is requested).

Table C.7.   ECAP error messages.*

Error message	Explanation and causes
BRANCH NO. XX IS MISSING	a. Branch serial numbers not sequential starting with 1. b. R, G, L or C data subgroup is missing from a branch card. c. Branch serial number repeated.
ERROR AC1 EXECUTION INHIBITED	a. Invalid self-inductance input data. b. Invalid mutual-inductance input data.
ERROR AC2 EXECUTION INHIBITED	a. Both nodes of a branch are connected to ground. b. Node data subgroup left off a branch card. This omission is allowed only for parameter modification data.
ERROR NO. XX CARD NO. = YY APPROXIMATE COLUMN NO. IS ZZ	XX is error code number. YY is position of card in deck starting with first command card. ZZ is blank suppressed column number.
FREQUENCY OF TIME STEP IS IMPROPERLY DEFINED FOR THIS PROBLEM	a. If AC analysis problem: FREQUENCY omitted or given zero value. b. If transient analysis problem: TIME STEP omitted or given zero value.
INPUT ERRORS MAKE EXECUTION IMPOSSIBLE XX ERROR(S) WERE DETECTED	A total of XX errors was detected during processing of the input data deck. Enter new ECAP job.

*Reproduced by permission from *The 1620 Electronic Circuit Analysis Program* (*ECAP*) (*1620-EE-02X*) *User's Manual.* © 1965 by International Business Machines Corporation, pp. 181–85.

Table C.7 (continued). ECAP error messages.

Error message	Explanation and causes
NODE NO. XX IS MISSING	a. Node numbers not sequential starting with 1. b. Comma not included in a node data subgroup.
SOLUTION NOT OBTAINED TO DESIRED TOLERANCE AT TIME .XXXXXXXXE $\pm$ yy	The sum over all the nodes of the absolute unbalance of currents at each node exceeds IERROR. Possible source of difficulty is: a. IERROR too small for the circuit b. TIME STEP inappropriate c. parameter values cover too wide a range at one node

Error code	Error	Remarks
1	Program or machine failure. Retry problem.	Control is returned to System/360
2	More than five time-dependent voltage sources have been entered.	
3	More than five time-dependent current sources have been entered.	
4	Card number xx is blank.	Ignored
5	Character in column indicated is not (but should be) one of the following: B, T, M, S, E, or I.	Check for C (comment) in other than column 1, or for command, solution control, or output specification statement starting before column 7.
6	The serial number of a time-dependent source has not been found before column 6.	
7	The serial number of a time-dependent source extends beyond column 5.	
8	The total number of entered values for one time-dependent source has exceeded 126.	
9	The command card MODIFY not allowed for a transient analysis.	
10	The two consecutive characters starting in the column indicated do not form a valid ECAP command, solution control, or output specification statement.	The allowable character combinations are: DC, AC, TR, MO, EX, CO*, RE*, EN†, 1U, 2U, 3U, PR, PU*, TY*, SE, WO, ST, CH*, FR, OU, TI, 1E, 2E, 3E, IN, FI, SH, OP

*Not used in System/360 version.
†Optionally used after the last EXECUTE statement of the job.

Table C.7 (continued). ECAP error messages.

Error code	Error	Remarks
11	The two consecutive characters starting in the column indicated do not form a valid output block indicator in an output specification statement.	The allowable character combinations are: NV, CA, BV, BA, BP, CV, SE, WO, ST, MI, VO,‡ CU‡
12	The two consecutive characters starting in the column indicated do not form a valid ECAP command, solution control, or output specification statement.	
13	The two consecutive characters starting in the column indicated do not form a valid ECAP command, solution control, or output specification statement.	
14	The serial number of a B, T, M, or S card has not been found before column 6.	
15	The serial number of a B, T, M, or S card extends beyond column 5.	
16	The number of B cards has exceeded 200.	
17	The number of T cards has exceeded 200.	
18	The number of M cards has exceeded 25.	
19	The number of S cards has exceeded 200.	
20	The first character of a data subgroup is not one of the following: N, B, R, G, E, I, L, C.	Check formulation of this and preceding data subgroups. Compare to allowable forms.
21	A data subgroup not completed on a single card.	
22	The number of passive element data subgroups ($R, G, L,$ or C) has exceeded one on a B or M card; or, the number of BETA or GM data subgroups has exceeded one on a T card.	
23	The number of voltage source data subgroups on a B card has exceeded one.	
24	The number of current source data subgroups on a B card has exceeded one.	

‡VO (for voltage) and CU (for current) can be used interchangeably with NV and CA, respectively.

Table C.7 (continued). ECAP error messages.

Error code	Error	Remarks
25	s cards are not applicable to a DC or AC analysis.	
26	The number of initial condition (EO or IO) data subgroups has exceeded one per B card.	
27	Initial condition data subgroups do not apply to job defined with the command card DC ANALYSIS or AC ANALYSIS.	
28	The character following the column indicated is not one of the following: $+-.0123456789E$,) (/	Check formation of data subgroup containing this character. Compare to allowable forms.
29	The total number of branches affected by all switches exceeds 200.	
30	The number of command cards entered before an EXECUTE command has exceeded four.	
31	Initial condition of a switch improperly specified (OFF or ON).	Check for numeric zero where letter O should have been entered.
32	One of two node numbers in an $N(n_1, n_2)$ data subgroup exceeds 50.	Check for omitted comma between *from* and *to* node numbers.
33	L or C data label encountered for job not defined with the command card AC or TRANSIENT.	
34	More than 50 parameters modified in one MODIFY group.	
35	Incorrectly formed data subgroup.	
36	$B(b_1, b_2)$ data subgroup appears on a B card.	
37	L data subgroup (M card) appears on a T or S card.	

APPENDIX D

SUMMARY
OF
ECAP
INPUT
CARD
FORMATS[1]

[1]*The 1620 Electronic Circuit Analysis Program* (*ECAP*) (*1620-EE-02X*) *User's Manual* (White Plains, N.Y.: International Business Machines Corp., 1965), pp. 171–80.

There are five classifications of input cards

- Command
- Comment
- Output specification
- Solution control
- Data

The card formats for each classification are summarized in this appendix. The formats pertaining to the DC analysis program are shown first in Table. D.1.

Table D.1.   ECAP DC analysis input card formats.

Card type		Card columns		Reference page
		1 2 3 4 5	6	7.............................72	
Command			DC[ANALYSIS]	16
			EX[ECUTE]	16
			⌐MO[DIFY]⌐ The user-compiled programs, 1U[SER] called 1USER, 2USER, and 2U[SER] 3USER, should be used only ⌐3U[SER]⌐ by those thoroughly familiar with the ECAP System Manual	32
Data	Branch Bnn		$N(n_{\text{initial}}, n_{\text{final}}), \begin{Bmatrix} R = d \\ G = d \end{Bmatrix}[, E = d][, I = d]$	16
	Dependent current source Tnn		$B(b_{\text{from}}, b_{\text{to}}), \begin{Bmatrix} \text{GM} = d \\ \text{BETA} = d \end{Bmatrix}$	19
Solution control			⌐SE[NSITIVITIES] ⌐ WO[RST CASE] ST[ANDARD DEVIATION] ⌐1ERROR $= p$ ⌐	21 21 21 41
Output specification			$\text{PR[INT]} \left[, \begin{Bmatrix} \text{VO[LTAGES]} \\ \text{NV} \end{Bmatrix} \right] \left[, \begin{Bmatrix} \text{CU[RRENTS]} \\ \text{CA} \end{Bmatrix} \right]$ [, CV][, BV][, BA][, BP] [, SE[NSITIVITIES][, WO[RST CASE]] [, ST[ANDARD DEVIATION]] [, MI[SCELLANEOUS]]	344 16 26 21 21 26
Comment		C any text		16

The formats used in the AC analysis program are next (Table D.2), and those used in the transient analysis program are last (Table D.3).

Braces and brackets are heavily used in the presentation of the formats. Braces indicate that one of the entries contained within the braces must appear

Table D.2.   ECAP AC analysis input card formats.

Card type			Card columns	Reference page
			1 2 3 4 5 6 7 72	
Command			AC[ANALYSIS]	54
			EX[ECUTE]	54
			┌MO[DIFY]┐ The user-compiled programs, 1U[SER] │ called 1USER, 2USER, and 2U[SER] │ 3USER, should be used only └3U[SER]┘ by those thoroughly famil- iar with the ECAP System Manual	57
	Branch	Bnn	$N(n_{\text{initial}}, n_{\text{final}}), \begin{Bmatrix} R = d \\ G = d \\ C = d \\ L = d \end{Bmatrix} [, E = d][, I = d]$	54
Data	Dependent current source	Tnn	$B(b_{\text{from}}, b_{\text{to}}), \begin{Bmatrix} \text{GM} = d \\ \text{BETA} = d \end{Bmatrix}$	66
	Mutual inductance	Mnn	$B(b_1, b_2), L = d$	60
Solution control			┌FR[EQUENCY] = d┐ └1ERROR = p ┘	54 62
Output specification			PR[INT]$\left[, \begin{Bmatrix} \text{VO[LTAGES]} \\ \text{NV} \end{Bmatrix} \right]\left[, \begin{Bmatrix} \text{CU[RRENTS]} \\ \text{CA} \end{Bmatrix} \right]$ [, CV][, BV][, BA][, BP] [, MI[SCELLANEOUS]]	344 54 54 54
Comment		C any text		16

Table D.3.   ECAP transient analysis input card formats.

Card type	Card columns	Reference page
	1 2 3 4 5 6 7 72	
Command	TR[ANSIENT ANALYSIS]	76
	EX[ECUTE]	76
	┌1U[SER]┐ The user-compiled programs, 2U[SER] │ called 1USER, 2USER, and └3U[SER]┘ 3USER, should be used only by those thoroughly familiar with the ECAP System Manual	—

Table D.3 (continued). ECAP transient analysis input card formats.

Card type		Card columns	Reference page
		1 2 3 4 5 | 6 | 7................................72	
Data	Branch Bnn	$N(n_{\text{initial}}, n_{\text{final}}), \begin{Bmatrix} R = d \\ G = d \\ C = d \\ L = d \end{Bmatrix} [, E = d][, I = d]$	76
		$\left[, \begin{Bmatrix} E0 = d \\ I0 = d \end{Bmatrix}\right]$	91
	Dependent current source Tnn	$B(b_{\text{from}}, b_{\text{to}}), \begin{Bmatrix} \text{GM} = d \\ \text{BETA} = d \end{Bmatrix}$	86
	Mutual inductance Mnn	$B(b_1, b_2), L = d$	79
	Switch Snn	$B = b_s, (b_1, b_2, b_3, \ldots, b_n), \begin{Bmatrix} \text{ON} \\ \text{OFF} \end{Bmatrix}$	85
	Time-dependent voltage source or current source $\begin{Bmatrix} \text{Enn} \\ \text{Inn} \end{Bmatrix}$	$\begin{Bmatrix} (k), p_0, p_1, p_2, \ldots, p_n \\ \text{P}(k), p_0, p_1, p_2, \ldots, p_n \\ \text{SIN}(t_p), V_1, V_0, t_0 \end{Bmatrix}$	82
Solution control		TI[ME STEP] $= p$	76
		OU[TPUT INTERVAL] $= p$ ⎤ Note:	76
		1E[RROR] $= p$ The 3ERROR	79
		2E[RROR] $= p$ feature is of	87
		3E[RROR] $= p$ no practical	—
		IN[ITIAL TIME] $= p$ use and is not	91
		FI[NAL TIME] $= p$ discussed in	76
		SH[ORT] $= p$ the text.	90
		OP[EN] $= p$	90
		EQ[UILIBRIUM] ⎦	93
Output specification		$\text{PR[INT]}\left[, \begin{Bmatrix} \text{VO[LTAGES]} \\ \text{NV} \end{Bmatrix}\right] \left[, \begin{Bmatrix} \text{CU[RRENTS]} \\ \text{CA} \end{Bmatrix}\right]$	344
		$[, \text{CV}][, \text{BV}][, \text{BA}]$	76
		$[, \text{BP}][, \text{MI[SCELLANEOUS]}]$	76
			76
Comment	C any text		16

on the corresponding card. Brackets imply that none of the entries contained need appear on the card unless desired. In any case, a card must not contain more than one of the entries within one pair of brackets or one pair of braces.

In a number of instances, a pair of brackets contains only one entry. This simply means that the entry can be omitted or included. For instance, the first command card, shown in Table D.1, must contain the entry DC, but the word ANALYSIS is optional. In the branch data card of Table D.1, either $R = d$ or $G = d$ must appear on the card (but not both). On the other hand, $E = d$ is optional, as is $I = d$.

A lower case d appearing on the right-hand side of an equality (for instance, the entry $R = d$ in the branch data card) indicates that an appropriate data subgroup format should be substituted. The data subgroup formats used in the DC analysis program are found in Table D.4. The formats for the AC analysis

Table D.4. ECAP DC analysis data subgroup formats.

(a) Nominal solutions or parameter modification solutions.

d	p_1	p_2	p_3
p_1	Nominal		
$p_1(p_2, p_3)$	Nominal	Minimum	Maximum
$p_1(p_2)$	Nominal	Per cent variation expressed as a decimal fraction	

(b) Parameter modification solutions only.

d	p_1	p_2	p_3
$p_1(p_2)p_3$	First	Total number of values excluding p_1	Last

program and the transient analysis program are given in Table D.5 and Table D.6, respectively.

A lower case p (Table D.3, for example) should be replaced by a numerical entry.

The numbers above the card are the column numbers that define the data fields. Command cards, solution control cards, and output specification cards have just one data field—columns 7 through 72. The required entries can be made anywhere in this field. The data card has three fields: columns 1 through 5, column 6, and columns 7 through 72.

Similarly, the data field for the comment card is columns 2 through 72; a C must appear in column 1.

Table D.5. ECAP AC analysis data subgroup formats.

(a) Nominal solutions or parameter modification solutions.

$x = d$	p_1	p_2	FREQ	R	G	C	L	E	I	BETA	GM
							x				
$x = p_1$	Nominal		√	√	√	√	√	√	√	√	√
$x = p_1/p_2$	Magnitude	Phase angle						√	√		

(b) Parameter modification solutions only.

$x = d$	p_1	p_2	p_3	p_4	p_5	FREQ	R	G	C	L	E	I	BETA	GM
										x				
$x = p_1(p_2)p_3$	First	Total number of values excluding p_1	Last				√	√	√	√		√	√	√
$x = p_1(p_2)p_3$	First	Multiplier	Last								√			
$x = p_1(+p_2)p_3$	First	Total number of values excluding p_1	Last								√			
$x = p_1(p_2)p_3/p_4(p_2)p_5$	First (magnitude)	Total number of values excluding the first (p_1 or p_4)	Last (magnitude)	First (phase angle)	Last (phase angle)								√	√

Table D.6. ECAP transient analysis data subgroup formats.

$x = d$	p_1	p_2	x									
			R	G	C	L	E	I	E0	I0	BETA	GM
$x = p_1$	Nominal		✓	✓	✓	✓	✓	✓	✓	✓	✓	✓
$x = (p_1, p_2)$	Nominal	Switched value	✓	✓	✓	✓	✓	✓			✓	✓

APPENDIX E

DIFFERENCES IN ECAP PROGRAMS FOR THE IBM 7094 AND 1620 COMPUTERS

Although this book has been written primarily for use with the System/ 360 version of ECAP (to be run on Model 40 or higher computers), it is expected also to be of value in the use of the IBM 7094 and IBM 1620 versions. It is the purpose of this appendix to point out the differences between these versions and the 360 version.

E.1. 7094 DIFFERENCES

The 7094 (or 7090, 7044, or 7040) ECAP version differs from the 360 version in the following ways.

E.1.1 General

1. No double precision is used, as contrasted to the fact that many 360 version calculations are done in double precision. The main result is a tendency toward larger current unbalances, requiring the user to maintain the impedance ratios at a node at less than about 10^5 (rather than the 10^6 or 10^7 limitation specified in comment 1 of Sec. 2.12).

2. The floating point number extremes are $1E \pm 38$.

3. There are a few less error code diagnostics.

E.1.2 DC Analysis

There are no significant differences in the DC analysis program.

E.1.3 AC Analysis

No 1ERROR check is made.

E.1.4 Transient Analysis

1. Only the NV, CA, and MI are available printouts.

2. There is no M card capability.

E.2. 1620 DIFFERENCES

The 1620 version differs from the 360 version in all of the ways that the above-described 7094 version differs (except that the 1620 has $1E \pm 99$ as its floating-point extremes), and in addition the differences in the 1620 version compared with the 360 version include the following.

E.2.1 General

Problem capacity is limited to 20 nodes, 60 branches, and 10 T cards. Only 20 parameters may be modified in a single modify routine. To save execution time when using the modify capability, one complete modify routine may be placed before the first execute card for a DC or AC analysis problem. That is, the first execute card of any DC or AC problem with one or more modify routines may be omitted.

E.2.2 DC Analysis

1. There is no 1ERROR check made.

2. Because completely new solutions are not made for a DC modify routine, no parameter should be modified by more than four orders of magnitude, and no problem should contain more than five modify routines.

3. Worst case analysis node voltage solutions are computed using the magnitude of the node voltage partial derivatives. When parameter tolerances are large, this technique will often lead to results with appreciable error since the partial derivatives generally are not constant for all values of each parameter.

4. The specification of node numbers on the WO card has no effect on the worst case solution since a worst case analysis will be performed on all nodes if the WO card is included.

E.2.3 AC Analysis

1. Only the multiplicative frequency iteration can be made. That is, the plus (+) sign before the factor in parentheses on the FREQUENCY card is ignored.

2. There is a five M card limitation.

E.2.4 Transient Analysis

1. Only nonperiodic time-dependent voltage sources are allowed, and all must have the same multiple (i.e., the same number in parentheses). Each source may be specified by a maximum of 21 points.

2. Only 20 switches (s cards) are allowed.

ANSWERS

TO

PROBLEMS

2.1. BP $= 0.2669$ mw

2.2a.

```
              DC
          B1  N(0,1),R=3,E=10
          B2  N(1,0),R=2
        C
          B3  N(0,2),R=2
          T1  B(2,3),BETA=6
        C
          B4  N(3,0),R=2
          T2  B(1,4),GM=-2
        C
          B5  N(0,4),R=2,I=12
        C
              PR,NV,CV,CA,BA
              EX
```

2.2b. $NV2 = NV3 = -NV4$

$$CA3 = (CV3)(Y_3) + \left(\frac{BETA}{R_2}\right)CV2 \simeq 0$$

$$CA5 = (CV5)(Y_5) \simeq 12$$

2.3. Node 0 is missing.

2.4.

Circuit	$NV3_{min}$	$NV3_{max}$
5	$R_2 = 4, \beta_1 = 5$	$R_2 = 1, \beta_1 = 20$
	13.33	80.00
6	$R_3 = 8, \beta_1 = 8$	$R_3 = 2, \beta_1 = 20$
	16.00	80.00

2.5.

```
              MO
          B2  R=4
          T1  BETA=3
              EX
```

Circuit	NV2
Original	18.0
Modify	12.0

2.6.

Circuit	NV$_{min}$	NV$_{nom}$	NV$_{max}$
Original	4.0	8.0	9.0
MODIFY 1	4.0	7.2	9.0
MODIFY 2	—	6.0	—
	4.0	8.0	9.0
MODIFY 3	—	9.0	—
	—	9.6	—
	8.0	10.0	10.0

2.7. Because the matrix elements do not properly reflect the range of parameter values, the results derived from these matrices can be expected to be inaccurate. Current unbalance is a frequent result. The lesson to be learned is to observe the nodal impedance rule.

2.8. CA3$_{min}$ = −1.0
CA3$_{max}$ = 2.3636

2.9. CV2$_{min}$ = 2.4
CV2$_{max}$ = 8.727

2.10.

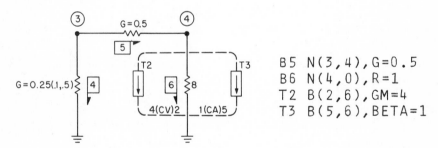

B5 N(3,4),G=0.5
B6 N(4,0),R=1
T2 B(2,6),GM=4
T3 B(5,6),BETA=1

2.11. Sensitivity $(\mu a/\Omega) = \dfrac{\partial NV4}{\partial R_4} = -0.048765 \ \mu a/\Omega$;

$R_x = 100 \ \Omega$, $I_G = 79.48 \ \mu a$; $R_x = 1000 \ \Omega$, $I_G = 0.0 \ \mu a$

2.12. b. $V_c(Q_1) = -7.57$ v $V_c(Q_2) = -23.02$ v
$V_e(Q_1) = -0.539$ v $V_e(Q_2) = -6.83$ v
$I_c(Q_1) = 1.8$ ma $I_c(Q_2) = 7$ ma

c. 31.2

d. -9.524 v $\leq V_c(Q_1) \leq -5.971$ v
-26.99 v $\leq V_c(Q_2) \leq -18.687$ v
-0.6855 v $\leq V_e(Q_1) \leq -0.4024$ v
-8.713 v $\leq V_e(Q_2) \leq -5.315$ v

e. Critical parameters R_2, R_5, R_7, R_8, V_{cc}, V_{be}

$V_{ce} = 30 \text{ v} \pm 3 \text{ per cent}$

$V_{be} = 0.7 \text{ v} \pm 50 \text{ mv}$

$R_{2,5,7,8} = R_{\text{nom}} \pm 2 \text{ per cent}$

f. $R_T = 600 \; \Omega$, $E_T = -1.2609 \text{ v}$

2.13. $V_c(Q_1) = -16.17 \text{ v}$ $V_c(Q_2) = -15.718 \text{ v}$

$V_e(Q_1) = -0.488 \text{ mv}$ $V_e(Q_2) = -13.256 \text{ v}$

$i_c(Q_1) = 1.62 \; \mu\text{a}$ $i_c(Q_2) = 14.28 \text{ ma}$

3.1. BP $= 0.7814 \; \mu\text{w}$

3.2.

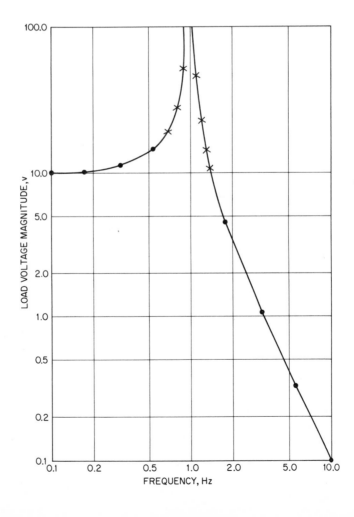

3.2 (*continued*).

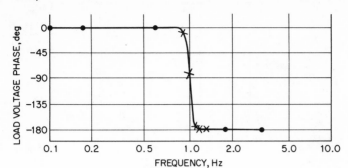

3.3. $M = (U/L)^{1/(n-1)}$

3.4.   BP $= 97.65$ mw

3.5. Errors: a. Two B3 cards; b. coupling coefficient > 0.999995; c. FREQ $= 0$; d. BETA specified without a resistive *from*-branch; e. parameter iteration requested in nominal solution; and f. different numbers are specified for source magnitude and phase iteration steps.

3.6.

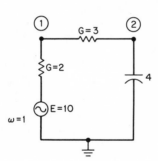

$x \equiv NV1$

$= 4.79 \underline{/-20.18}$

3.7.

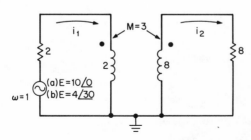

a. $u \equiv i_2 = 0.9023 \underline{/15.746}$ for $I = 10 \underline{/0}$

b. $u \equiv i_2 = 0.3609 \underline{/45.746}$ for $I = 4 \underline{/30}$

3.7 (*continued*).

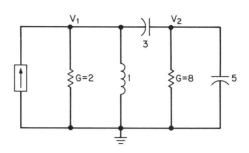

c. $u \equiv v_2 = 0.9026 \,\underline{/15.779}$

d. $u \equiv v_2 = 0.36104 \,\underline{/45.779}$

3.8. At the frequency extremes, the impedance of either L_2 or C_1 is several orders of magnitude less than the next smallest impedance at its nodes, and the range of numbers required for accurate solutions is greater than the system can handle. Removal from the equivalent circuit of unreasonably small impedances is often of practical value.

3.9. $Z_{AB} = 2.99993 \,\underline{/27.153} \; \Omega$ at $\omega = 1$

3.10. $L \simeq 20 \, \text{h}$

3.11.

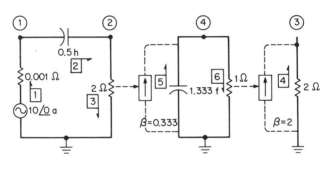

FREQ	ECAP-determined NV3/NV2
0.00159	$\dfrac{0.066 \,\underline{/88.6}}{0.100 \,\underline{/89.4}}$
0.159	$\dfrac{2.82 \,\underline{/-8.08}}{7.07 \,\underline{/45.0}}$
15.9	$\dfrac{0.050 \,\underline{/-90.0}}{9.99 \,\underline{/0.573}}$

3.12. See Fig. a3.12 on next page.

3.13. See Fig. a3.13 on next page.

3.14. See Fig. a3.14 on p. 366.

3.15. $Z|_{f=0.01} = 2.005 \,\underline{/-1.8}$; $Z_{f=1} = 0.486 \,\underline{/-5.0}$;

$Z|_{f=10} = 0.4996 \,\underline{/-0.47}$

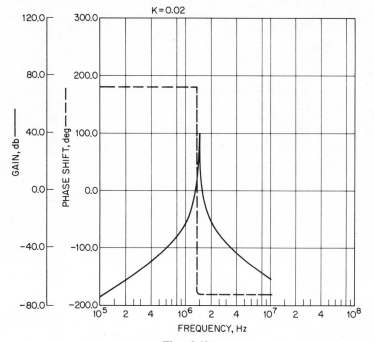

Fig. a3.12

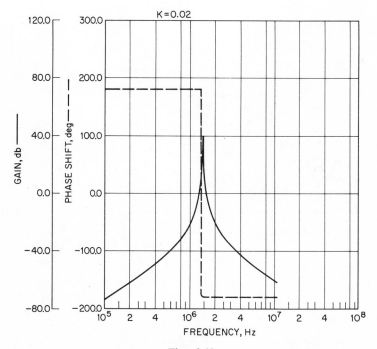

Fig. a3.13

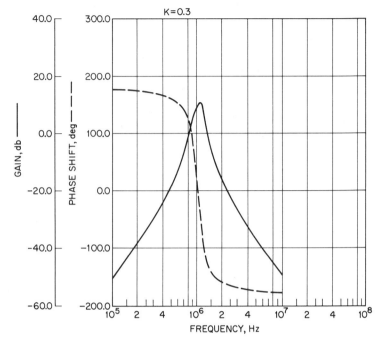

Fig. a3.14

3.16. See Fig. a3.16 on next page.

3.17. See Fig. a3.17 on p. 368.

4.1. Add:

```
B2 N(1,0),R=(1E4,1E-4)
S1 B=2,(2),ON
```

4.2.

```
I143 P(5),0, 3.83, 7.07, 9.24, 10.0, 9.24, 7.07, 3.83, 0,
  *        -3.83,-7.07,-9.24,-10.0,-9.24,-7.07,-3.83, 0
```

4.3. See Fig. a4.3 on p. 369.

4.4. See Fig. a4.4 on p. 369.

4.5. a. no switch action

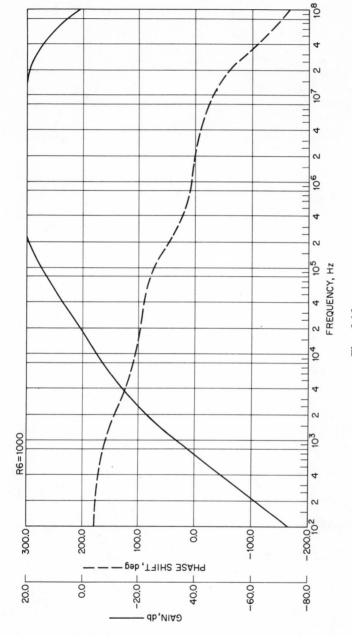

Fig. a3.16

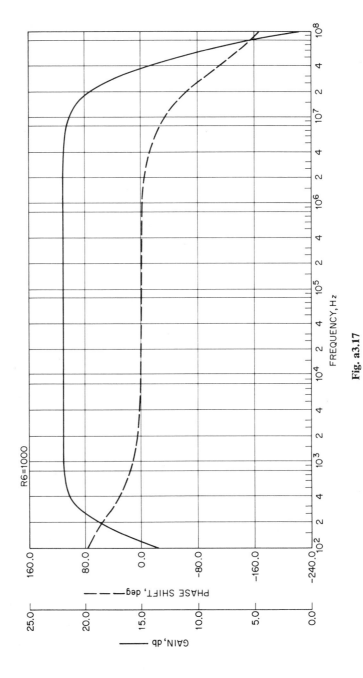

Fig. a3.17

```
        TR
 C   PART A
     B1  N(1,0),R=1
     I1  (10),-1.25,8.75
     S1  B=1,(2),OFF
     B2  N(0,2),R=1,E=(24,50)
 C
 C   PART B
     B3  N(0,3),R=1
     E3  P(5),0,5,10,0
     B4  N(0,3),R=1,E=(9.999,.001)
     S2  B=4,(4,5),ON
     B5  N(0,4),R=1,E=(12,18)
        TI=1
        FI=30
        PR,NV
        EX
```

Fig. a4.3

```
        TR
 C   START OF MODEL
     B1  N(1,0),G=.1,E=15
     B2  N(1,0),G=(1E-6,.1),E=-5
     B3  N(1,0),G=(1E-6,.6),E=-10
     B4  N(1,0),G=(1E-6,1.2),E=-12.5
     S1  B=2,(2),OFF
     S2  B=3,(3),OFF
     S3  B=4,(4),OFF
 C   END OF MODEL
     B5  N(0,1),R=.001
     E5  (20),0,20
 C
        TI=1
        FI=20
        PR,NV,CA
        EX
```

Fig. a4.4

b. no switch action

c. switch turns off at $t = 0.0$ and remains off

4.6. a. B2 is simultaneously a *from-* and a *to*-branch

   b. GM is used with a reactive *from*-branch

   c. TI is not specified (elicits error message)

4.7. $(TI/2^n) \leq 2ERROR \cdot TI$

   $n \geq 3.32 \log (2ERROR)^{-1}$

4.8.

$$t = 3.0 \qquad \left\{ \begin{array}{l} \left(\dfrac{1}{R_1} + \dfrac{1}{SHORT}\right) e_1(3) = \dfrac{E_1}{R_1} \\[3mm] \left(\dfrac{1}{R_3} + \dfrac{1}{OPEN}\right) e_2(3) = \dfrac{E_3}{R_3} \end{array} \right.$$

SHORT $= 0.01$

OPEN $= 5E3$

$$t = 5.375 \begin{cases} \left(\dfrac{1}{R_1} + \dfrac{1}{\Delta t/C}\right)e_1(5.375) = \dfrac{E_1}{R_1} + \dfrac{e_1(5)}{\Delta t/C} \\[2ex] \left(\dfrac{1}{R_3} + \dfrac{1}{2L/\Delta t}\right)e_2(5.375) = \dfrac{E_3}{R_3} - \left[J_L(5) + \dfrac{\Delta t}{2L} \cdot e_2(5)\right] \end{cases}$$
$\Delta t = 0.375$

$$\begin{aligned} t = 5.375+ \\ \text{SHORT} = 0.01 \\ \text{OPEN} = 5\text{E}3 \end{aligned} \begin{cases} \left(\dfrac{1}{R_1} + \dfrac{1}{\text{SHORT}}\right)e_1(5.375+) = \dfrac{E_1}{R_1} + \dfrac{e_1(5.375)}{\text{SHORT}} \\[2ex] \left(\dfrac{1}{R_3} + \dfrac{1}{2L/\Delta t}\right)e_2(5.375+) = \dfrac{E_3}{R_3} - J_L(5.375) \end{cases}$$

$$t = 6.0 \begin{cases} \left(\dfrac{1}{R_1} + \dfrac{1}{\Delta t/C}\right)e_1(6) = \dfrac{E_1}{R_1} + \dfrac{e_1(5.375+)}{\Delta t/C} \\[2ex] \left(\dfrac{1}{R_3} + \dfrac{1}{2L/\Delta t}\right)e_2(6) = \dfrac{E_3}{R_3} - \left[J_L(5.375+) + \dfrac{\Delta t}{2L} \cdot e_2(5.375+)\right] \end{cases}$$
$\Delta t = 0.625$

4.9. a. $\text{NV}_1(t) = 4.0$ for every t.
  b. Only the steady-state solution would be calculated if C_3 were removed since the circuit would contain no reactive elements or time-dependent sources.

4.10. a. $\Delta t_{\min} = 0.0625$
  b. 6.4375

4.11.

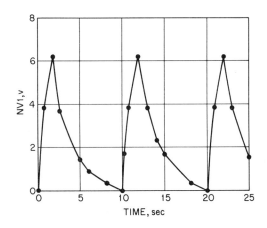

4.12. See Fig. a4.12 on next page.

4.13. See Fig. a4.13 on next page.

4.14. See Fig. a4.14 on p. 372.

4.15. See Fig. a4.15 on p. 372.

4.16. See Fig. a4.16 on p. 373.

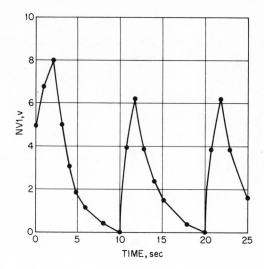

Fig. a4.12

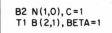

B2 N(1,0), C=1
T1 B(2,1), BETA=1

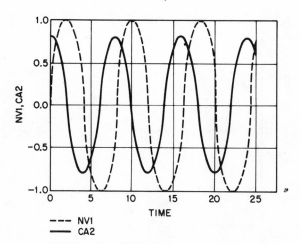

--- NV1
— CA2

Fig. a4.13

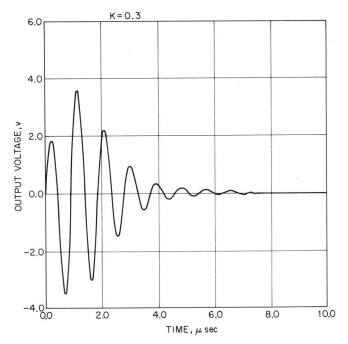

Fig. a4.14

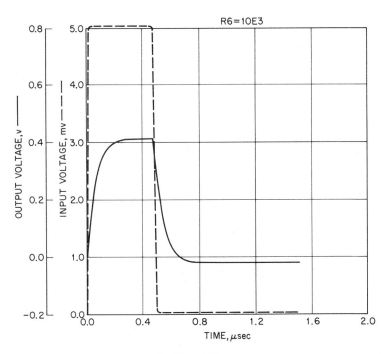

Fig. a4.15

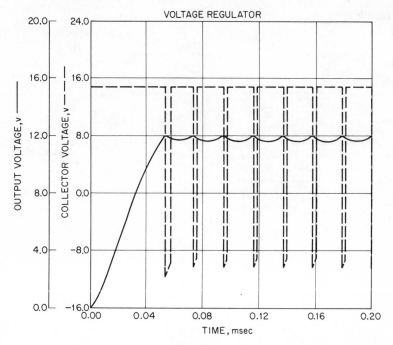

Fig. a4.16

5.1.

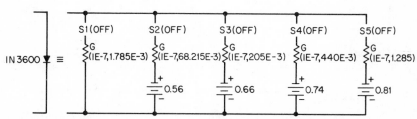

5.2.

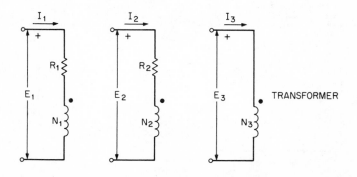

5.2 (*continued*).

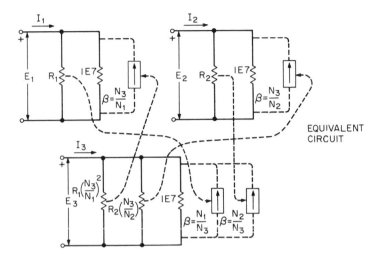

5.3.

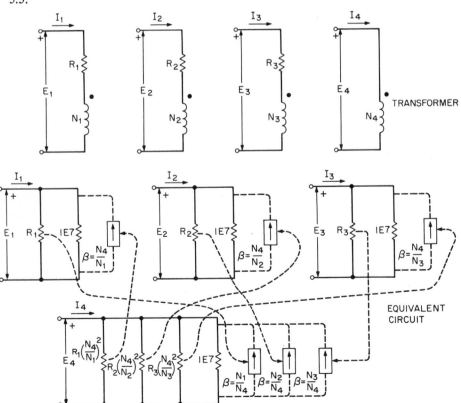

5.4.

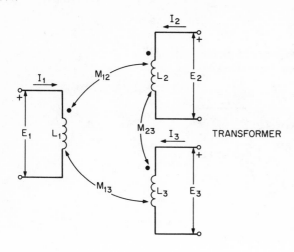

TRANSFORMER

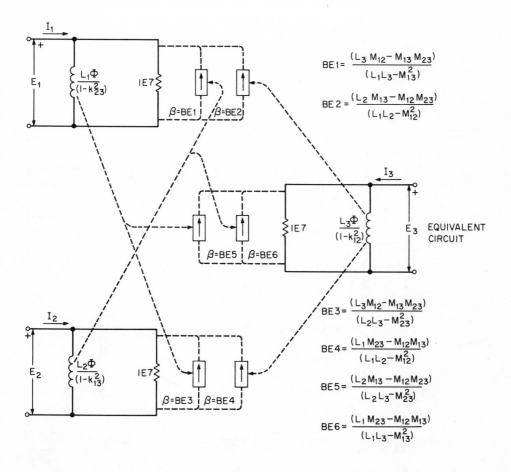

$$BE1 = \frac{(L_3 M_{12} - M_{13} M_{23})}{(L_1 L_3 - M_{13}^2)}$$

$$BE2 = \frac{(L_2 M_{13} - M_{12} M_{23})}{(L_1 L_2 - M_{12}^2)}$$

EQUIVALENT
CIRCUIT

$$BE3 = \frac{(L_3 M_{12} - M_{13} M_{23})}{(L_2 L_3 - M_{23}^2)}$$

$$BE4 = \frac{(L_1 M_{23} - M_{12} M_{13})}{(L_1 L_2 - M_{12}^2)}$$

$$BE5 = \frac{(L_2 M_{13} - M_{12} M_{23})}{(L_2 L_3 - M_{23}^2)}$$

$$BE6 = \frac{(L_1 M_{23} - M_{12} M_{13})}{(L_1 L_3 - M_{13}^2)}$$

For verification:

$$\frac{L_1\Phi}{(1 - k_{23}^2)} = \frac{L_2\Phi}{(1 - k_{13}^2)} = \frac{L_3\Phi}{(1 - k_{12}^2)} = 0.356 \text{ mh}$$

$$\text{BE1} = \text{BE2} = \text{BE3} = \text{BE4} = \text{BE5} = \text{BE6} = 0.4285$$

5.5.

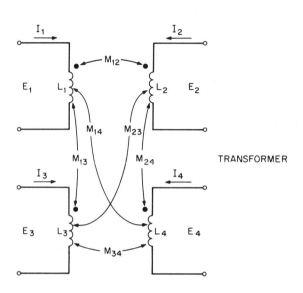

TRANSFORMER

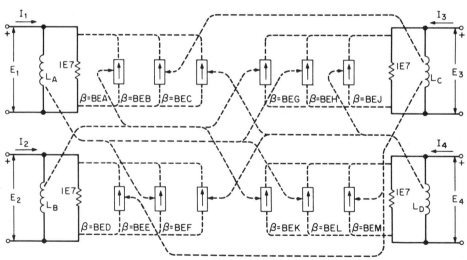

$$L_A = \frac{L_1\Phi}{(1 + 2k_{23}k_{34}k_{24} - k_{24}^2 - k_{34}^2 - k_{23}^2)}$$

$$L_B = \frac{L_2\Phi}{(1 + 2k_{13}k_{14}k_{34} - k_{13}^2 - k_{14}^2 - k_{34}^2)}$$

$$L_C = \frac{L_3\Phi}{(1 + 2k_{12}k_{14}k_{24} - k_{12}^2 - k_{24}^2 - k_{14}^2)}$$

$$L_D = \frac{L_4\Phi}{(1 + 2k_{12}k_{13}k_{23} - k_{12}^2 - k_{13}^2 - k_{23}^2)}$$

$$\Phi = (1 + 2k_{12}k_{13}k_{23} + 2k_{12}k_{14}k_{24} + 2k_{13}k_{14}k_{34} + 2k_{23}k_{24}k_{34}$$
$$+ k_{12}^2k_{34}^2 + k_{13}^2k_{24}^2 + k_{14}^2k_{23}^2 - k_{12}^2 - k_{13}^2$$
$$- k_{14}^2 - k_{23}^2 - k_{24}^2 - k_{34}^2 - 2k_{12}k_{13}k_{24}k_{34}$$
$$- 2k_{12}k_{14}k_{23}k_{34} - 2k_{13}k_{14}k_{23}k_{24})$$

$$BEA = \frac{M_{12}(1 - k_{34}^2)L_3L_4 + M_{13}M_{34}M_{24} + M_{14}M_{23}M_{34} - L_3M_{14}M_{24} - L_4M_{13}M_{23}}{L_1L_3L_4(1 + 2k_{13}k_{14}k_{34} - k_{13}^2 - k_{14}^2 - k_{34}^2)}$$

$$BEB = \frac{M_{13}(1 - k_{24}^2)L_2L_4 + M_{14}M_{23}M_{24} + M_{12}M_{24}M_{34} - L_4M_{12}M_{23} - L_2M_{14}M_{34}}{L_1L_2L_4(1 + 2k_{12}k_{14}k_{24} - k_{12}^2 - k_{14}^2 - k_{24}^2)}$$

$$BEC = \frac{M_{14}(1 - k_{23}^2)L_2L_3 + M_{12}M_{23}M_{34} + M_{13}M_{23}M_{24} - L_2M_{13}M_{34} - L_3M_{12}M_{24}}{L_1L_2L_3(1 + 2k_{12}k_{13}k_{23} - k_{12}^2 - k_{13}^2 - k_{23}^2)}$$

$$BED = \frac{M_{23}(1 - k_{14}^2)L_1L_4 + M_{13}M_{14}M_{24} + M_{12}M_{14}M_{34} - L_1M_{24}M_{34} - L_4M_{12}M_{13}}{L_1L_2L_4(1 + 2k_{12}k_{14}k_{24} - k_{12}^2 - k_{24}^2 - k_{14}^2)}$$

$$BEE = \frac{M_{12}(1 - k_{34}^2)L_3L_4 + M_{23}M_{34}M_{14} + M_{24}M_{13}M_{34} - L_3M_{14}M_{24} - L_4M_{13}M_{23}}{L_2L_3L_4(1 + 2k_{23}k_{34}k_{24} - k_{24}^2 - k_{34}^2 - k_{23}^2)}$$

$$BEF = \frac{M_{24}(1 - k_{13}^2)L_1L_3 + M_{13}M_{14}M_{23} + M_{12}M_{13}M_{34} - L_1M_{23}M_{34} - L_3M_{12}M_{14}}{L_1L_2L_3(1 + 2k_{12}k_{13}k_{23} - k_{12}^2 - k_{13}^2 - k_{23}^2)}$$

$$BEG = \frac{M_{23}(1 - k_{14}^2)L_1L_4 + M_{12}M_{14}M_{34} + M_{13}M_{14}M_{24} - L_1M_{24}M_{34} - L_4M_{12}M_{13}}{L_1L_3L_4(1 + 2k_{13}k_{14}k_{34} - k_{13}^2 - k_{34}^2 - k_{14}^2)}$$

$$BEH = \frac{M_{13}(1 - k_{24}^2)L_2L_4 + M_{12}M_{24}M_{34} + M_{14}M_{23}M_{24} - L_2M_{14}M_{34} - L_4M_{12}M_{13}}{L_2L_3L_4(1 + 2k_{23}k_{34}k_{24} - k_{24}^2 - k_{34}^2 - k_{23}^2)}$$

$$BEJ = \frac{M_{34}(1 - k_{12}^2)L_1L_2 + M_{12}M_{13}M_{24} + M_{12}M_{14}M_{23} - L_2M_{13}M_{14} - L_1M_{23}M_{24}}{L_1L_2L_3(1 + 2k_{12}k_{13}k_{23} - k_{12}^2 - k_{13}^2 - k_{23}^2)}$$

$$BEK = \frac{M_{24}(1 - k_{13}^2)L_1L_3 + M_{12}M_{13}M_{34} + M_{13}M_{14}M_{23} - L_1M_{23}M_{34} - L_3M_{12}M_{14}}{L_1L_3L_4(1 + 2k_{13}k_{14}k_{34} - k_{13}^2 - k_{14}^2 - k_{34}^2)}$$

$$BEL = \frac{M_{14}(1 - k_{23}^2)L_2L_3 + M_{12}M_{23}M_{34} + M_{13}M_{23}M_{24} - L_2M_{13}M_{34} - L_3M_{12}M_{24}}{L_2L_3L_4(1 + 2k_{23}k_{34}k_{24} - k_{24}^2 - k_{34}^2 - k_{23}^2)}$$

$$BEM = \frac{M_{34}(1 - k_{12}^2)L_1L_2 + M_{12}M_{14}M_{23} + M_{12}M_{13}M_{24} - L_1M_{23}M_{24} - L_2M_{13}M_{14}}{L_1L_2L_4(1 + 2k_{12}k_{14}k_{24} - k_{12}^2 - k_{24}^2 - k_{14}^2)}$$

For verification:

$$L = L_A = L_B = L_C = L_D = 0.32839 \text{ mh}$$
$$BEA = BEB = BEC = BED = BEE = BEF = BEG = BEH = BEJ$$
$$= BEK = BEL = BEM = 0.302157$$

5.6. See Fig. a5.6 on next page.

5.7. See Fig. a5.7 on next page.

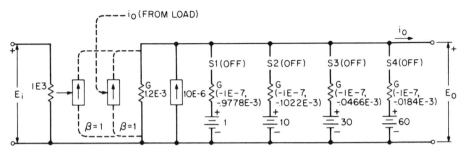

Fig. a5.6

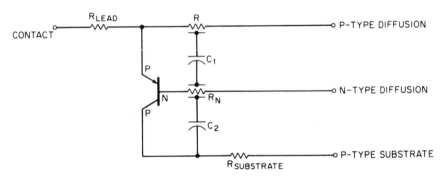

Fig. a5.7

5.8.

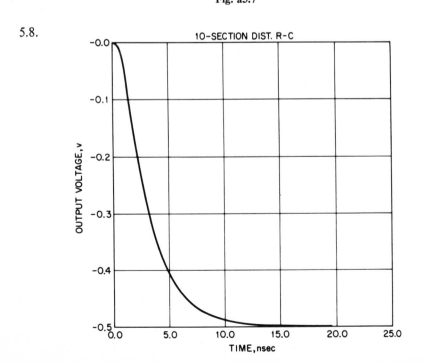

5.9.

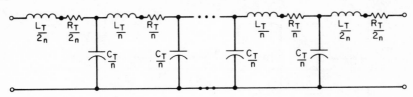

5.10.

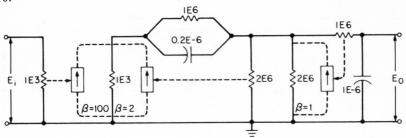

5.11.

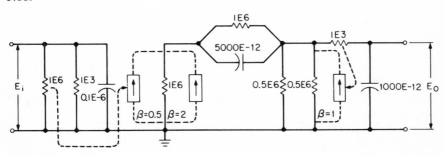

5.12. $I = 3.4\ \mu a$

5.13. See Fig. a5.13 on next page.

5.14. See Fig. a5.14 on next page.

5.15. See Fig. a5.15 on pp. 381–82.

5.16. See Fig. a5.16 on pp. 382–83.

6.1. See Fig. a6.1 on p. 383.

6.2. See Fig. a6.2 on p. 383.

   Note: Output circuit must provide a path to ground.

6.3. See Fig. a6.3 on p. 384.

6.4. See Fig. a6.4 on p. 384.

6.5. See Fig. a6.5 on p. 385.

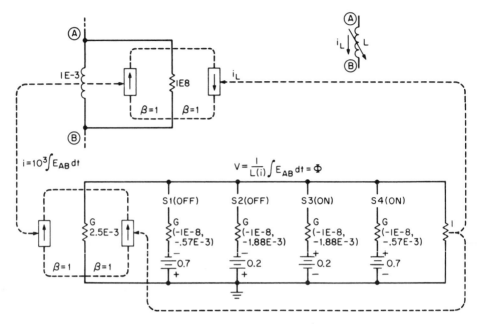

Fig. a5.13

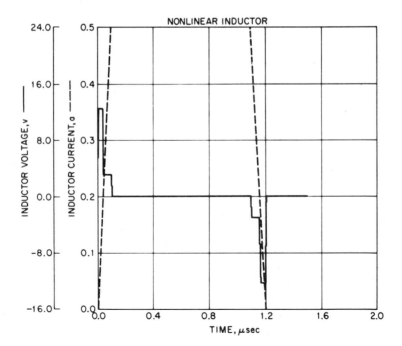

Fig. a5.14

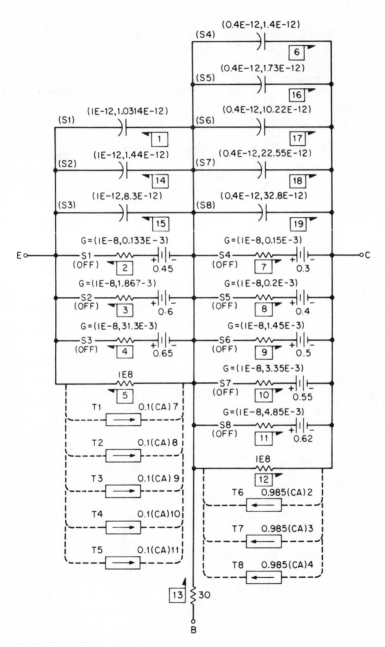

Fig. a5.15 (*continued on next page*)

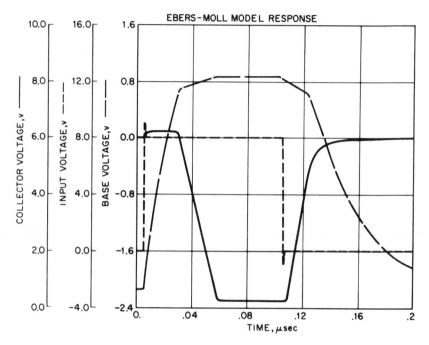

Fig. a5.15 (*continued*)

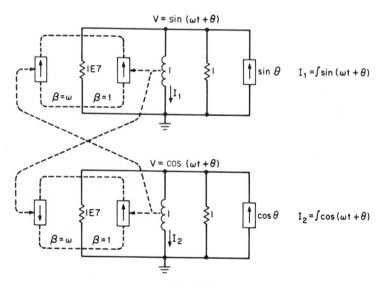

Fig. a5.16

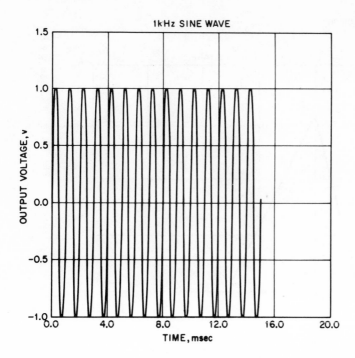

Fig. a5.16 (*continued*)

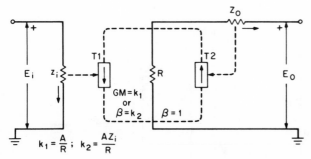

Fig. a6.1

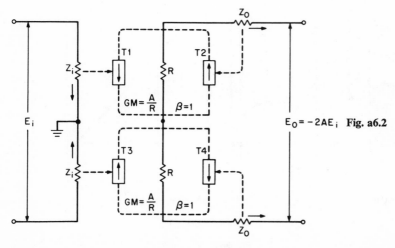

Fig. a6.2

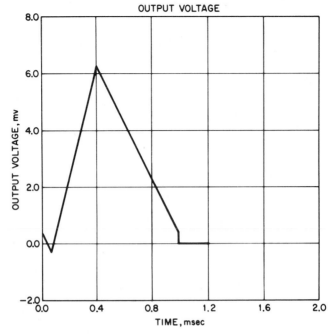

Fig. a6.3

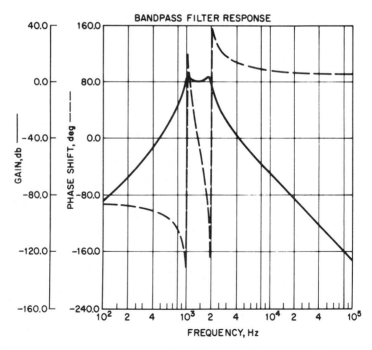

Fig. a6.4

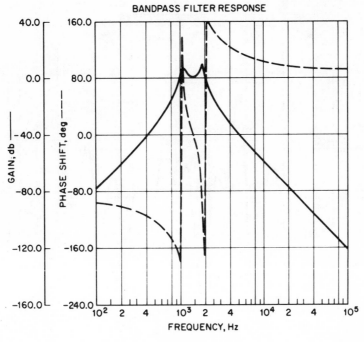

Fig. a6.5

6.6.

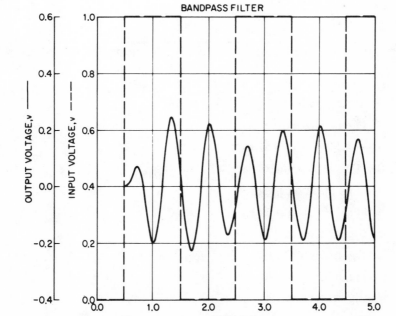

6.7.

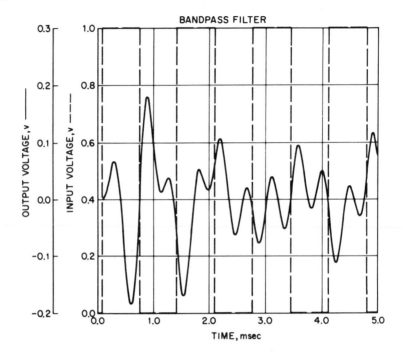

6.8.

R_i, ohms	V_i, volts	I_{OUT}, ma
300	5.0	187.70
100	10.0	533.38
400	10.0	246.86

6.9. See Fig. a6.9 on next page.

6.10. $T_S = 4.27$ nsec to 8.52 nsec

6.11. See Fig. a6.11 on next page.

6.12. See Fig. a6.12 on p. 388.

6.13. See Fig. a6.13 on p. 388.

7.1. See Fig. a7.1 on p. 389.

7.2. See Fig. a7.2 on p. 389.

7.3. See Fig. a7.3 on p. 390.

7.4. See Fig. a7.4 on p. 390.

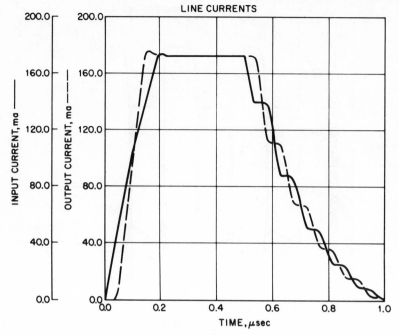

Fig. a6.9

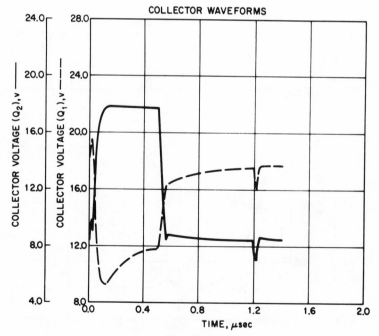

Fig. a6.11

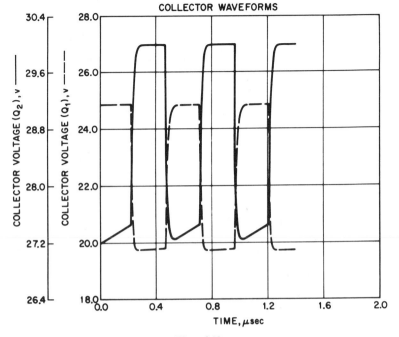

Fig. a6.12

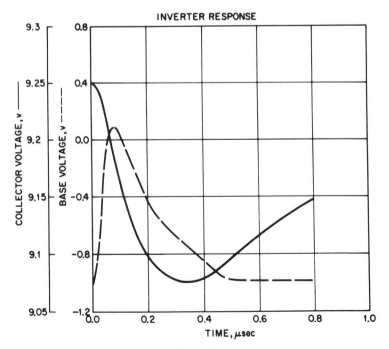

Fig. a6.13

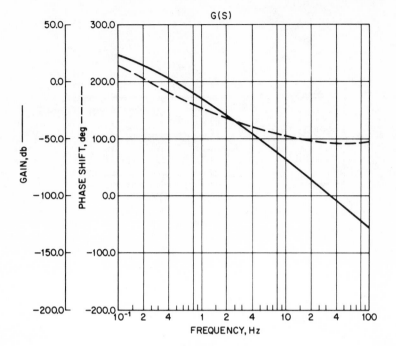

Fig. a7.1

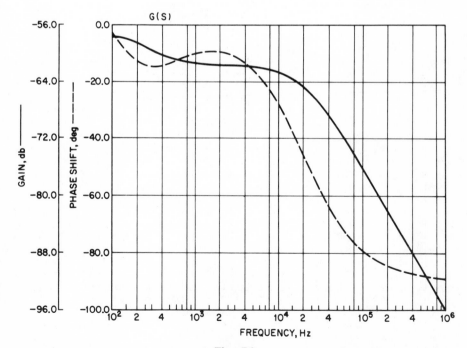

Fig. a7.2

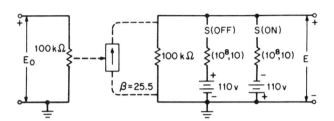

Fig. a7.3

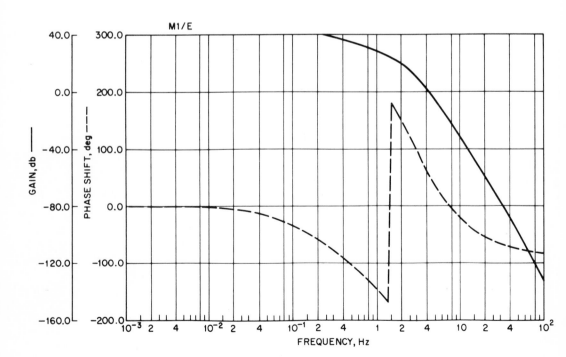

Fig. a7.4

7.5.

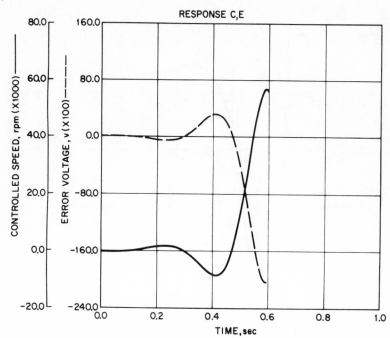

7.6.

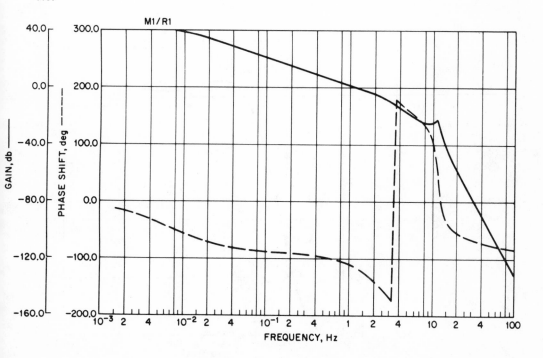

7.7.

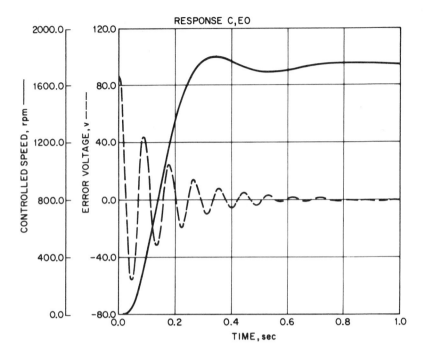

7.8.

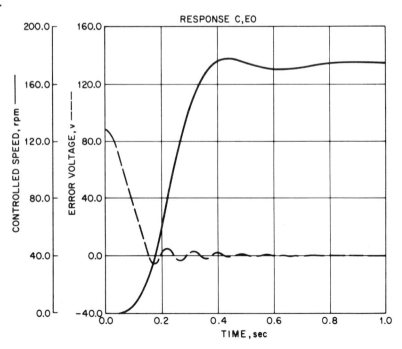

7.9.

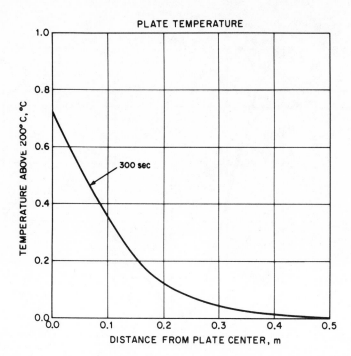

7.10.

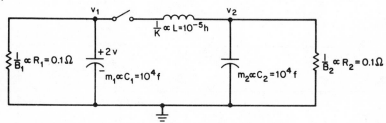

7.11.

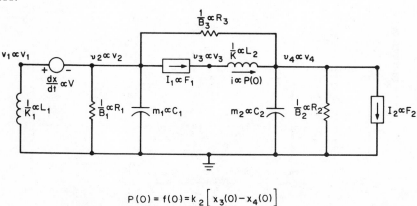

$$P(0) = f(0) = k_2 \left[x_3(0) - x_4(0) \right]$$

7.12.

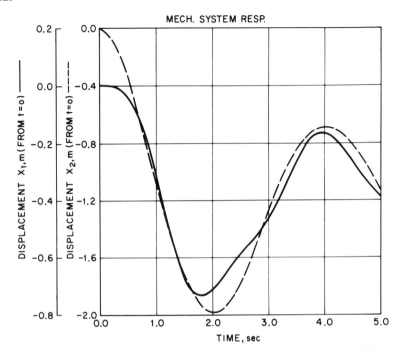

7.13.

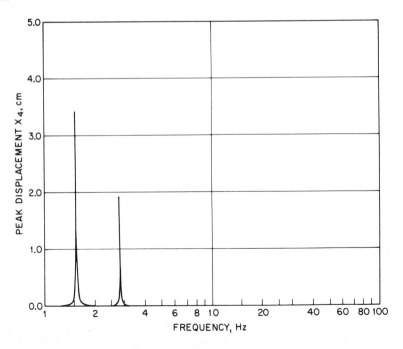

7.16.

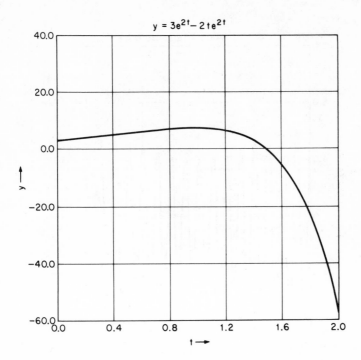

$y = 3e^{2t} - 2te^{2t}$

7.17. See answer 7.16.

7.18.

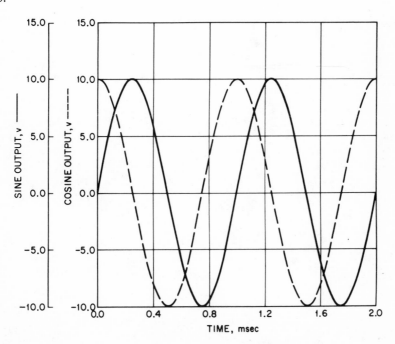

7.19.

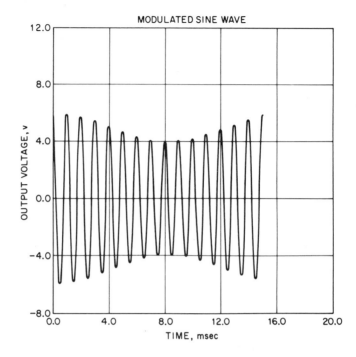

INDEX